GENERAL BIOLOGY

Dr. Uma Devi Koduru

Professor Retd. Andhra University

Visakhapatnam (AP) India

KHANNA BOOK PUBLISHING CO. (P) LTD.

Publisher of Engineering and Computer Books

4C/4344, Ansari Road, Darya Ganj, New Delhi-110002

Phone : 011-23244447-48   **Mobile:** +91-9910909320

E-mail : contact@khannabooks.com

Website : www.khannabooks.com

Copyright © Khanna Book Publishing Co. (P) Ltd.

No part of this publication may be reproduced, stored in a retrieval system or transmitted, in any form or by any means, electronic, mechanical, photocopying, recording or otherwise without prior permission of the publisher.

This book is sold subject to the condition that it shall not, by way of trade, be lent, re-sold, hired out or otherwise disposed of without the publisher's consent, in any form of binding or cover other than that in which it is published.

Disclaimer: The website links provided by the author in this book are placed for informational, educational & reference purpose only. The Publisher does not endorse these website links or the views of the speaker/ content of the said weblinks. In case of any dispute, all legal matters to be settled under Delhi Jurisdiction only.

ISBN: 978-93-91505-02-8

GENERAL BIOLOGY
by **Dr. Uma Devi Koduru**

First Edition: 2024

Published by:
Khanna Book Publishing Co. (P) Ltd.
CIN: U22110DL1998PTC095547

Visit us at: www.khannabooks.com
Write us at: contact@khannabooks.com

To view complete list of books,
please scan the QR Code:

Dedicated to

MY MENTOR PROFESSOR

PSRL NARASINGA RAO

With Love, Gratitude, Respect & Appreciation

PREFACE

Biology, the study of living organisms, is intriguing, since it includes us. Anything makes sense, when it relates to us. Knowledge of the living world and how it functions, is cardinal to understand the principles of life and its unifying features. The awareness acquired through a systematic study of the subject will be pluralistically useful. It is a welcome step taken by AICTE (All India Council of Technical Education), to introduce a course in 'General Biology' for all specializations of undergraduate studies in engineering in India, on par with the already offered courses in physical sciences. Design of living forms and the execution of life activities, are perfect and beyond improvement. They serve as exemplary models to emulate in engineering design. The syllabus has been aptly devised by AICTE, driving towards this end.

This book covers in toto, all the topics in the prescribed syllabus, and an additional topic on applied biology–biotechnology (included in the syllabus modified by APSHE (Andhra Pradesh State Higher Education) for the colleges in the state of AP). The choice of the content in each topic, and the extent to which it is dealt, is meant to achieve the learning objectives intended in the syllabus.

Topics dealt with are: an introductory chapter on the importance of Biology followed by chapters dealing with organization underlying the bewildering biodiversity prevailing in microscopic to gregarious large forms and their classification; working knowledge on inheritance pattern; physical and chemical nature of the genetic material and mechanism of its expression; genetic predetermination and the level of its flexibility and its role in disease; the architecture and functioning of various kinds of biomolecules that are at the helm of organizing the structure of life forms, and running life processes; the mechanism of basic metabolic activities and energy transfers, that are in tune with the laws of thermodynamics; and finally, the outstanding and exciting applications of biology – genetic engineering, cloning, industrial applications of biomolecules, recombinant vaccinces, fermentation technology, and highly sensitive and extremely useful miniature devises like biosensors and biochips. Interesting and exemplary experiments have been detailed where relevant, to draw attention to the simplicity, precaution and the conceptual design involved in an elegant scientific enquiry and nurture scientific temper.

The book will also be useful for all under and post graduate students in life sciences and to the +2 level study in life sciences, to acquire a comprehensive understanding of the topics. A pedantic dealing of the topic has been avoided; detailing only enough to know the essentials and breadth of each topic to create interest for further study. All topics have stand alone illustrative figures to facilitate easy understanding.

ACKNOWLEDGEMENTS

I owe my interest in biology to the teachers who introduced the subject to me in such an inspiring way, that study of biology became an avocation all through my professional life as a researcher and a teaching faculty in the university. "General Biology" is the outcome, of my more than four-decade romantic journey in the university (Andhra University, Visakhapatnam, India), as a student, learner, researcher and teacher of biology. I am thankful to my teachers, my alma mater and my students for the knowledge I gained to be able to pen this book.

"Ask Google" has become a popular adage. I express my sincere thanks to Google, which helped me get access in seconds, with the click of a button, to various resources on the topics I looked for, as a back-up and for factual details while writing the book. The several resources that I used, are mentioned in the reference section of the book.

I thank my research students, Dr. J. Padmavathi, Professor V. Sridevi (ANITS, Visakhapatnam), Dr. S. Suneetha and Dr. M. Prajna for helping me in various stages of writing the book. My special thanks to my research student Dr. G. Sandhya, who perused all the chapters and made useful suggestions. I am obligated to my family for encouraging me and my son, N. Sasi Mallik for drawing the figures that I conceptualised.

Author

CONTENTS

Chapter 3: Genetics   43

Chapter 4: Biomolecules   75

Chapter 5: Enzymes   91

Chapter 8: Laws of Thermodynamics at Play in Metabolism   167

Chapter 9: Microbiology   191

Chapter 10: Biotechnology   221

1. Biology – An Exciting Science; Biomimicry

We must learn to think not only logically but bio-logically. ~ **Edward Abbey**

Biology is the study of complicated things which give the appearance of having been designed for a purpose. ~ **Richard Dawkins**

1.1 BIOLOGY – SOME EXCITING ASPECTS OF THIS SCIENCE AND WHY WE NEED TO STUDY IT

Biology is the study of living organisms. This includes us – the humans. Nothing makes sense unless it relates to us. Biology helps to understand what we are, why we are as we are, how we carry on life activities and fight or succumb to disease. Knowing about oneself is the first and necessary step to conduct life in a conscious and sensible way.

Living beings are not just us; there are more than eight million species inhabiting the earth as estimated to date. We have only discovered a fifth of them and given them a name. Even with this modest coverage of the living world, the extent of biodiversity in the living world is astounding.

There is no place on earth without living beings, be it a deep ocean vent, blistering desert, pitch dark cave, freezing Antarctic ice, or steaming hot springs. Such is the adaptability of living beings!

Given the immense biodiversity in form and habitat, the magnificence of the biological world is that, the language of life in all living beings is one and the same! The blue print which specifies the form, function and behaviour of each and every living organism is written using the same chemical material – the DNA (Deoxyribo Nucleic Acid) (**Figure 1.1**). The DNA is composed of four types of nucleotides (A, T, G, C, for nucleotides Adenine, Thymidine, Guanine and Cytosine triphosphates) several in number, (e.g. 3 billion pairs in humans) which constitute its subunits. The blueprint is composed in the form of a code referred to as the genetic code. The genetic code is in a triplet (sequence of three nucleotides) module and means the same in all organisms. In such incomparable creatures as a microscopic bacterium, an eerie slime mould, a barnacle or a star fish, a sea weed or the corn plant, the mighty banyan tree or the moss, a fish or a human – the genetic code used to execute the organism's blue print is essentially identical! (**Figure 1.1**). The manner in which living beings duplicate their genetic material (DNA) which determines who they are, is fundamentally the same.

The energy currency through which life (metabolic) activities are transacted in all living organisms is the same! It is the power molecule – ATP (Adenosine Tri Phosphate).

What is true for a bacterium is so for a human too! This is indeed a humbling perspective of life.

The world of biology is incredibly simple yet complex, phenomenally interactive and reactive, self-regulating, and, intricately designed. A perusal of living beings with incisive tools to delve into their functioning at molecular level, leads one to an affirmative conclusion that an overall improvement in a biological system is not likely – if not impossible. Following are some examples to agree with this conclusion.

Our bone marrow produces 260 billion red blood cells (RBCs) and 135 billion white blood cells (WBCs) per day. Our eyes can identify 10 million colours and the nose can remember 50,000 different scents. Messages travel along the nerves from the human brain at ~ 200 miles an hour. We take in about four gallons of air per every minute – all our life. Our heart pumps blood continuously – all our life time, at a speed of 3 to 4 mph. Our skeletal system is a great architectural structure; to cite an example: the thigh bone (femur) that supports the weight of our body during walking is more powerful than a solid concrete of the same weight.

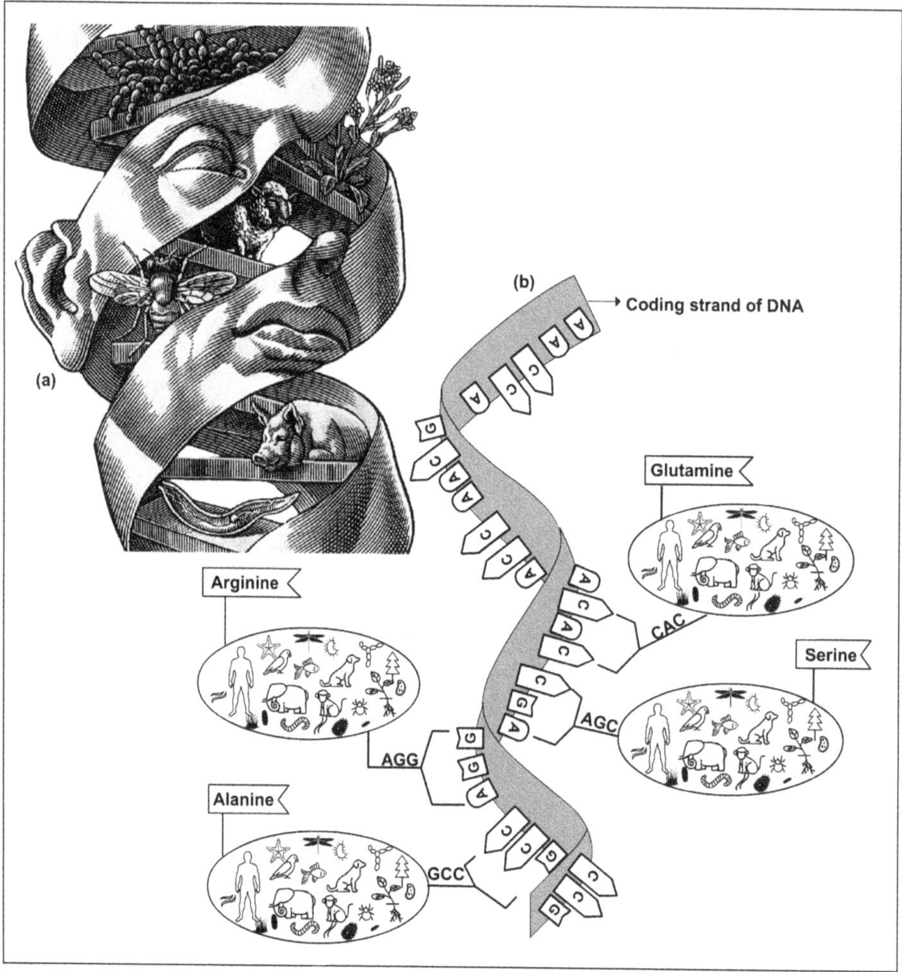

Figure 1.1: Universal language of life. a. *DNA is the material of the blue print of all living organisms.* (Picture credit *https://fineartamerica.com/featured/dna-in-allliving-things-artwork-bill-sanderson.html*). ***b.*** *The code of the blue print which is a triplet sequence of nucleotides in the DNA strand are read the same in all living organisms. The genetic code is universal: translated into the same amino acid in all living beings. Alanine, arginine, glutamine and serine are four of the twenty amino acids that make up the proteins.*

The nucleus in the cells of the human body with an average approximate diameter of six microns (0.006mm) harbours 6 feet long DNA! What is more – it looks like a haphazard bundle of threads in the nucleus, but, within a short notice, the exact file (gene) with the information required by an organism for that moment to carry out a particular function is retrieved to be copied so that information in the copy is processed and the requisite function is executed! During cell division, the 3 billion nucleotide long DNA in the nucleus of a cell in humans is copied nucleotide by nucleotide at an average rate of assembling million nucleotides per hour within one replication unit, with several such replicating units copying parallelly at the same time such that the process is completed in about eight hours! Given the speed of copying, it has high fidelity with very few errors! These very few errors are simultaneously corrected through 'proof reading' during the copying process so that the final daughter DNA is an exact replica of the original. As for reading the genetic code – about 6 to 9 codes are translated (equals to six to nine amino acids assembled) per second during the synthesis of the amino acid chain (polypeptide) for manufacture of a protein.

The information storage capacity of human brain is a million gigabytes with a very complex mechanism of memory storage! The human eye has a resolution of about 576 megapixels. The resolution of the most sophisticated digital cameras today (year 2021) is about 100 megapixels.

With this level of compactness, architecture and speed of operations that run non-stop with very rare errors, what more improvement in biological systems can one think of?

Life is still the most complicated than anything that man has built to date. Emulating structures and phenomena in living organisms – reverse engineering of biological models, is the best bet to come up with designs for energy efficient, elegant, perfect tools and gadgets. Scientists working on brain and neurons are helping nerds in neural networks (artificial intelligence) in the field of bio-robotics to develop mechanical systems that mimic the way biological systems process information.

Avocation in the fields of biology such as genetic improvement through breeding and genetic engineering of crops and animals is what keeps hunger at bay and feeds the burgeoning human population. About three quarters of all medicines used are derived from chemicals of plant origin.

Biology is a discipline at the forefront of key ecological issues and universal challenges like global warming and climate change. A study of biology helps to understand how life flourishes and what destroys it, ways and means to preserve life on the planet, for a friendly coexistence with all living beings, through judicious use of natural resources.

A study of biology is therefore to be pursued for becoming a self- and socially- conscious individual; to thrill and marvel at the wondrous community of living beings and, perhaps, copy a few cheat codes to invent marvellous devices that facilitate a healthy and eco-friendly life. An understanding of the subject of biology results in a deep-seated sympathetic and appreciative, profound knowledge that no life is mean – ant is no inferior to us – humans, and is as important a living creature in the ecosystem as we. Thus, it inculcates the attitude of 'live and let live' all fellow living organisms, which results in a fine balance in nature.

Biology is thus, no less a discipline of science than physics and chemistry; it deserves an equal if not higher status as them. The boundaries between different sciences are becoming obscure with inter-disciplinary approach to understanding phenomena. Psychological research has demonstrated that previous knowledge influences how one organizes and links new information in dealing with a problem. A basic knowledge in biology is sure to be handy for a future engineer – for that matter, anyone.

1.2 OBSERVATIONS OF 18TH CENTURY IN BIOLOGY WITH PROFOUND IMPACT THAT CASCADED INTO SEVERAL DISCOVERIES

Simple observations in living organisms have led to proposing insightful axioms in physics and chemistry. The discovery of Brownian movement and the law of conservation of energy – the first law of thermodynamics were both made in living systems.

1.2.1 Brownian Movement

Robert Brown, a Scottish botanist, while investigating the phenomenon of fertilization in a then newly discovered plant – *Clarkia pulchella* commonly called ragged robin/ pink fairy, observed a strange phenomenon. He intended to study, the mechanism by which, pollen grains that carry the male germ cells impregnate the ovule which harbours the egg. While viewing the pollen grains of this plant suspended in water under the microscope, he observed tiny (~ 5microns long) oblong particles ejected from the pollen grains making jittery movements – what he described as "a constant rapid oscillatory movement". He was convinced that this motion was neither caused by currents in the water in which the pollen grains were suspended, nor due to its evaporation. He concluded that the tiny particles were plant equivalent of sperm and they were jiggling because they were alive. Next, he checked to see if the pollen of plants dead a century ago (fossils) showed similar movement. And they did! Brown called this a "very unexpected fact of seeming vitality (life) being retained by these 'molecules' so long after the death of the plant." He further tested tiny fragments of fossilized wood (of eon age – dead centuries ago) and found similar movement. To confirm that this movement was tied to particles of organic origin, be it from long dead organisms, he looked at the suspended particles from chips of glass and granite and smoke under the microscope. He observed the random rapid motion in these samples as well. Finally, as an inarguable support of the non-living nature of this phenomenon, he demonstrated it in the fluid-filled vesicles in a rock from the Great Sphinx. Brown could not explain the reason for the movement of the microscopic particles that he observed in his investigations in 1857.

Eight years later, in 1905, a satisfactory explanation to Brown's observation was provided by Albert Einstein, a theoretical physicist. Einstein reasoned that if tiny but visible particles were suspended in a liquid, the invisible atoms in the liquid would bombard the suspended particles and cause them to move rapidly and randomly. This explanation served as a convincing evidence for the existence of atoms and molecules – the basis for the modern atomic theory. Einstein's proposition was experimentally verified by Jean Perrin in 1908 for which he was awarded the Nobel prize in physics in 1926.

The kinetic theory of gases which explains pressure, temperature and volume of gases is also based on the Brownian motion model of particle movement. Based on observations of Brownian movement, Einstein derived that the number of units such as atoms, molecules, or ions in one mole (equivalent to molecular weight) of any substance is 6.022×10^{23} which is now called the Avogadro number/constant. Avogadro constant is used to convert mass of a substance to number of particles (molecules) or vice versa.

Named as Brownian movement in the honour of Robert Brown who first reported it, it is defined as the uncontrolled or erratic movement of particles in a fluid due to their constant collision with other fast-moving molecules. The rapid, oscillatory, jittery movement of particles christened Brownian movement or motion, is described variously as 'random walk', 'drunken sailor walk' or pedesis (Greek word meaning leaping) (**Figure 1.2**). Usually, the random movement is observed to be stronger in smaller sized particles, in less viscous liquid and at a higher temperature.

Brownian motion is the random movement of particles in a fluid due to their collisions with other atoms or molecules. The impact of the tiny fast-moving atoms and molecules in the medium can move particles much bigger than them. Brownian motion may be considered a macroscopic (visible) picture of a particle influenced by many microscopic random effects. Brownian motion prevents particles from settling down, leading to the stability of colloidal solutions.

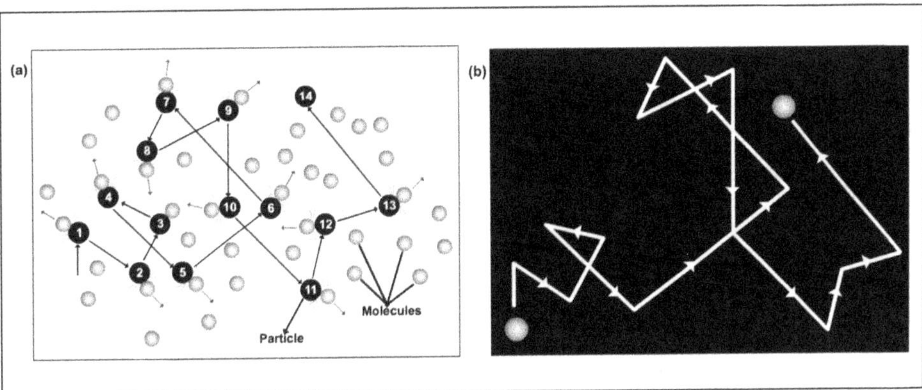

Figure 1.2: Brownian motion. a. *Rapid, oscillatory movement of particles in a liquid. The movement is driven by the collision of the particles with the molecules in the liquid.* **b.** *The path that a particle takes due to the random Brownian movement is haphazard and appears as an uncontrolled confused walk of a drunken sailor.*

1.2.2 Law of Conservation of Energy (1st Law of Thermodynamics)

Julius Robert Mayer voyaged to East Indies as a physician on a Dutch freighter carrying cargo in 1840. His observations during his year's stay as a physician in Jakarta in Java – a tropical place south of the equator, led to the discovery of the most important physical law – the law of conservation of energy. While letting blood by lancing the veins of sick sailors, he observed that it was far too red than what he observed in his patients back in Europe. In the first instance he thought that he hit an artery – arterial blood carries more oxygen than venous blood and is therefore darker. Then he found that it is normal in tropics. He pondered over this seemingly insignificant fact. He reached the conclusion that the cause must be due to the lesser amount of oxidation of food required to maintain the body temperature in the hot climate of the tropics. Just as food fuels our power output, Mayer realized that it also fuels heat supply. Led by this reflection to consider the body as a machine dependent on outside forces for its capacity to act, he stepped into a novel realm of thought, which led to the discovery of the mechanical theory of heat, and to the first full and comprehensive appreciation of the great law of conservation of energy (chemical energy of the food being converted to heat energy). Thus, the profound physical law was formulated by Mayer from his insight into a biological observation. There are accounts of his other observations that led him to the intuition – about putting his hand in the sea water being churned by the swift waves propelled by the winds of a storm during the voyage and noticing that the water was warmer than during normal weather when the waves are slower and calmer. Some say the trip had nothing to do with his vision of indestructability of energy and that he first noticed a temperature shift due to motion (change of kinetic energy to heat energy) when watching a horse stir a cauldron of paper pulp. Many quote that it was his observation of the blood samples that led to the intuition of conservation of energy.

Whatever be the source of his insight, Mayer reported in his 1842 scientific paper about the interchangeability of mechanical and heat energy, indestructibility of energy and mentioned: "Energy is neither created nor destroyed." Mayer also came up with a rough estimate of what it would take to bring a certain amount of water up a certain temperature – the unit which is now popularly known as Joule – a unit of heat. It brings out the relationship between calorie (a unit of energy) to joule (a unit of heat). One calorie equals 4.184 joules.

Mayor later made yet another important observation – that source of the energy radiated by the sun was the result of constant bombardment by high-energy meteors. He stated that the sun is the ultimate source of energy for both plants and animals and that when absorbed, plants convert this light energy to chemical energy through the process of photosynthesis. He wrote: "The plants take in one form of power (light); and produce another power – chemical difference." He used the term 'power' for light energy and 'chemical difference' for chemical energy. Thus, Mayer concluded from this observation that one form of energy is converted to another, but never lost.

Mayer's great doctrine of conservation of energy was not appreciated by his peers, just as any novel and revolutionary thought is received in science. The truth and the importance of the doctrine dawned upon scientists when the concept was better expressed by a physicist, James Prescott Joule.

1.3 SCIENCE VS. ENGINEERING

The ruins excavated of Mohenjo-daro, one of the world's earliest major cities of the Indus valley civilization, located in the Sindh province of Pakistan, dating back to 2600 BC, contemporary with the

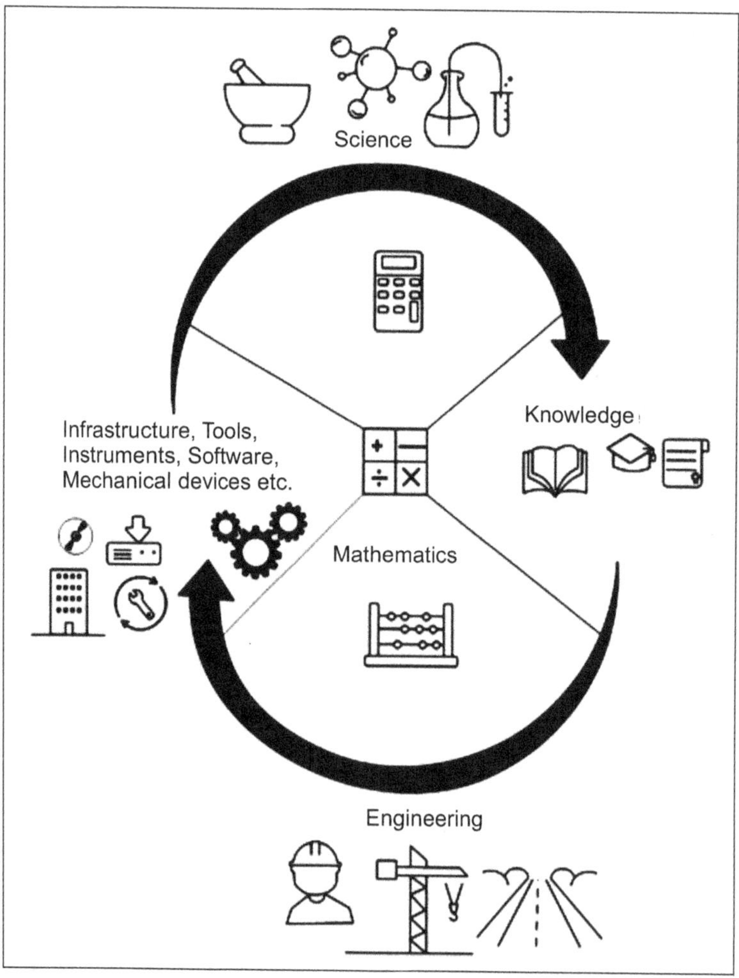

Figure 1.3: Relationship between science and engineering. *Both fields are supportive to each other – one begets another. Both require mathematics.*

civilizations of ancient Egypt – Mesopotamia; indicate that it is based on remarkably sophisticated civil engineering and urban planning. The earliest records of scientific investigation dates back to 1600 BC to the ancient Egyptian text – 'Edwin Smith Papyrus', a surgical treatise on trauma. Thus, engineering preceded organized scientific investigation by ten centuries. Engineering does not always require the conscious use of scientific knowledge. But for a modern engineer, science serves as an important tool to solve complicated engineering problems. Science and Engineering are mutually supporting fields: advances in science are possible due to design of incisive engineering tools, and, insightful scientific discoveries are very useful and handy for engineering design (**Figure 1.3**). Science and engineering are a winning combination indeed!

The crux of science is inquiry and that of engineering is design. Scientific investigation follows a bottom-up approach; engineering adopts a top-down tactic.

A scientific method involves: **1.** observation; **2.** asking questions; **3.** looking up for existing answers; **4.** predicting an answer to the question, and from it, formulating a hypothesis; **5.** planning and carrying out experiments to test the hypothesis; **6.** analysing the data; **7.** if the data support the hypothesis, carrying out further alternate experiments to test if the hypothesis can be rejected; **8.** if the results from these experiments still support the hypothesis, accepting it "for now" and reporting it. The accepted and reported hypothesis becomes knowledge and it is open to be tested further and rejected if new evidence against the supported hypothesis is found. Thus, scientific method has rigour and courage to stand the test of time and has no problem of painstakingly gathered data to be falsified in the light of new evidence. This is the bottom-up approach.

Engineering creatively applies scientific knowledge to analyse events, develop materials, innovative tools and construct objects that are useful to human kind and benefit society.

All science is measurement; any design process involves estimates and calculations. Therefore, mathematics is the supporting pillar for both science and engineering avocations.

The differences in attitude and style between science and engineering are tabulated in **Table 1.1**. It might be apt to sum up with the quote of Theodore von Kármán: "Scientists study the world as it is, engineers create the world that has never been".

Table 1.1: Science vs. Engineering

	Science	Engineering
1	Involves observation, questioning and inquiry through an axiomatic methodology.	Entails creation of a design as a solution to a problem.
2.	Examines unknown, seeks knowledge.	Shuns unknown, uses knowledge.
3.	Information flows from matter to mind.	Information flows from mind to matter.
4.	Obtains information through instruments.	Applies information to make tools and devices.
5.	Shapes its description to fit the physical world.	Shapes the physical world to fit its description.
6.	Science seeks unified one correct option and if more than one options are available, then, only one must be right.	Seeks for several working designs, and, if many of them work, success is assured.
7.	Attempts to get all possible information about a phenomenon or process.	Focusses on relevant and limited information required to solve a problem on hand.

Two examples how knowledge from biological science has been applied in engineering to develop useful devises are described below. One is biomimicking of the human eye in the design and creation of a camera, the other, is the modelling of an aircraft from the knowledge of body structure and flight pattern of birds.

1.4 HUMAN EYE AND CAMERA

Eyes are the most important and complex of the sense organs, as we form ~ 70% of our impressions through them. They are the physical portal of information to the brain.

The human eye, a biological organ is the result of ~540 million years of evolution. The cameras – the electromechanical devices that we use today, are a result of 200 years of design improvisation. The first photograph was taken in 1826 with a wooden camera – a pinhole type. It was however, conceptualised much earlier, in the fifth century, by Mo-Ti, a Chinese philosopher.

The camera bears an uncanny resemblance to the human eye. The anatomy and functioning of the eye were a subject of investigation from the ninth century onwards, with dozens of great treatises on ophthalmology published between ninth and fourteenth century. The design and construction of the present-day camera should therefore have been based on the knowledge of the framework of human eye. Just as the many man-made designs inspired from living organisms, a camera is far from being equal to its original! Many improvisations are being attempted to enhance the image forming ability of a camera to bring it to the efficiency level of an eye. A camera set up equivalent to a human eye would theoretically cost 35 million US$ according to the estimate of Caler Ward (2015). It is yet to be constructed though! And still it wouldn't match a human eye because the image seen by an eye is just not the result of its optical system, but, more due to its co-ordination with the brain, i.e., it is a subjective device. A camera can only record an image but not show an interpretation of it. Whatever be its optical power and resolution, a camera will remain to be an absolute objective appliance that captures images.

To draw a parallel between the human eye and a camera, the anatomy of the eye and the components of a camera and the working pattern of both are described below.

1.4.1 Human Eye

1.4.1(a) Structure

Human eye is approximately 23mm of height and 24mm long. The anterior segment which constitutes $1/6^{th}$ of it is made up of cornea, iris and lens. The posterior semi-spherical part that makes $5/6^{th}$ of the eye, contains the vitreous, lined by the retina. The retina connects at the posterior most end to the optical nerve that extends to the brain (**Figure 1.4**).

The cornea is transparent and jelly like, being the only tissue in the human body without blood vessels. It is curved. The pigmented circular structure concentrically surrounding the black pupil in centre of the eye is the iris. The amount of light entering the eye is controlled by the size of the pupil, which is adjusted by the iris and sphincter muscles. The minute architectural details of the iris exhibit variations which are very unique to an individual and are more accurate than fingerprints to serve as an individual's identification. Behind the pupil, there is a transparent lens with a spherical curvature. (**Figure 1.4**).

The retina is a light-sensitive layer in the back of the eye (**Figure 1. 4**) that contains highly evolved cells called rods and cones. There are 7 million cones and 100 million rods. Rods help to see better in the dark. Cones are of three types differing in their sensitivity to colours. The human eye only sees three colours with each type of cone being specific to one of the colours – red, blue or green. These three types of cones work together to sense combinations of light waves that help us see millions of other colours and shades. Rods cannot perceive colour, just – black, white and grey. However, they are very sensitive and tell us the shape of something. The retina has a mesh work of blood vessels which is very intricate and unique to an individual. Retinal scans can serve as a person's identification mark. They can serve to differentiate even between identical twins. The posterior most part of the retina is called the macula; there is a small involution of the retina near the macula which is termed the fovea (**Figure 1.4**).

The iris has 256 unique characteristics compared to 40 that a fingerprint has. Therefore, iris scans are increasingly being used as a person's id. A retina is even more useful identification mark of an individual. A retinal scanner uses infrared light to map the unique pattern of blood vessels on a person's retina.

1.4.1(b) Working

When we look at an object, the light it generates enters the eye through the cornea which begins to focus the light. From the cornea, light passes through the pupil. The size of the pupil changes to regulate the amount of light entering the eye. The light then passes through the lense. The lens shape is changed by the ciliary muscle for near focus. The light refracted through the lens is focussed onto the retina. Photons of light falling on the light-sensitive cells (cones and rods) form an image on the macula, the region of the retina where the rods and cones are most densely distributed. The image is converted into electrical signals that are transmitted by the optic nerve to the visual centre of the brain. It is then that vision of the object happens.

1.4.2 Camera

1.4.2(a) Components

The camera consists of a body which is a light proof box enclosing the camera shutter, a convex plastic or glass lens (fixed or removable and interchangeable), lens aperture, shutter and the image sensor (**Figure 1.4**). The image sensor has a grid with millions of microscopic light information gathering elements called pixels. Each megapixel consists of one million pixels. The LCD (Liquid Crystal Display) screen of the camera is for previewing and then viewing the captured image.

1.4.2(b) Working

When shutter release button of the camera is pressed, it opens. The light from the object flowing into the camera lens passes through its aperture through the lens and opens the shutter to the image sensor. (**Figure 1.4**). The lens focusses light to create an image. The lens aperture has different size openings which are referred to as F stops operated by the iris (**Figure 1.4**). The size of the lens aperture (to which it has been set) determines the amount of light that reaches the image sensor. The amount of time the light is exposed to the image sensor is determined by the set shutter speed – the time the shutter remains open. Shutter speed is expressed as seconds or fractions of a second. The colours and characteristics of the light that the image sensor is exposed to, is recorded by it and then saved to the camera's memory card. The information in the memory card can then be reproduced on the camera's LCD screen, or a computer screen, or printed on photo paper.

The human eye and a camera have similar components with comparable anatomical similarity and technical similarities in the optical working mechanism (**Figure 1.4**). The similarities and differences between a human eye and a camera are given in **Table 1.2**.

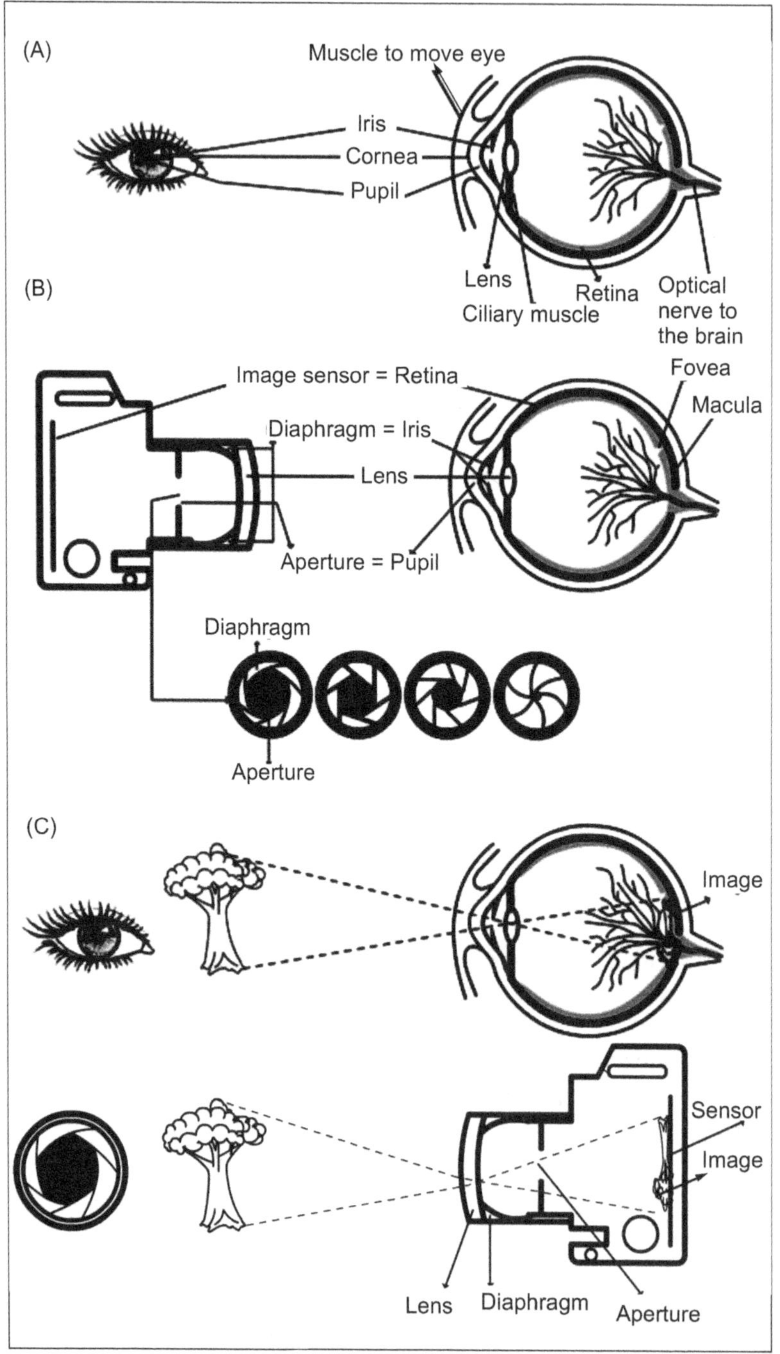

Figure 1.4: Human eye and the Camera. A. *The external view and internal anatomy of human eye.* **B.** *Comparison between parts of a camera and a human eye.* **C.** *Mechanism of image formation in an eye and a camera – the light is refracted by the lens and an inverted image is formed on the macula of the retina in the eye and the sensor in a camera.*

Table 1.2: Similarities and differences between human eye and a camera

	Human eye	Camera
	Similarities	
1.	**Eye lens = Camera lens**	
	a. The lens of the eye and a camera (made of glass/plastic) are both transparent with a spherical curvature.	
	b. The convex curvature of the eye lens together with the cornea and the camera lens, allow both, to view in addition to the front, a limited area to the right and the left (though not in focus). They both function by refracting light and focussing it on the retina or image sensor.	
2.	**Iris and Pupil = Diaphragm and Aperture**	
	The size of the pupil (which is opening to the eye lens) is managed by the contracting and expanding of the iris. When the iris contracts, the pupil becomes smaller and the eye takes in less light. When the iris widens in darker conditions, the pupil becomes larger, so it can allow in more light.	The size of the aperture is controlled by the diaphragm. Different aperture sizes, denoted by F number/stops can be set by adjusting the diaphragm (**Figure 1.4**). Larger aperture (lower F stop value) lets in more light than a smaller aperture (higher F number).
3.	**Retina = Film/Image sensor**	
	The retina collects the light reflected from the surrounding environment to form the image.	The same task is performed either by a film or a sensor in digital cameras.
	The retina is almost the same size (32mm in diameter) as the sensor on a full frame camera (35mm in diameter).	
4.	**Focus**	
	Both the eye and camera have the ability to focus on one single object and blur the rest, whether in the foreground (shallow depth of field) or off at a distance. Likewise, the eye can focus on a larger image, just as a camera (greater depth of field) and capture a large scape.	
	Dissimilarities	
1.	**Position of the lens**	
	Lens is seated deep inside the eye covered by the pupil and cornea.	Lens is the outermost part of the optical system in a camera. It is visible just in front of the device.
2.	**Retina vs. Film or Silicon image sensor**	
	a. The shape of the retina is curved.	The film or the silicon image sensor is flat.
	Because of the curvature of the retina, its edges are about the same distance from the lens as its centre. On a flat sensor the centre is closer to the lens and the edges are further away from the lens. The image formed on the retina has therefore more sharpness in its corners than that formed on the sensor of a camera.	
	b. Retina has an equivalent of 576 million megapixels (photoreceptor cells called rods and cones). The cones are colour receptive and of three types – one each sensitive to green, red and blue. The rods are sensitive to white, black and grey.	The most advanced prototype sensor envisioned by Canon will only have 250 megapixels. Present (year 2020) cameras have the highest of 200 megapixels. All pixels are of one type and colour receptive. The red, blue and green colours are recognized through a Bayer array filter fixed over the sensor which pairs two of the three colours at a time. Black and white receptivity is not existent in the camera pixels.

	c. Distribution of the photoreceptor cells (pixels) in the retina is not uniform. The densest concentration of photo receptors is in a small central area, about 6mm across called the macula. Cones which are the colour sensing cells are densely packed in the central region of the macula which is called the fovea. The rest of the macula around 'colour only' fovea area contains both rods and cones.		Distribution of the pixels is uniform. Each pixel is set out in a regular grid pattern. Every square millimetre of the sensor has exactly the same number and pattern of pixels.
	Due to much higher number of pixels in the retina, the eye forms better resolution pictures than a camera. Also, the eye can capture more clearer pictures in the dim light because of the presence of black and white receptive rod cells. The camera gives blurred pictures in low light.		
3.	**Pupil vs. Camera aperture**		
	Pupil has several pinholes through which light enters the lens.		Aperture is a single hole through which light enters.
4.	**Focal length and focussing**		
	Has a fixed focal length of 22mm. Therefore it has a fixed visual scope.		Focal length can be changed simply by adjusting the distance of the lens and sensor, or by changing the lens for a one with different focal length. Therefore, visual scope can be varied.
	The pupil adjusts its size while focusing but, in a camera, the lens moves to change focus. The focal length of eye lens can be changed by a change in its shape brought about by the action of ciliary muscles.		
5.	Has stereoscopic vision and captures 3D images as well.		Camera is only able to capture 2D images so far.
6.	The human eye has blind spots. Hence, the focus diminishes farther away from the centre of the eye.		Cameras do not have blind spots; the focus of the image stays constant.
7.	**Image formation**		
	a. The eye does not really capture a full, highly-detailed image in a single instant of time. The detailed scene that we think we see is actually a construct of the brain, which builds it up in part, from the information the eye is delivering over time as a constant video feed and in part, from several "assumptions" that are effectively built into the system.		The camera shows the actual image captured at a particular moment on its sensor.
	b. The eye has ~570 million photo sensors(pixels) in the retina, but the optic nerve that carries the signals from the sensor to the brain has only 1.2 million fibres. So only less than 10% of the retina's data is passed on to the brain at any given instant.		The camera sends every pixel's data from the sensor to a computer chip for processing into an image.
	c. What is seen through an eye is a video.		The camera creates still images.
	d. The images seen by the eye may not be permanently and vividly stored in the memory of the brain.		Digital cameras don't lose their memory unless purposefully deleted.

The camera is thus like a robotic eye. As we have seen, the human eye is obviously more efficient than any camera that has been produced as yet. The present technology is far from producing any type of imagery system that's better than the human eye.

1.5 BIRD FLYING AND AIRCRAFT

Insects that appeared on earth 325 million years ago were the first animals to have been endowed with the power to fly. Among vertebrates, the ability to fly arose in birds that originated on earth much later than their invertebrate insect cousins – about 60 million years ago but much before man appeared (which was only ~ 300 thousand years ago). Flying is a rapid mode of transport to explore new and unknown territories – it is believed that insects owe their widest distribution in every environmental niche on earth due to their power of flight.

Nature is an inspiration for many tools and devises designed by man. The Hindu epic Ramayana, composed around 5th century BCE (~1,500 years ago) describes 'Pushpaka vimana' – a chariot which could fly in the air (through the power of thought!) carrying people. Various descriptions of the Pushpaka vimana are found in literature but it has no semblance to the modern-day aeroplanes. Greek mythology has a story of Daedalus and his son, Icarus, escaping from the prison by flying with bird like wings that they crafted with feathers.

Today's airplanes are an outcome of biomimicry of the birds. The attempts to mimic birds for an airborne transport vehicle began in 1486 with the design of 'ornithopter' by Leonardo da Vinci after a keen observation of the anatomy and flight of birds. Flying however materialised much later – after five centuries in 1903, with the first aeroplane designed based on a pigeon, by the Wright brothers. The physical principles that Wright brothers used have since been applied to every airplane, balancing two sets of opposing forces – the upward force called lift, opposing the downward acting force of gravity, and the forward acting force called thrust, opposing the backward acting force called drag (**Figure 1. 5**). The helicopter equipped with two oppositely spinning rotors, that can take off and land vertically, fly both forward and backward and also hover in the air, was designed in 1936 taking the inspiration from the mode of flight of the humming bird.

The real first aeroplane came into operation in 1909, when the engineers mimicked the structure – wings, body shape and weight and mode of flying of a bird, to come up with an aerodynamically sustainable design for a vehicle that is to be airborne. The birds did not just serve as an inspiration for flying – they provided the design for the aeroplanes. Described below is the similarity in the design of an aeroplane to that of a bird. The mimicry of a bird resulted in a commendable achievement – an aeroplane that renders the bird a pygmy when compared in size, and ability – being capable of carrying hundreds of humans for thousands of kilometres at a stretch across continents over the mighty oceans.

1.5.1 Structure of the Bird and Mode of its Flight

Birds fly with their wings. But everything about a bird's body is designed for flight. They have a very low body weight. They are covered with light but strong smooth and glossy water proof feathers which perfectly insulate their body and help in flight. The birds have hollow pneumatic bones connected to the pulmonary system that allows air circulation thereby increasing skeletal buoyancy. Both the feathers and bones have unusual combinations of structural features organized hierarchically from nano- to macroscale that enable a balance between lightweight and bending/torsional stiffness and strength. Birds have a unique muscular system that facilitates flight. Their breast muscles are extremely enlarged with skeletal modifications to accommodate them. They have a special pulley system that allows a muscle under the wing to raise it. The wings have a curved air foil (aerofoil)shape which helps in flight. The overall shape of the bird is streamlined. Its streamlined shape is achieved through special arrangement of feathers which greatly facilitates flying.

1.5.1(a) Principles of Flying

There are four forces in flying: **1.** lift; **2.** weight (gravitational force); **3.** thrust and **4.** drag. Lift and drag are considered aerodynamic forces because they exist only during the flight movement. Lift acts opposite to weight, and, thrust, opposite to drag. Lift is a mechanical aerodynamic force created by the motion of a flying object. When the lift is strong enough, the object overcomes weight or gravity and it is held in the air. Thrust is the forward movement of the flying object and drag is caused due to difference in the velocity of the flying object and air. Drag acts opposite to thrust slowing down motion during flight (**Figure 1.5**).

1.5.1(b) Mode of a Bird's Flight

To fly, a bird flaps its wings. This results in two things – it lifts, and simultaneously propels the bird forward. As the wings push downward and backward, more air is moved below the wing than above it. This difference in the amount of air, or air pressure, causes lift, resulting in upward and forward movement of the bird. Thus, birds overcome drag and produce lift through movement of their wings. Lift allows the bird to glide for considerable time when again it flaps its wings to continue flying.

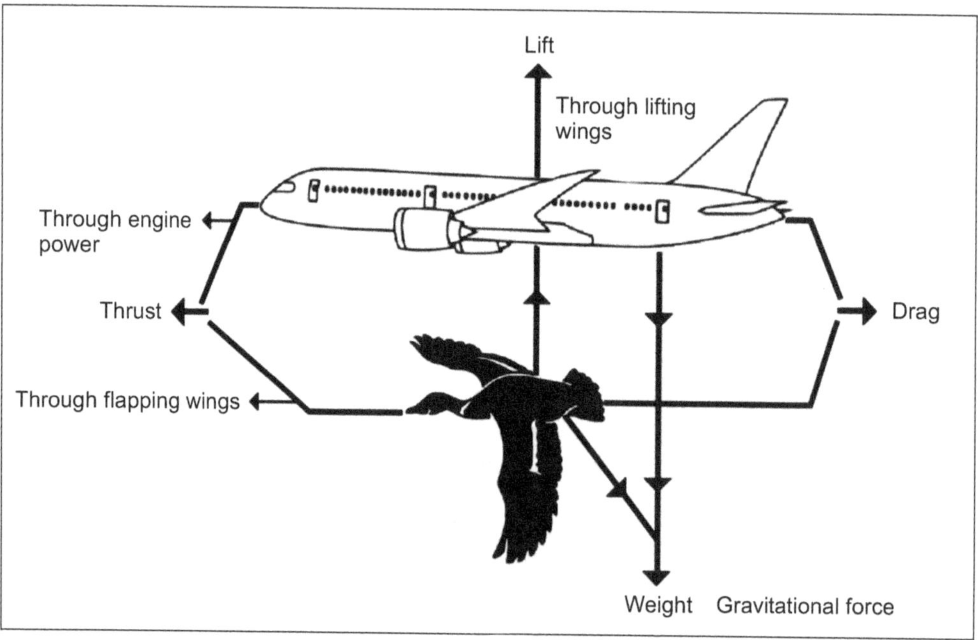

Figure 1.5: The four forces in flight. *Weight is the gravitational force that pulls an object down. Lift is an opposite force which the bird and aeroplane achieve overcoming weight through lifting their wings. Thrust and drag operate in opposite directions. Drag is caused due to the difference in the velocity of the flying object and the surrounding air. Thrust, the forward movement is possible by overcoming drag. Thrust is achieved by the bird through flapping of its wings with the help of its strong chest muscles. An aeroplane thrusts forward using its engine power.*

An aeroplane is an exemplary example of biomimicry of a bird. The similarities and differences between a bird and an aeroplane are listed in **Table 1.3**.

Table 1.3: Comparison of a bird and an aeroplane

	Bird	**Aeroplane**
	Structural similarities	
1.	**Body shape, Texture and Built**	
	Both have a streamlined shape and a smooth surface to prevent air resistance (**Figure 1.6**). The birds achieve smoothness through their glossy feathers and the aircraft has polished surfaces. Both are light weight. The bird has air filled bones and light feathers. An air craft is made mostly with alloys of light weight metals like aluminium and titanium; steel is used only in parts which require a lot of strength like the landing gears. Light weight composite materials are also used in the fabrication of an aircraft.	
2.	**Wings**	
	Wings are the most important for flying. The wing of an aeroplane is designed based on the aerofoil shape of the bird's wing (**Figure 1.6**). It tapers to the outside. Both wings have different thicknesses as they move away from the aeroplane body. The tail and propeller blades of the plane also have aerofoil shape.	
	Similarities in mechanism of flight	
1.	Because the aeroplane is designed to have similar structure of the body and wings and body size to weight ratio like that of a bird, it follows the same principles of flight – creating lift and thrusting forward overcoming weight and drag (**Figure 1.5**).	
2.	Birds take off by raising their wings and creating a low pressure by which the body goes up through the process of 'lift'. Aeroplanes achieve lift through the same process (**Figure 1.6**).	
3.	The aerodynamics while gliding is exactly similar in birds and aeroplanes. The wings in both work by bending the air downwards and exerting a force equivalent to the force that the weight of the bird or aircraft exerts on the air.	
	Structural dissimilarities	
1.	Wings are flexible and they can be flapped.	Wings are fixed. Flapping the wings is not possible.
2.	Have feathers which also help in flight.	No structures equivalent to feathers are there on the body of the aeroplane.
	Dissimilarities in mode of achieving flight	
1.	Birds generate thrust through flapping their wings with the help of their strong chest muscles. When the wings go down, the feathers tend to rise and come into contact with their neighbours. The air is thus trapped creating a huge thrust force for the bird. With flapping, the wing goes back slightly in a rowing motion, which increases the thrust.	Airplanes have fixed wings. Therefore, they achieve thrust by moving forward quickly using engine power.
2.	Birds control pitch (ascending or descending) by forcing their rear end up or down resulting in the wings pointing more to the up or down.	The horizontal stabilizer on the rear end of the aeroplane (**Figure 1.6**) controls pitch.
3.	Birds manoeuvre the right and left orientation (technically called yaw) through twisting their horizontal tail in one direction or the other.	Aeroplanes manage yaw with the help of the rudder which is a hinge on the vertical tail at its rear end (**Figure 1.6**).
	Birds manage both pitch and yaw with their tail, aeroplanes have a horizontal stabilizer for manipulating pitch and a vertical tail for yaw.	

The aeroplane thus, is an impressive case of biomimicry of the bird. However, an aeroplane is far from what some birds can do with their flexibility and in several other attributes of flight. Migratory birds like the frigate birds and swifts are known to fly months together in air without touching the ground resting for only a brief span during flight with one eye open – half brained uni-hemispheric slow wave sleep. Bar tailed godwit birds, tracked through satellite tags, were found to fly from Alaska to New Zealand, non-stop without food or drink, a distance of 11,500 Kms in nine days! Humming birds can hover and remain in the same place in still air as long as they wish! Birds have no navigation aid for their flight! Birds from Siberia fly 4,000 kms to South India and return back without any external navigation equipment, just with star maps coded in their brains and their ability to perceive earth's magnetic field or subsonic noise of the ocean waves. Nocturnal prey birds like the owls fly and land in silence (with no sound) enabled by the serrated feathers on their wings and legs. Birds use less energy to fly – they purposefully enter clouds particularly the fluffy, white cumulus clouds to hitch a ride without using their own energy. A pilot avoids clouds, especially the cumulus clouds, as they would cause turbulence and icing of the aircraft. For all the abilities and elegance that birds possess, branding a less intelligent (stupid) human as having "Bird brain" is an insult to the birds after all!

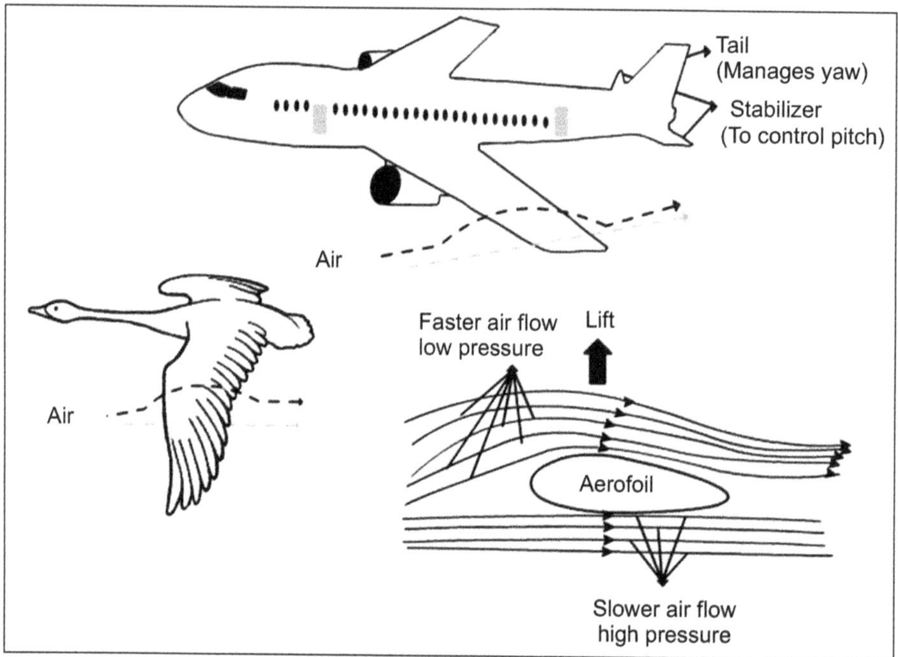

Figure 1.6: Mechanism of flight in the bird and an aeroplane. *Both the bird and the aeroplane lift in the air by raising their wings. In both, the wings are aerofoil in shape. The air flow is faster above the wing and slower below it. Due to this, a pressure difference is created between the top and bottom of the wing which results in lift. In the aeroplane, the pitch (ascending and descending) is controlled by the horizontal stabilizer and yaw (turning left or right) is managed by the rudder which is a hinge on the vertical wing (tail). A bird manages both pitch and yaw with its tail.*

Engineers are engaged in improvising the design of aeroplanes to combine all these magical features in the flight patterns of different bird species in the next generation aircraft to make air travel more energy efficient, travel farther without requirement to land for fuel, achieve more speed and less noisy and more sustainable. Several designs are in the pipe line like – **1.** morphing wings which can change shape to increase speed; **2.** wings with flapping motion; **3.** using ultra-light weight materials in the fabrication of aeroplane body; **4.** using solar energy and turbines while the aeroplane is in flight; **5.** wing and tail structures with individually controlled feather-like tips which help better control of flight through minimising drag; **6.** a smooth wing root – the point where the wing joins the fuselage – designed to imitate the graceful, aerodynamic arch of an eagle or falcon; **7.** fringed wings that muffle the noise of the wind passing over the wings and covered landing gears (inspired from owls) for reducing noise.

Aircraft companies are even planning to schedule flights in such a way that several planes travelling in the same direction form a V shape like that formed by flying flock of snow geese (**Figure 1.7**). Through this 'V' formation during flight, the follower geese expend less energy by surfing on the air movement created by the leader bird. The geese thus benefit from a free lift, which enables them to stay aloft with minimal fatigue over long distances. The birds change positions taking turn to lead the flight so that all in the flock benefit. Airbus fello'fly demonstrator project aims to prove the viability of this flight technique—known as "wake-energy retrieval" for commercial aircraft. Preliminary studies of Airbus fello'fly team have shown that this collaborative activity of flights destined to fly in the same direction to be scheduled at the same time, results in 5-10% savings in fuel per fello'fly trip.

Biological systems are the most successful control systems being the result of billions of years of evolution. The present-day species represent the most competent ones with the less smart ones having become extinct. There are quite a few remarkable bioinspired innovations: the ubiquitous, convenient sticker like the Velcro copied from the burr needles of the spiny amaranth plant, modelling the front of a bullet train as the beak of a kingfisher bird, mimicking cat eyes in the design of simple but highly utilitarian road studs that guide road travel during night.

Figure 1.7: Flock of snow geese in flight in a 'V' shape. *The birds at the back expend less energy as they surf on the air movement created by the other birds in front. The birds change position during flight with the ones at the back coming to the front taking turns to lead the flock. Through this strategy they succeed in flying farther.*

Biology indeed is a science to become familiar with, be it a layman, scientist, politician, or an engineer!!

1.6 SUMMARY

1. Biology is an interesting science about the living beings including us – the humans. The living world has a tremendous diversity in form, structure and habitat. Yet, the material which dictates their form and function (DNA – the genetic material or the blue print), the language used for dictating (genetic code) and the mechanism of duplication of the blue print are the same in all living beings. The energy currency in all living organisms is the same chemical molecule – ATP (Adenosine tri phosphate). The speed and mechanism of various functions executed by living organisms are wondrous and awe inspiring. They have a perfect design beyond much improvement. The living organisms serve as perfect models for design of efficient tools in the physical world. Study of biology helps in understanding how life operates and it dawns on one

the universality in life processes and the essential similarity between all living organisms. It also helps one, comprehend the ecological aspects of the living world and how the physical world is affected by natural forces – both physical and biological. Therefore, a study of biology is essential for everyone, whatever their avocation.

2. Observations in living organisms have led to propounding profound physical and chemical laws. Brownian movement – the random, rapid movement of particles which led to the proposition of kinetic theory of gases and computation of Avogadro number (number of molecules in one mole of a substance) was first observed by Robert Brown in the pollen grains of a flowering plant suspended in water. The insight about the conservation of energy and transformation of energy from one form to another, popularly known as the first law of thermodynamics, happened from the observation of the difference in blood colour between people in temperate and tropical climates by a Danish physician, Julius Robert Mayer.

3. Engineering existed much before logical scientific studies were undertaken. However, scientific knowledge greatly helps in devising better tools and structures. Engineering and science are interdependent. Advances in one serve to further improvements in another. Science studies things as they exist. Engineering designs things which were never before existent.

4. Several wonderous devises have been designed on biological models which is the art of biomimicry. Two impressive examples are the camera and an aeroplane. Camera was designed based on the structure of a human eye with mechanical components imitating their equivalent functional units in the eye. The mechanism of image capture in the human eye and the camera which is modelled on it is essentially similar. The structural form of an aeroplane was modelled mimicking a bird, the flight dynamics in both are one and the same. In both cases, the designed forms are much less efficient than the originals from which they have been adopted.

1.7 SAMPLE QUESTIONS

1.7.1 Subjective Questions

Q.1. Why is the study of Biology important and useful? List some intriguing aspects of living organisms.

Q.2. Explain how an aeroplane has been designed using the bird as a model. Discuss the similarities and differences between them. Summarise the aspects in which an aeroplane still does not match a bird's flight ability and whether it can be improvised and if so, how?

Q.3. How were the kinetic theory of gases and first law of thermodynamic emerge from observations made in Biology?

1.7.2 Objective Questions

Q.1. The genetic code is:
- (*a*) Universal and identical in all organisms.
- (*b*) Different in different organisms.
- (*c*) Different in plants and animals.   (*d*) Similar in organisms.

Q.2. The speed at which DNA is duplicated is very high. Therefore:
- (*a*) There are frequent mistakes in copying.

(*b*) Very few mistakes while copying which are carried to the daughter DNA.

(*c*) Very few mistakes while copying which are simultaneously corrected so that the daughter DNA is an exact replica of the original.

(*d*) There are no mistakes while copying.

Q.3. Science and Engineering are

(*a*) Interdependent for striding forward in both fields.

(*b*) Not related. (*c*) Mutually exclusive.

(*d*) Discrete fields.

Q.4. Brownian movement happens because of:

(*a*) Collision of molecules. (*b*) Floating of molecules.

(*c*) Collision of particles with molecules.

(*d*) Interaction between molecules.

Q.5. The aperture in the camera is equivalent to:

(*a*) The iris of human eye. (*b*) The pupil of human eye.

(*c*) The ciliary muscle of the eye. (*d*) The diaphragm of the eye.

Q.6. The image formed by a human eye is subjective because

(*a*) It is formed on the retina. (*b*) It is captured by the corneal lens.

(*c*) Light is focussed on the macula.

(*d*) The information is carried to the brain which interprets it.

Q.7. The forward movement during the flight of an aeroplane is achieved by

(*a*) Overcoming weight. (*b*) Lifting the wings.

(*c*) Using the horizontal tail bar. (*d*) Engine power.

ANSWERS

1. (*a*) **2.** (*c*) **3.** (*a*) **4.** (*c*) **5.** (*b*) **6.** (*d*) **7.** (*d*)

2 Classification and Model Organisms

At the beginning of its existence as a science, biology was forced to take cognizance of the seemingly boundless variety of living things for no exact study of life phenomena was possible until the apparent chaos of the distinct kinds of organisms had been reduced to a rational system. Systematics and morphology, two predominantly descriptive and observational disciplines took precedence among biological sciences.

~ Theodosius Dobzhansky

There is a finite number of species of plants and animals – even of insects – upon the earth. The universality of the genetic code, the common character of proteins in different species, the generality of cellular structure and cellular reproduction, the basic similarity of energy metabolism in all species and of photosynthesis in green plants and bacteria and the universal evolution of living forms through mutation and natural selection all lead inescapably to a conclusion that although diversity may be great, the laws of life, based on similarities, are finite in number and comprehensible to us in the main even now.

~ H. Bentley Glass

2.1 INTRODUCTION

The earth is the only planet, as we know today, which has life. It is the home for more than 8 million species, with new species being constantly added, as they are discovered, specially, in the tropics. Living organisms are diverse. There is an overwhelming biodiversity on earth. Living beings exist everywhere on earth – land and water, in the myriad terrains and water bodies with infinite ecological niches. They adapt to the surrounding conditions so well that, even the most seemingly harsh unsuitable environments appear worth habitable. They survive against all odds of environmental severity and continue their species through reproduction, followed by diligent parenting in some animal species. Microscopes have opened up a world of an awe-inspiring even more overwhelming picture of the enormous variety of microbes that exist on earth.

From fossil records, it is estimated that life originated on earth about 3.5 billion years ago. Of course, the living beings have constantly changed over time. About 99.9 % species that once inhabited the earth have now become extinct. It is only the microbes which have survived from the beginning of life on earth to the present day. Darwin's theory of evolution that is substantiated by vast evidence is the tenet of biology. Life is a constant process of evolution through blind experimentation; survival of the fittest is the law that prevails in the living world.

Understanding life processes in all the innumerable life forms on earth is a task beyond imagination. But why understand the living beings? It is through this knowledge, that man can make his living more comfortable and learn to live in balance with nature, appreciating the importance of the various life forms in maintaining the living planet – earth. Living beings show similarities and differences – miniscule to enormous. If organisms with close similarity are sorted out and grouped together – the task called classification, the number of groups would be far fewer than the number of species. Such classified groups can be characterized as one unit and all individuals of that group can be approximated to be of that type. Study of living beings thus becomes easier and faster.

Classifying organisms, a field called taxonomy, began from 1700's with the seminal work of Carl Linnaeus. While Linnaeus' principles still form a framework for hierarchical classification of living organisms, more criteria have since been added for assessing similarity between organisms, the latest being genetic features. Thus, the present-day classification is described as molecular taxonomy and reflects the evolutionary (phylogenetic) relationships between different groups.

2.2 PURPOSE OF CLASSIFICATION

Classification is a means by which study of ~8.7 million species recognized at present on earth becomes manageable. Classification involves grouping organisms based on some level of similarity. Therefore, if one or a few representative members of the group are studied in detail, it encompasses the study of all the organisms in that group. Classification helps to identify an organism. It enables to understand the relationship between organisms of different groups. It helps to study the evolutionary history of living organisms. Classification allows to identify, group and properly name organisms through a standardized system. The conventional means of giving a universal name for an organism is the 'binomial nomenclature' – giving two names to an organism: the first is the name of genus and the second, the species name. Thus, humans are named *Homo sapiens*. Such naming avoids confusion about the reference of an organism called differently in different languages.

2.3 HIERARCHY OF LIVING FORMS

Hierarchical structures are highly stable because they are organized. Simple systems evolve into complex forms. On this premise, it is but implied that complex systems are hierarchies. Living organisms are no exception to this. They are highly organized and structured in a hierarchical manner.

Lower levels of organization are progressively integrated to make up higher levels. Each hierarchical level has a higher organizational complexity than its previous lower level. Yet another significant feature of this organization is that each higher level has novel features not found in its immediate lower level. These are described as 'emergent' properties that arise due to the assemblage of the units of lower level of hierarchy. The whole is thus not a sum of its parts.

2.3.1 Levels of Hierarchy in the Living World

The living world is organized in twelve hierarchical levels starting with atom to the biosphere. It extends from: atom – molecule (– macromolecule) – cell organelle – cell – tissue (– tissue system) – organ – organ system – organism – population – community – ecosystem and biosphere (**Figure 2.1**).

Atoms are the smallest and basic fundamental blocks of matter. They form the base of the organizational structure of both living organisms and non-living things.

Atoms such as carbon, oxygen, hydrogen, nitrogen come together to form molecules. Molecules are chemical structures that have at least two atoms but can have more of them held together by one or more chemical bonds. For example, a molecule of water is made up of one atom of oxygen bonded to two atoms of hydrogen, with each atom arranged in a very precise way to form a bond angle of 104.5°. Most biologically important molecules – the biomolecules are macromolecules. Macromolecules are large molecules that are usually formed by polymerization of several monomers. For example, the biomolecule Deoxy ribonucleic acid (DNA) is a polymer of four kinds of nucleotide monomers each of which in turn is made of several kinds of molecules. A protein is a macromolecule formed by polymerization of several monomers constituted by 20 different kinds of amino acids.

Macromolecules like lipids and carbohydrates aggregate into membranes that form organelles by enclosing several other macromolecules within them. Organelles are small structures that exist within cells. For example, mitochondria and chloroplasts are cell organelles that carry out indispensable functions for life. Mitochondria, referred to as power houses of cells, are the organelles of cellular respiration generating power for the cell. In green plants, chloroplasts are the chief organelles for photosynthesis that trap solar energy to be stored as chemical energy in the sugars that are synthesized by them. Several such organelles assemble together to form a cell (**Figure 2.1**).

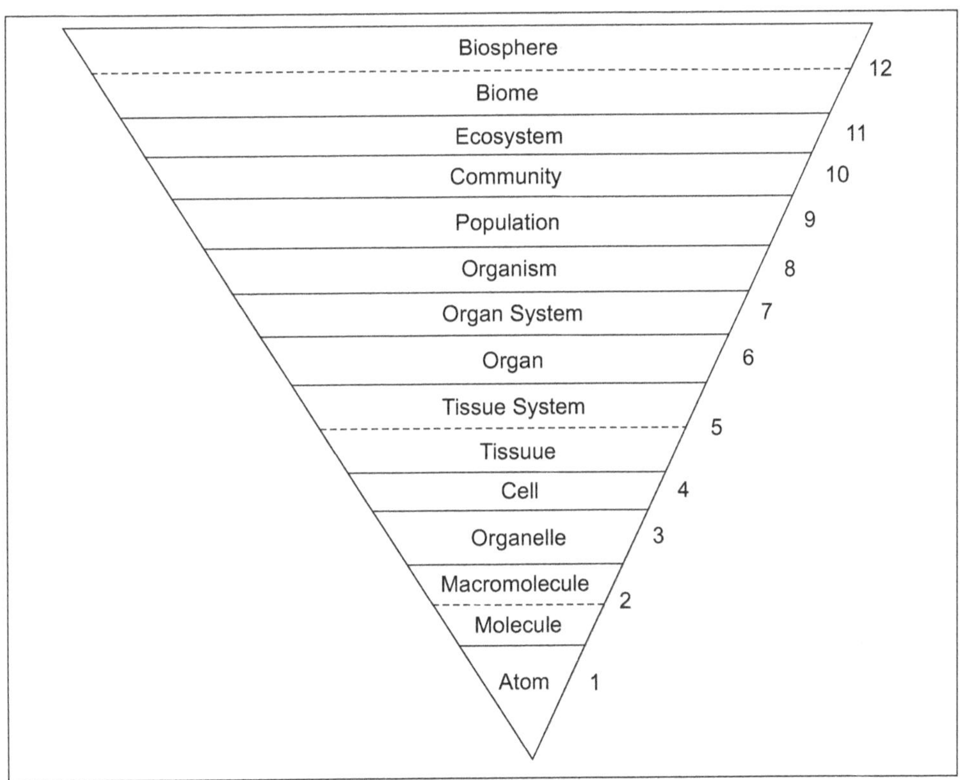

Figure 2.1: *Hierarchical organization in the living world.*

An organism can be unicellular or multicellular, simple or complex. In multicellular organisms, the cell is the smallest fundamental unit of structure and function. Cells that have similar structure aggregate to form tissue. Tissues designed to do a similar function organize into tissue system. There are four different types of tissues in animals: connective, muscle, nervous, and epithelial. In plants, tissues are divided into three types: vascular, ground, and epidermal. Tissues organize to form organs like the skin, liver, kidney, heart, etc. that perform specific functions. The organs that work together to perform a specific task e.g.

excretion, digestion etc. form an organ system. For instance, the circulatory (cardiovascular) system transports blood throughout the body and to and from the lungs; it includes organs such as the heart and blood vessels comprising of the arteries and veins. Organ systems performing different functions like the nervous system, digestive system, excretory system, circulatory system etc. together constitute an organism (**Figure 2.1**).

Organisms are individuals of a species, which can grow and multiply. A group of organisms of the same species living within a specific defined area constitute a population. For example, all the foxes in a forest constitute a fox population. Several different populations living in a particular location form a community. For example, population of foxes, tigers, lions, deer, bears, cheetahs, wild dogs, hyenas, sisham, teak, babul and pine trees etc. in a forest constitute a community. Several different communities interacting with each other and with the physical world (climate, soil and water) around them constitute an ecosystem. The vegetation and the animal and microbial communities living within a forest constitute the forest ecosystem. An ecosystem consists of all the living things in a particular area together with the abiotic, non-living parts of that environment such as soil and rain water. There are several types of ecosystems – pond ecosystem, river ecosystem, marine ecosystem, desert ecosystem etc. Ecosystems with similar climatic conditions constitute a biome. A congregation of all the biomes on earth constitute the top of the pyramid of hierarchy which is the biosphere. It the highest level of organization of living organisms constituting all the living organisms on earth – its surface, below its surface and in the atmosphere (**Figure 2.1**).

Hierarchy extending from an atom to the biosphere is detailed with an example in **Table 2.1**. A systematic exploration of the hierarchy in biological systems would illuminate on the question of what is life.

Table 2.1: Hierarchical organization in the biological world

Level of Hierarchy	Example	Hierarchy Type
12	All living organisms on and below the surface of land, water and atmosphere of earth.	Biosphere
	All terrestrial ecosystems of tropical zone.	Biome
11	Terrestrial	Ecosystem
10	Human, plant, animal and microbial populations in a particular location	Community
9	Human population	Population
8	Man	Organism
7	Nervous system	Organ system
6	Brain	Organ
5	Nervous tissue, connective tissue	Tissue system
	Nerve tissue	Tissue
4	Nerve cell (neuron)	Cell
3	Nucleus	Organelle
2	Deoxyribo Nucleic Acid (DNA)	Macromolecule
	Adenine, Thymidine, Guanine, Cytosine, Deoxyribose, Phosphate	Molecules
1	C, N, O, H, P	Atoms

2.3.2 Taxonomic Hierarchy

Just as the hierarchal organization of the biological world, classification of the living forms is patterned in a hierarchy. There are eight taxonomic units starting from the species to the domain (**Figure 2.2**) The smallest recognized taxonomic unit is the species. Individuals of the same species can mate. The next higher taxonomic level is the genus which may have two or more species. Species within a genus are reproductively isolated i.e., they cannot mate – they are reproductively incompatible. Several related genera are placed in one family, related families are grouped into an order. A class consists of orders with some level of similarity. Several classes with some features in common are constituted into a phylum. Related phyla are clustered into a kingdom. The domain is at the highest level in taxonomic hierarchy (**Figure 2.2**)

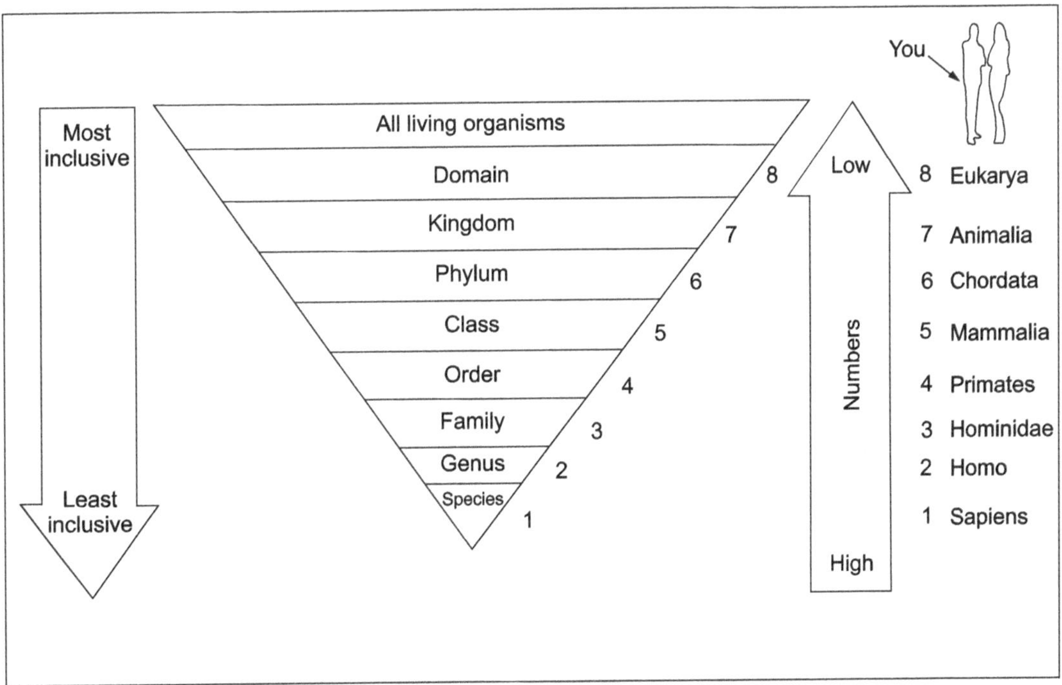

Figure 2.2: Currently followed taxonomical hierarchy (slightly modified form of Carl Linnaeus). *The higher hierarchical group includes more than one of the previous group. The number of members within a hierarchical unit increases with the decrease in the level of hierarchy. The number of taxonomic units at each hierarchical level increases from the top of hierarchy downwards.*

In taxonomical hierarchy, the number of different organisms in a unit decreases from the lowest level to the highest (**Figure 2.2**). There are a greater number of species than genera and more families than orders and so on. The number decreases because the upper taxonomic unit includes all the members in the lower taxonomic level, i.e., the taxonomic units become more inclusive from lower to higher levels (**Figures 2.2, 2.3**).

Domain Eukarya	Man Chimpanzee	Gorilla Orangutan	Monkey Baboon Langur	Lion Bear Tiger	Dog Cat Mouse	Fish Frog Parrot Snake	Insects Worms Molluscs Jelly fish	Fern Pine Rose Oak	Bacteria Amoeba Euglena Fungus
Kingdom Animalia	Man Chimpanzee	Gorilla Orangutan	Monkey Baboon Langur	Lion Bear Tiger	Dog Cat Mouse	Fish Frog Parrot Snake	Insects Worms Molluscs Jelly fish		
Phylum Chordata	Man Chimpanzee	Gorilla Orangutan	Monkey Baboon Langur	Lion Bear Tiger	Dog Cat Mouse	Fish Frog Parrot Snake			
Class Mammalia	Man Chimpanzee	Gorilla Orangutan	Monkey Baboon Langur	Lion Bear Tiger	Dog Cat Mouse				
Order Primates	Man Chimpanzee	Gorilla Orangutan	Monkey Baboon Langur						
Family Hominidae	Man Chimpanzee	Gorilla Orangutan							
Genus Homo	Man								
Species Sapiens	Man								

Figure 2.3: Inclusiveness of living forms decreases from higher order of taxonomic hierarchy to the lower one. *Demonstrated taking man as an example.*

2.4 CLASSIFICATION OF LIVING ORGANISMS

Living organisms can be classified in various ways taking a single feature of interest depending on the purpose of investigation. However, authentic taxonomic classification of organisms takes into account as many features as possible relating to morphological, ecological, biochemical and most importantly, genetic features for categorizing organisms into groups. Below are discussed the several ways living organisms can be classified.

2.4.1 Classification Based on Cellularity

Anton van Leeuwenhoek's observations with the microscope he devised, opened the doorway to the world of single-celled organisms. In 1674, he described several single-celled organisms which he called 'animalcules'. They were the protozoans (amoeba). With the previous (1655) discovery of cells by Robert Hooke, it was realized that cell is the basic structural and functional unit of all living organisms. It soon became evident that a substantial number of living organisms are single-celled. These single-celled organisms have great diversity. Examples of single-celled organisms are the bacteria, yeast, *Chlamydomonas*, *Spirulina*, amoeba, *Plasmodium*, *Paramecium*, *Euglena* etc. Since they are very small and microscopic, they are described as microbes. Their study is called microbiology. Single-celled organisms carry out all the functions required for life. The single-celled organisms evolved into multicellular forms. These evolutionary steps can be traced in the algal group – Volvocales (**Figure 2. 4**). In this group, some species like *Chlamydomonas* are single-celled. Species like *Eudorina* and *Pandorina* are colonial forms with a group of *Chlamydomonas*-like-cells coming together in a mucilaginous sheath. The next level of sophistication is seen in *Volvox*. *Volvox* is a colonial organism with division of labour among the cells in the colony. It is about as simple as a multicellular organism can be. It is a spheroid with about 95% of its volume made of transparent extracellular matrix. It consists of only two kinds of cells – 2000 to 4000 (depending on the size of the colony) small, biflagellate vegetative or somatic cells on the surface of the spheroid, and about 16 large, potentially immortal reproductive cells just internal to the somatic cell layer. The *Volvox* colony is visible to naked eye being up to 2mm in diameter and is seen rolling in fresh water bodies. (**Figure 2.4**).

Multicellular organisms can be very simple like hydra (animals), sea anemones or sea weed (macro alga with no differentiation of stem, root and leaves), to highly complex organisms like birds, mammals, man, the oak, mango and neem trees. In multicellular organisms, the cells organize into tissues and show hierarchical organization as discussed above. The cells in multicellular organisms have less stress because they do not need to perform all functions of life. A lower stress means a longer life span. However, being multicellular makes the organism take a longer time to mature and reproduce.

Based on this feature of cellularity, the entire gamut of living organisms can be classified into: **1.** single or unicellular, and **2.** multicellular organisms.

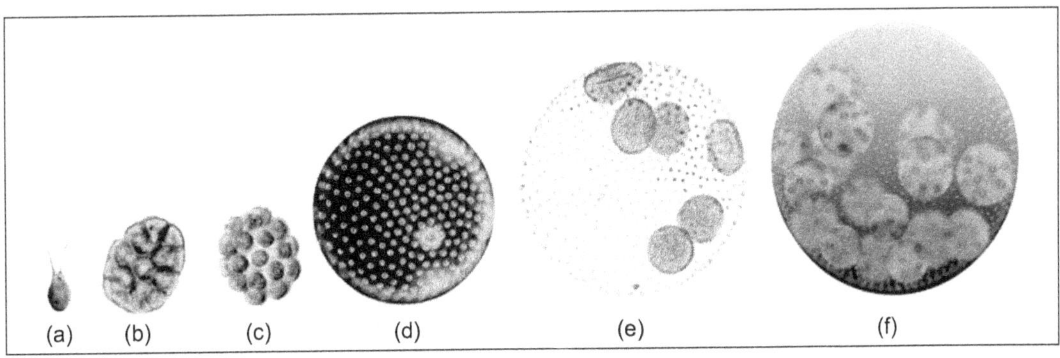

Figure 2.4: Evolution of single-celled organism to a colonial form with the simplest division of labour among the cells of the colony observed in Volvocales. a. *Chlamydomonas, a single-celled organism.* **b & c**. *Pandorina and Eudorina – colonial forms made of several single-celled organisms each leading its independent life except for staying together.* **d to f**. *Volvox, a macroscopic colonial form with up to 4000 individual Chlamydomonas-like-cells. The cells in the colony are of two types: majority of the cells are vegetative cells which enclose ~16 cells that take part in reproduction called the germ cells. The division of labour among members of the Volvox colony is the first step towards evolution of multicellular organisms.*

2.4.2 Classification Based on Ultrastructure of the Cell

The internal organization of the cell unfolded after the installation of electron microscope in mid 1940's. It became clear that there are two types of cell organization. Minute, single-celled organisms, commonly called the bacteria, were found to have a simple cell organization. The cells of larger single-celled organisms and those of the several larger plants and animals had a complex organization. Based on the organization of their cellular structure, all living organisms can be classified into two groups: Prokaryota and Eukaryota. The difference between the two is profound – far greater than that between an amoeba, a plant and an animal. The differences between them are listed in **Table 2.2.**

The classification of living organisms into two broad categories of Pro- and Eu- karyota was introduced in the late 1930's and universally accepted by biologists until recently. Prokaryotes have a simple cell structure. The cell is bound with a membrane enclosing a nucleoid – a thread folded like a spaghetti which constitutes its genetic material (naked DNA). The eukaryotic cell has its genetic material enclosed in a double membrane body called the nucleus and a few bacteria like structures in addition to a few membrane bound bodies and a network of membranes, filaments and fibres. Besides difference in the cell organization, the pro- and eu-karyotes differ in their gene structure and organization and in their biochemistry (**Table 2.2**). The structure of a cell of a prokaryote and a eukaryote are shown in **Figure 2.5.**

Table 2.2: Differences between prokaryota and eukaryota

S. No.	Prokaryota	Eukaryota
1.	Small cell size of 0.1 to 5 microns.	Larger cells of 10 to 100 microns.
2.	Large cell surface to volume ratio and have faster metabolic rate, higher growth rate and a shorter generation time (of less than an hour).	Cells are voluminous and have a longer generation time lasting a day, week to years.
3.	Are unicellular organisms.	Some are unicellular organisms but a large majority are multicellular organisms.
4.	Cell wall is made of peptidoglycan called murein.	Cell wall is present only in plant cells made up of cellulose and other carbohydrates. Other eukaryotes including animals do not have a cell wall and only enclosed by a cell membrane – the plasma membrane.
5.	Genetic material is constituted by a circular DNA within the cell. DNA is not closely associated with proteins and therefore described as naked. The region in which it is present is termed the nucleoid.	DNA is densely packed in association with proteins to form chromatin which is enclosed within a double layered membrane. This membrane bound structure is called the nucleus.
6.	Have a simple cell skeleton with two kinds of proteins organized as a circular filamentous ring in the centre of the cell.	Have an elaborate cytoskeleton made of three structural elements: microtubules, microfilaments and intermediate filaments.
7.	The cell has no membranous structures.	The cell encloses several membrane bound structures called organelles: Endoplasmic reticulum Mitochondria Golgi apparatus Lysosomes Peroxisomes Chloroplasts (in photosynthesizing organisms) Vacuoles (not frequent in animal cells) Centriole (in animal cells)

8.	Ribosomes are of 70S type and have 23S, 16S and 5S rRNA.	Ribosomes are 80S and have 28S, 18S, 5.8S and 5S rRNA.
9.	The genome size is small ranging from 500kb to 12Mb.	Bigger genome size ranging from 10 Mb to > 100,000 Mb.
10.	Most of the DNA is coding.	A large part of DNA in majority of the eukaryotes has no known function.
11.	Genes related to one function are together and expressed as a unit called the operon.	Genes related to one function are highly dispersed and expressed individually.
12.	Most genes are uninterrupted and entirely coding.	Most genes are interrupted by non-coding regions.
13.	Non-essential genes are harboured in small structures called plasmids.	Plasmids are rare in eukaryotes. Present only in single-celled eukaryotes like yeast.

kb = kilo (1,000) base (nucleotide) pairs, Mb= million base pairs of DNA.

There is evidence that the first life forms were the prokaryotes with a simple cell structure; they existed on earth since at least 3.8 billion (380,000,000) years. The eukaryotes evolved from the prokaryotes much later, about 2.7 billion years ago. The basic divergence in cellular structure represents the greatest single evolutionary discontinuity to be found in the living world. In 2012, a single-celled organism named *Parakaryon myojinensis* was discovered in the deep sea, off the coast of Japan which was reported to have intermediate characters between a prokaryote and a eukaryote. It is more than 100 times the size of the bacterium *E. coli* with naked DNA enclosed in a single membrane within the cell. The cell of this organism has some bacteria like inclusions but no cell organelles.

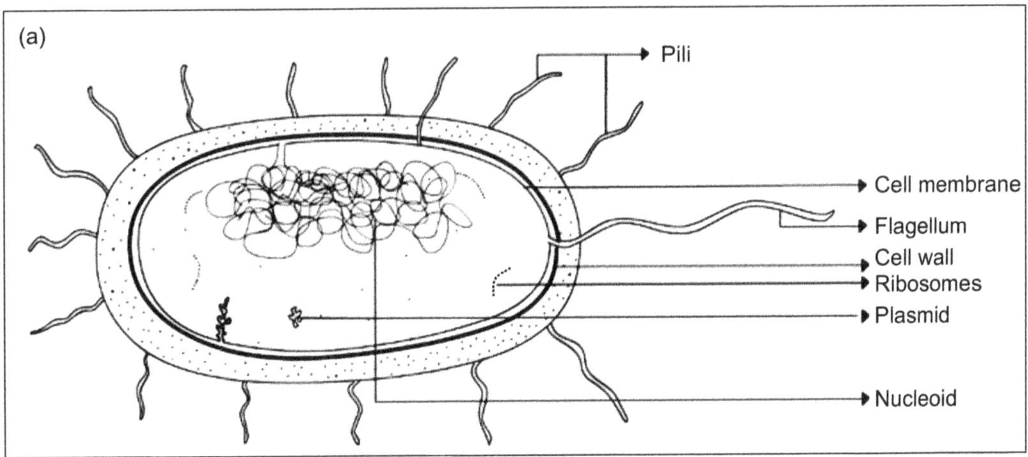

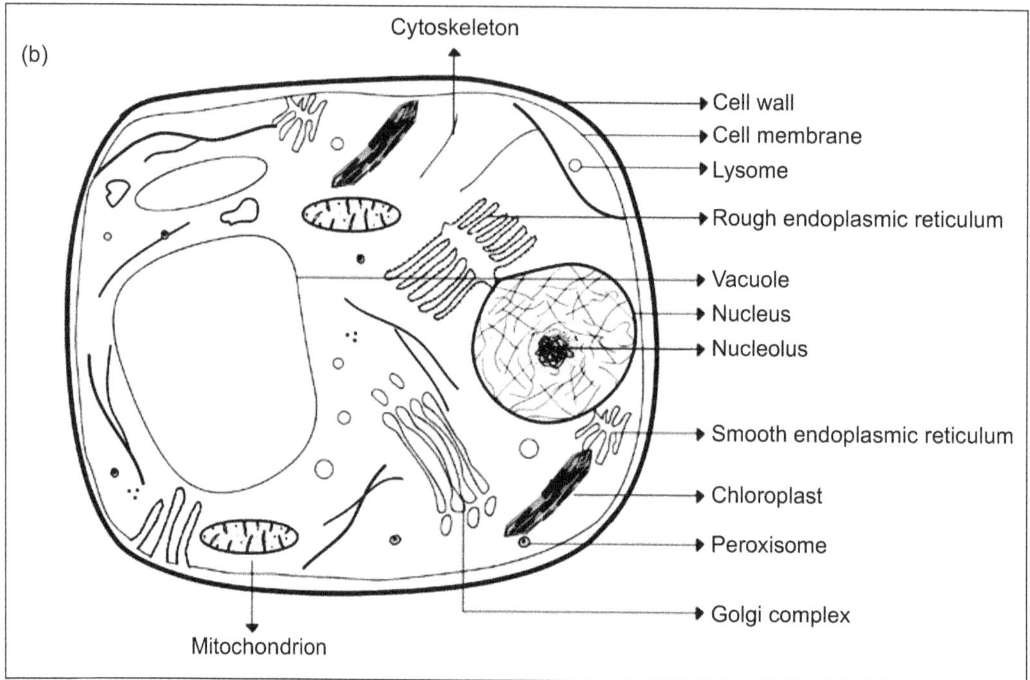

Figure 2.5: Diagrammatic representation of a prokaryotic and a eukaryotic cell. a. *Prokaryotic cell with a nucleoid.* **b.** *Eukaryotic cell (plant cell) with a membrane bound nucleus and several membrane bound organelles.*

2.4.2(a) Origin of Eukaryote from Prokaryote – The Endosymbiont Theory

In late 1960's, Lynn Margulis proposed the endosymbiont theory to explain the origin of a eukaryotic cell. Though initially it was not taken, at present there is overwhelming evidence in favour of it, to accept it as the most plausible route through which a eukaryote evolved from a prokaryote. The hypothesis of the endosymbiont theory is as follows:

The origin of the eukaryote happened in colonies of prokaryotes with different types of prokaryotes. Though these prokaryotes were capable of living independently, organizing into a colony endowed them perhaps more protection and also served as a means of obtaining more energy.

Living in proximity, the members within the colony went one step further in cooperation for mutual benefit (symbiosis). An anaerobic prokaryote which had become larger due to some enclosed membranes formed due to invagination of its cell membrane, engulfed (through endocytosis) an aerobic bacterium. The engulfed bacterium was not digested but retained as an inclusion within the larger anaerobic prokaryote. The engulfed bacterium has two membranes – the outer membrane is the cell membrane of the anaerobic bacterium and the inner membrane is that of its own (**Figure 2.6**).

When the larger prokaryotic cell which engulfed the smaller prokaryote divided, the smaller cell within it also divided so that both the daughter cells had a copy of the engulfed prokaryotic cell. When this cell lived in conditions where oxygen was available, the relationship between the two became symbiotic. The larger prokaryote is anaerobic and oxygen is toxic for it. The smaller prokaryote harboured within it is aerobic, and used up the oxygen for its respiration thereby protecting the anaerobic prokaryote from the surrounding oxygen which is toxigenic to it. In turn, the aerobic prokaryote is provided a protective environment to inhabit within the anaerobic prokaryotic cell. Eventually, the smaller prokaryotic cell lost its capability to live independently and became fully committed to a symbiotic life, and in the process, it specialised to perform a single function – of energy generation through oxygenic respiration. The aerobic prokaryotic cell that started off as an endosymbiont metamorphosed into the mitochondrion. A eukaryotic cell originated thus with a nucleus, a mitochondrion and more compartmentalisation with the internalised membrane system (**Figure 2.6**).

At a later stage, a photosynthetic prokaryote (a cyanobacterium – a bacterium which can photosynthesise due to the presence of light capturing pigments) established itself as an endosymbiont in the cell with a nucleus, mitochondrion and internal membrane system in a route similar to that taken earlier by the aerobic prokaryote. This new endosymbiont dedicated itself to the function of photosynthesis and transformed into the chloroplast (**Figure 2.6**). Thus, two types of eukaryotic cells arose: the plant cell with an organelle for synthesis of food (chloroplast) and one for providing energy (mitochondrion) and the animal cell with only a mitochondrion (**Figure 2.6**).

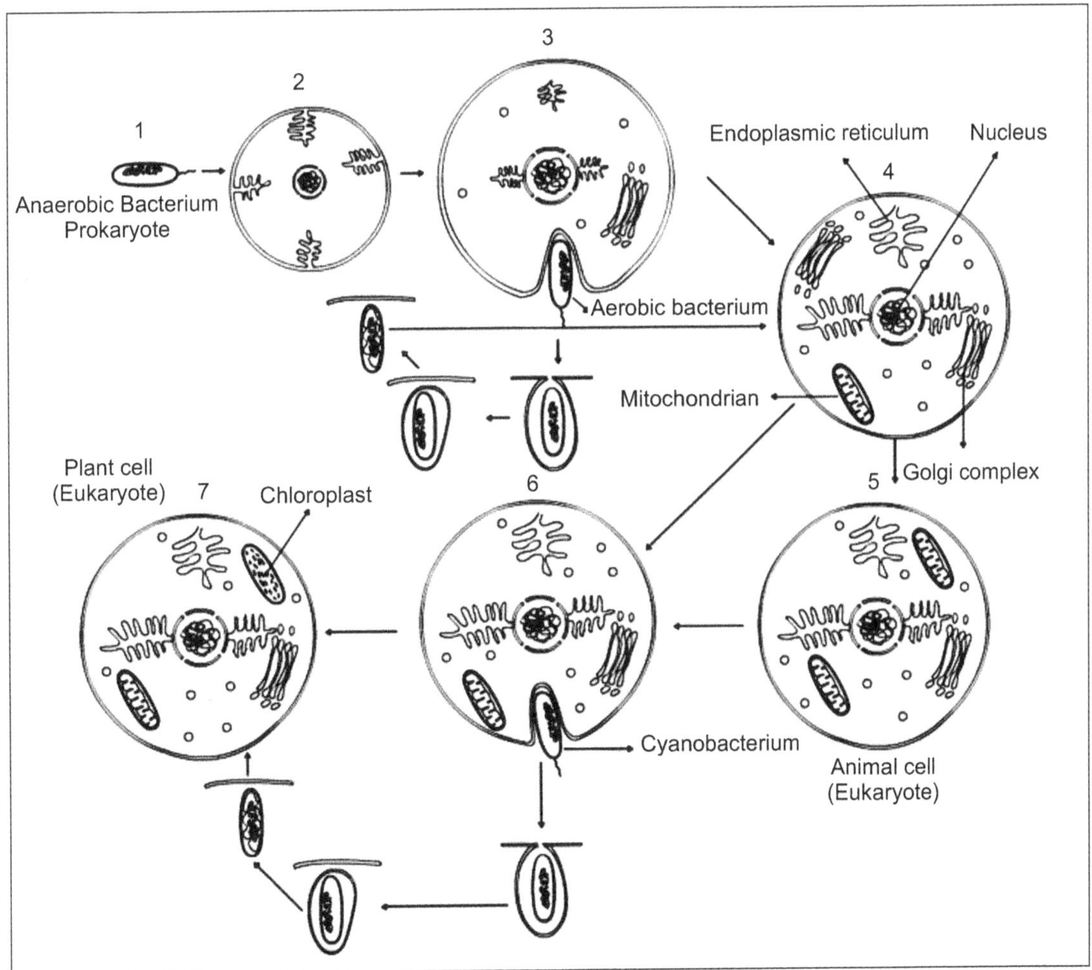

Figure 2.6: The Endosymbiont theory explaining the origin of eukaryotic cell due to symbiotic association between prokaryotes: an anaerobic bacterium that developed an internal membrane system with an aerobic bacterium and a cyanobacterium. *The aerobic bacterium transformed into the mitochondrion and the cyanobacterium, into the chloroplast in a plant cell.*

There are multiple independent lines of evidence for the endosymbiont theory about the origin of a eukaryote from a prokaryote. The mitochondrion and chloroplast in the eukaryotic cell bear an uncanny resemblance to a prokaryote. The similarities between these two important cell organelles – mitochondrion and chloroplast of a eukaryotic cell with a prokaryote are listed in **Table 2.3.**

Table 2.3: Similarities between mitochondria and chloroplasts of a eukaryotic cell and a prokaryote

Feature	Similarity
Size	Both mitochondria and chloroplasts are similar in size to a bacterium (prokaryote).
Multiplication	Mitochondria and chloroplasts divide through a process of binary fission similar to a prokaryote.
DNA	Mitochondria and chloroplasts have naked circular DNA just like a prokaryotic cell.
Ribosomes	Mitochondria and chloroplasts have 70S ribosomes as in a prokaryotic cell.
Sensitivity to antibiotics	Mitochondria and chloroplasts are affected by the antibiotics targeted to prokaryotic bacteria.
Membrane composition	The membrane composition of mitochondria and chloroplasts is more similar to prokaryotic membranes than to eukaryotic membranes.

2.5 CLASSIFICATION BASED ON ENERGY AND CARBON UTILIZATION

Energy is a prerequisite for all activity. Life is a manifestation of energy. Based on the type of nutrients that living organisms utilize for obtaining energy, all living organisms can be classified into two groups: Autotrophs and Heterotrophs. The autotrophs use inorganic nutrients and carbon di oxide and synthesise sugars which are the source of energy. The heterotrophs feed upon the autotrophs for their nutritional requirements. The autotrophs are able to synthesise organic compounds from inorganic substances through either photosynthesis or chemosynthesis. Autotrophs which have chlorophyll utilize sunlight as the source of energy to convert CO_2 and water to carbohydrate through a process called photosynthesis. Some bacteria can utilize inorganic substances like elemental sulphur, hydrogen sulphide, molecular hydrogen, ammonia, manganese, or iron for obtaining energy to convert CO_2 into carbohydrate. The energy they utilize is chemical energy released through breaking of bonds in the inorganic compounds. This process is termed chemosynthesis. The classification based on nutrition can be further refined using three parameters together – the source of: **a.** energy (sun/chemical: photo/chemo), **b.** carbon (organic/inorganic: auto/hetero) and **c.** electrons (organic/inorganic: organo/litho) (**Table 2.4**). There are four subclasses each in autotrophs and heterotrophs (**Table 2.5**).

Table 2.4: Classification of living organisms based on mode of nutrition

Source of:			Group
Energy	**Carbon**	**Electrons**	
Sunlight (phototroph)	Inorganic (lithotroph)	Inorganic (autotroph)	Photolithoautotroph e. g. All plants
		Organic (heterotroph)	Photolithoheterotroph
	Organic (organotroph)	Inorganic (autotroph)	Photoorganoautotroph
		Organic (heterotroph)	Photoorganoheterotroph

	Inorganic (lithotroph)	Inorganic (autotroph)	Chemolithoautotroph (Exclusively prokaryotes)
Chemical (chemotroph)		Organic (heterotroph)	Chemolithoheterotroph (Exclusively prokaryotes)
	Organic (organotroph)	Inorganic (autotroph)	Chemoorganoautotroph
		Organic (heterotroph)	Chemoorganoheterotroph e.g. Herbivorous and carnivorous animals, decomposing bacteria, fungi and amoeba

Table 2.5: Subdivisions within the major nutritional divisions of autotrophs and heterotrophs.

1. Photolitho
2. Photoorgano
3. Chemolitho
4. Chemoorgano

This classification does not aptly classify organisms because organisms sometimes switch their nutritional mode depending on the availability of nutrients.

2.6 CLASSIFICATION OF ANIMALS BASED ON EXCRETORY PRODUCT

Animals can store carbohydrates and fats but cannot store proteins and amino acids. The amino acids that are released due to breakdown are deaminated; the amino group is excreted and the organic acid is utilized as an energy source or converted to carbohydrates and fats. The proteins, the amino groups and, nitrogen bases released during the metabolism of nucleic acids, constitute the nitrogenous waste that is excreted through urine and faecal matter. Depending on the form of the nitrogenous waste excreted, animals are classified into five groups three of which are major groups (**Table 2.6**), and two minor. The two minor groups are aminotelic and guanotelic. The aminotelic animals like some molluscs (e.g. snails and oysters), excrete amino acids as such. Spiders are guanotelic, they excrete guanine, a nitrogenous base component of nucleic acids (DNA and RNA).

Table 2.6: Classification of animals based on the nitrogenous excretory product (major groups)

Group type	Excretory product			Examples
	Type	Toxicity	No. of Nitrogen atoms per molecule (Water requirement for excretion)	
Ammoniotelic	Ammonia	Highly toxic	One (0.5 L /g ammonia)	Most aquatic animals, Earth worms.
Ureotelic	Urea	Less toxic	Two (0.05 L/g urea)	Mammals and Amphibians

Uricotelic	Uric acid	Least toxic	Four (0.01 L/ g uric acid)	Reptiles, Birds and Insects.

2.7 CLASSIFICATION BASED ON HABITAT

The earth basically has two main surfaces – land and water. Based on their habitat (environment in which an organism lives), living organisms can be broadly classified as terrestrial and aquatic. There are several types of terrestrial habitat like deserts, mountains, grasslands, forests etc. Similarly, aquatic habitats are of several types: ponds, lakes, rivers, streams, marshes, bogs, seas, oceans, estuaries etc. The aquatic habitat is broadly of two types: freshwater and marine (salt water). There are also regions of extreme conditions like deep ocean vents and cold seeps, hot springs, extremely saline ponds, Tundra and Taiga regions in the Arctic circle and the Antarctica. No region on earth is devoid of life. There are organisms which inhabit even the most hostile of environments. Such organisms (predominantly microbes) are resilient in these conditions because they have adapted to the environment. Living beings can thus be classified variously based on their habitats. Animals are classified as listed in **Table 2.7** and plants as in **Table 2.8.**

Table 2.7: Classification of animals based on habitat

Group type	Examples
Terrestrial	Humans, Cat, Dog, Tiger, Fox
Aerial	Birds, Insects
Arboreal	Monkeys, Lizards
Aquatic	Fish, Whale, Otters
Amphibious	Frogs, Toads

Table 2.8: Classification of plants based on habitat

Group type	Habitat	Examples
Terrestrial		
(I) Mesophytes	Land with sufficient water.	Neem, Mango, Shoe flower
(II) Xerophytes	Land with insufficient water.	
1. Psammophytes	Shifting sand in deserts, on sea shore and river banks	Sand fescue, Sharp leaved willows
2 .Lithophytes	Rock and gravel	Lichens, Algae, Moss
3. Halophytes	Saline soils	Mangroves
4. Psychrophytes	Cold soils	Lichens, Pine trees
5. Eremophytes	Deserts	Cacti, Succulents
6. Oxylophytes	Acid soils	Bog mosses
7. Sclerophytes	Bush lands	Acacia, Eucalyptus
(III) Hygrophytes	Excessive moisture	Cyperus, Insectivorous plants

Aquatic		
(IV) Hydrophytes	Fresh water	Water lily, Hydrilla
	Marine	Sea weeds, Sea grasses, Kelp

Animals and plants in aquatic habitat are classified as benthic and planktonic depending on whether they are attached to the bottom of the water body or free floating.

In all the various types of classifications of living organisms discussed above, organisms which group together in one classification can fall apart in another. For example, yeast and bacteria come into one group – unicellular organisms when classified based on cellularity. When classified based on ultrastructure of cell, bacteria belong to Prokaryota while yeast belongs to Eukaryota. When classified based on habitat, all microbes, plants and animals living in water come under one group – aquatic, while when classified by other criteria, for example, on their carbon source, they separate into different groups as heterotrophs and autotrophs. Thus, the above discussed classifications reflect similarity between organisms within a group with respect to only one main character.

2.8 CLASSIFICATION BASED ON GENETIC RELATEDNESS – MOLECULAR TAXONOMY: THREE DOMAIN CLASSIFICATION

In the early 1970's, scientists discovered prokaryotes that lived in extreme environments like hot springs, salt evaporation ponds, ocean vents, acid lakes and dead sea. These extremophilic prokaryotes showed more similarities with the eukaryotes than to the bacteria. Carl Woese tried to classify the living organisms to include them. He used molecular data – the sequence of the gene coding for rRNA of the ribosomal small subunit (16S/18S ribosomal RNA gene in the prokaryotes and eukaryotes respectively) for classification. This gene sequence served as a chronometer to trace evolutionary relationships between all living organisms. Based on the analysis of this gene sequence, Woese showed that the living organisms fall into three major groups. He gave the status of 'domain' to these groups. The position of different organisms in this three-domain system of classification as revealed by the analysis of small subunit rRNA gene is shown in **Figure 2.7**.

The three domains are: Bacteria, Archaea (mostly extremophilic prokaryotes) and Eukaryota. The bacteria are the true prokaryotes and therefore also referred to as Eubacteria. Some of the groups in Eubacteria are the Cyanobacteria (photosynthesizing bacteria or blue green algae), Spirochetes (Gram negative bacteria) and Actinobacteria (Gram positive bacteria). The cell structure of members of Archaea is prokaryotic in that there is no nuclear membrane but they have many resemblances to eukaryotes in the structural organization of the gene and several of their metabolic path ways. This phylogenetic classification shows evidence that eukaryote arose from the archaea (**Figure 2.7**). The archaea are the oldest of the living organisms and they can be distinguished from eubacteria superficially from their habitat – thriving in extreme conditions. Some of the groups in Archaea are: halophiles, thermophiles, thermoacidophiles and methanogens. Some archaea live under normal temperatures and salinity; they are also found in human gut. The cell membranes of archaea are different from bacteria and eukaryote. They contain branched or ringed hydrocarbons. The archaea are not affected by some of the antibiotics that kill bacteria. The Eukaryota includes the kingdoms of Plantae, Animalia, Fungi and Protista. Organisms which cannot be considered as plants or animals or belong to fungi (heterotrophs having chitin in the cell walls) are the Protists, e.g. amoeba, *Paramecium*, *Euglena*, slime moulds, green and red algae, diatoms etc. Several groups of Protista are chiefly single-celled organisms.

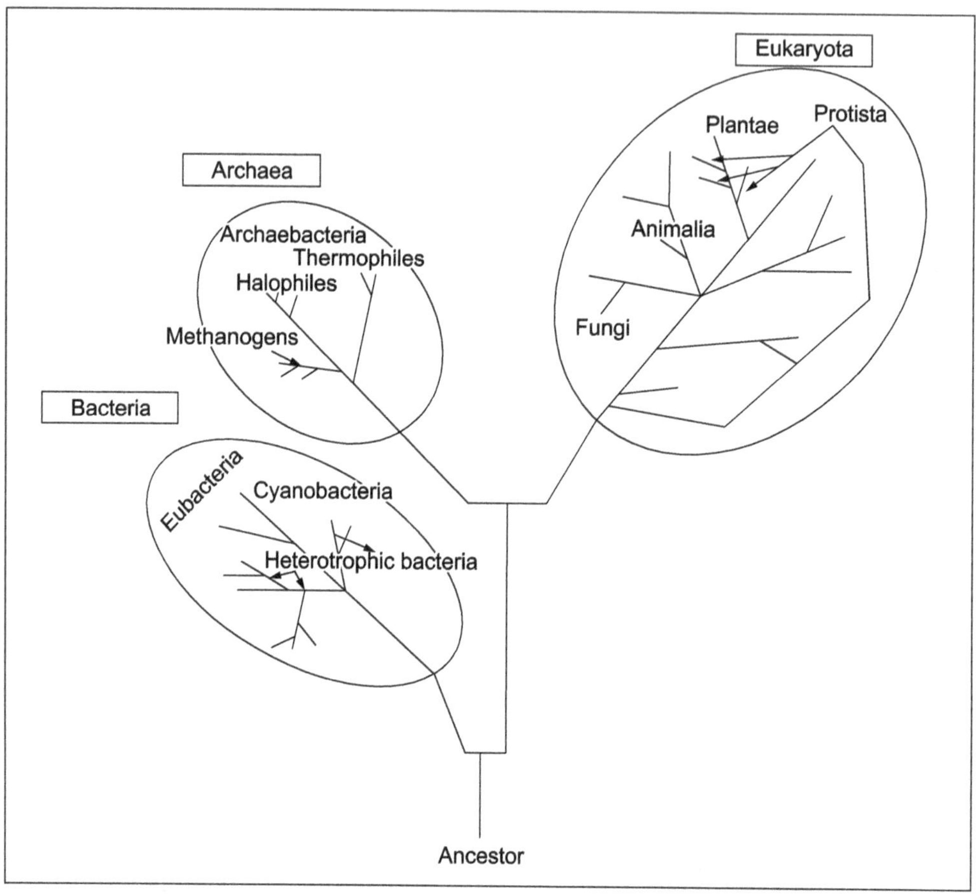

Figure 2.7: Phylogenetic classification of the living kingdom based on sequence of rRNA gene of the ribosomal small subunit (16S in bacteria and archaea and 18S in eukaryotes). *All living organisms can be grouped into three domains – Bacteria, Archaea and Eukaryota. Bacteria and Archaea are single-celled prokaryotes with the latter consisting of mostly extremophilic organisms (living in extreme environmental conditions). They include the kingdoms Eubacteria and Archaebacteria respectively. The Eukaryota is a large domain with wide diversity. It consists of uni- and multi-cellular organisms classified into four kingdoms – Plantae, Animalia, Fungi and Protista.*

The three-domain classification with six kingdoms (**Table 2.9**) is the most accepted taxonomic division of living organisms.

Table 2.9: Classification of living organisms based on rRNA gene coding for rRNA (16S/18S) of the ribosomal small subunit (Woese 1990)

	Domain		
	Archaea	**Bacteria**	**Eukaryota**
Kingdom	Archaebacteria	Eubacteria	Protists ,Fungi, Animalia, Plantae
Cell type	Prokaryotic	Prokaryotic	Eukaryotic
Cell membrane chemistry	Branched or ring hydrocarbons linked to glycerol	Peptidoglycan	Proteins and lipids
Cellular nature	Unicellular	Unicellular	Uni- and multi-cellular

2.9 MODEL ORGANISMS FOR BIOLOGICAL STUDIES

There is an overwhelming biodiversity on earth. There is a vast variety of size, colour, organization and life style among living beings. It is not possible to study all of them to understand the basic phenomena of life. When studies of the function of a cell at molecular level were initiated, it soon became clear that the basic process of duplicating, transferring and expressing the genetic blue print was similar in all living organisms starting from the simple bacteria to man – the most complex organism on earth. The basic metabolic reactions have similar pathways in all living organisms because of their shared ancestry. It soon dawned, that to understand the fundamental life processes, simple organisms can be studied and thus arose the concept of model organisms in biological research. To study the mechanism of heredity and develop improved fruit trees for the orchards of the monastery, Mendel did not experiment with fruit trees but instead used a very unrelated pea plant which is a very suitable material to study inheritance.

Yet another important field is biomedical research. Many diseases in humans like Parkinson's, Alzheimer's, cancer, immune-deficiency diseases have a genetic basis. To understand the pathways of development of disease, it is not possible for ethical reasons to experiment in humans. A small, simple fly (*Drosophila*) of 2 to 3 mm size and a 1 to 2 mm worm (*Caenorhabditis*), both of which have no physical resemblance to humans, have a lot of similarity to humans at genetic level. They were found to have genes causing several of the diseases mentioned above and are therefore used extensively for research on them. It has been found that mechanisms of immunity, function of defence genes and response to infection is similar across animals and plants. Many complex human behaviours, such as aggression, circadian rhythms, sleep, learning and memory and mating are also seen in simple animals like a worm or an insect. What is genetically and biochemically true in yeast, worms, flies and mice tends to be true in humans too.

Therapeutic drug discovery requires extensive screening of several potential chemical molecules for their efficacy and non-toxicity in living systems. Animals need to be used for this purpose. Today, there is increased concern about the welfare of animals in scientific research. Therefore, organisms which can be ethically used on a large scale (like flies and worms) serve as good models in these experiments.

To study the progression pattern of some diseases in humans through its natural course, it takes dozens of years. By inducing such a disease in an animal model with a short life span, the disease course becomes much shorter and hence the study time can be much shortened.

Organisms that are: **1.** small in size, **2.** have a short life cycle, **3.** easy to breed, **4.** with a large progeny size, **5.** less expensive to grow and maintain in a laboratory, **6.** with a small genome, **7.** in which genetic manipulation is easy and **8.** whose raising does not require licencing and bring up ethical concerns are usually chosen as model organisms. A large majority of higher eukaryotes have a large amount of genetic material (genome), most of which is unexpressed being repetitive in nature. For example, rice, maize and wheat – the major cereal crops, and humans have very large genomes of about three billion base pairs. Simpler eukaryotes have smaller amount of DNA and perform nearly all the functions of higher eukaryotes with a large genome. It is easier to sort out the genetic factors guiding a metabolic pathway and understand their mode of function in organisms with smaller genomes. Hence organisms with a small genome are used as model organisms. All the model organisms except the mammalian, have much smaller genomes of up to 140 million base pairs. Except for mice, no licencing is required to raise them for experiments in the laboratory, nor, to sacrifice the embryos.

Research on model organisms has contributed greatly to our knowledge of genetics, and biology in general with implications in medicine and agriculture. Knowledge gained through investigations with model organisms had such profound consequences that several Nobel prizes have been awarded for work done using them, e.g.: 12 Nobel prizes for work on the bacterium *E. coli*, three for yeast, five for fruit fly (*Drosophila melanogaster*), three for nematode worm (*Caenorhabditis elegans*) and 17 for the common house mouse. An account of the most popular model organisms is given in **Table 2.10.**

Table 2.10: Model organisms for studies in biology

Model organism	Domain/Kingdom Characteristics	Investigations done and knowledge contributed
Escherichia coli, Common gut bacterium Image credit: https://www.philpoteducation.om/	Bacteria/Eubacteria Minute, microscopic. Has a generation time of ~20 min. Millions can be cultured in a small flask in less than a day. Genome size is ~4.6 million base pairs.	Was useful in understanding fundamental concepts of molecular biology: gene replication and its expression, genetic code and mechanism of protein synthesis. Most widely used model organism, dubbed: "work horse of molecular biology". Used for synthesis of recombinant DNA proteins.
Saccharomyces cerevisiae, Baker's yeast Image credit: https://www.sciencedirect.com/science/article/ pii/B9780124200678000040	Eukaryota/Fungi Unicellular, microscopic simplest eukaryote. Generation time of 90 to 140 min. Easy to culture in flasks on simple media. Genome size is ~12 million base pairs.	Many fundamental aspects of eukaryotic DNA replication, transcription, translation, mutation, etc were understood through studies in yeast. ~20%of genes known to have a role in disease in humans have counterparts in yeast. Many genes important for human biology were discovered by studying their homologues in yeast. Cell cycle control genes (that have a role in cancer) were discovered in yeast. The dynamics of aging in humans is studied in yeast because both have similar processes of aging. Being used for studies on developmental pathways involved in the neurodegenerative Parkinson's disease. Pioneering model organism for study of functional genomics and systems biology.
Caenorhabditis elegans, Nematode worm Image credit: https://www.wsj.com/articles/science-wants-to-know-can-worms-swim-1435598011	Eukaryota/ Animalia An emblematic premier multicellular animal model. A nematode of 1.2mm length. Generation time of 3.5 days, produces a progeny of 300 at a time. Male and bisexual types consists of about 1000 somatic cells and 1000-2000 germ cells. Worm is transparent and all the cells are visible under a dissection microscope. Genome size of ~100 million base pairs.	Many of its genes have counterparts in humans. A study of their function and role in disease can be extrapolated to humans. A favourite model to study aging, apoptosis (programmed cell death) and developmental biology. The mechanism of gene silencing with small RNAs was unravelled from studies with this organism. Has been used for studying the development of neurons, biology of human pathogens and the mechanisms of host immunity and pathogen virulence. Extensively used in screening potential drug molecules.

Drosophila melanogaster, Fruit fly 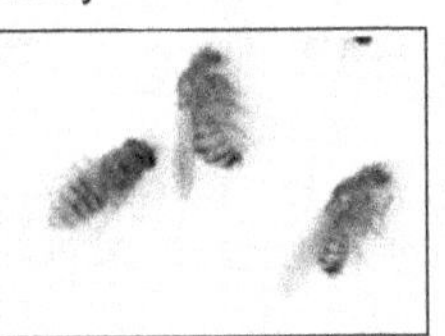 Image credit: https://www.sciencesource.com/	**Eukaryota/Animalia** A 3mm fly. Generation time of 12 days and a life span of 70 to 80 days. They can be easily cultured (hundreds of insects can be bred in a small bottle). Genome size of ~140 million base pairs. Most studied organism. Dubbed genetic "work horse".	Seminal discoveries in genetics: chromosomes as carriers of genes, linkage of genes etc. were made from studies with this fly. An estimated 75% of known human disease genes have a match in the fruit fly genome. Being used as a genetic model for several human diseases: Parkinson's, Alzheimer's, fragile X syndrome and Rett's syndrome. Used to investigate obesity-associated diseases, cardiovascular dysfunction and cancer. Useful in assessment at preclinical stage in nutrigenomics, a field that is at the intersection between diet, health, and genomics. Was helpful in studying microbiota interactions, organ regeneration, wound healing, learning, behaviour, and aging. Chemical screening for drug discovery has been successfully performed in *Drosophila* for several disorders of the central nervous system, kidney and metabolism. 'Drosophila avatars', consisting of patient-specific tumours modelled in transgenic flies, are very promising for personalized medicine.
Mus musculus, House rat  Image credit: https://www.leeser-will.de/schaedlinge/hausmaus-2/	**Eukaryota/Animalia** A small-sized mammal 2 to 4" long and ~30g in weight. A generation time of 10 weeks, short life span (one mouse year = 30 human years), large litters, ease of husbandry. Visible phenotypic variants. Most closely related to humans and therefore christened "pocket sized man'. Mice and men share about 97.5 per cent of their working (coding) DNA.	Best model for probing the immune, endocrine, nervous, cardiovascular, skeletal and other complex physiological systems that mammals share. Used in studies on diseases like cancer, atherosclerosis, hypertension, glaucoma, cancer, blood disorders, cataracts, obesity, seizures, respiratory problems, deafness, Parkinson's, Alzheimer's, cystic fibrosis, HIV and AIDs, muscular dystrophy and spinal cord injuries. Served as an ideal model for studies in developmental biology and aging. Mice are termed the 'Swiss army knife' of cancer research. "Humanized mice" are developed that carry a match of the tumour, blood system, and the immune system of humans, or a particular individual, for personalised treatment of cancer.
Arabidopsis thaliana, Rock cress  Image credit: https://www.naturepl.com/	**Eukaryota/Plantae** A small (~20cm tall) weed plant of the cabbage/mustard family with a short generation time of six weeks. Produces large number of small seed. Genome size of ~115 million base pairs. Unlike animal model organisms for which the preservation of lineages requires continuous attention, *Arabidopsis* seeds can survive for years at room temperature and even longer with refrigeration.	Greatly expanded our knowledge of plants and revealed key, common processes in diverse organisms beyond plants. 8,000 (25%) of *Arabidopsis* genes have homologues in the rice genome.

The genomes of all these model organisms have been sequenced (determining the sequence of nucleotides in their DNA) long ago and there is a thorough information on their genetic makeup at molecular level. Several mutants (knock down and knock in / loss- or gain- of gene function) to study various genetic conditions are available in all of them. Tools to tweak their genome and generate mutants or induce disease are available for all model organisms. All the wealth of information gathered through investigations on them over time is available in public data bases. Scientists working on these model organisms have easy access to the stocks of these organisms.

2.10 SUMMARY

1. The living organisms are very diverse. There is immense biodiversity on earth with an estimated number of more than eight million types of organisms. Studying all of them is impossible. To deal with this, grouping of organisms with similar features is done. This is called classification.

2. Most living organisms are complex and structured in a hierarchal manner which bestows them a stable form. They are built from simple entities that are integrated to form progressively more complex structures: atom–molecule–macromolecule–organelle–cell–tissue–tissue systems–organ–organ systems–organism–population–community–ecosystem–biome–biosphere.

3. Classification, technically called taxonomy, is also structured in a hierarchal pattern. Several units of the lower group are classed into one unit at the next level. There are eight taxonomic units. The bottom to top hierarchal taxonomic units are: species – genus – family – order – class – phylum – kingdom – domain.

4. In both the taxonomic and organizational level of hierarchy, the higher-level unit is inclusive of more than one of the lower ones. The higher level of organization has not exactly a sum total of the characteristics of its components but has newer features called the emergent features.

5. Classification of organisms can be done based on different criteria:

   (*a*) Based on cellularity they can be classified as unicellular and multicellular.

   (*b*) Considering the ultrastructure of the cell, they can be classified as Prokaryota (without an internal membrane system in the cell) and Eukaryota (with an elaborate membrane system and organelles within it). The difference between the pro- and eu-karyotes is even more than that between plants and animals. Eukaryotes are believed to have originated from prokaryotes. Symbiotic association between prokaryotic cells followed by division of labour resulted in a eukaryotic cell. This concept is popularly known as the endosymbiont theory and has strong evidence on several counts.

   (*c*) The ability to synthesise their own food or not, is yet another criterion used in the classification of living organisms. Organisms that can synthesise their own food are grouped as autotrophs while those which cannot, are grouped together as heterotrophs. They are further differentiated based on the source of carbon: organisms which utilize inorganic carbon (from carbon di oxide) are termed lithotrophs and those which obtain carbon from organic sources are termed organotrophs. Among the autotrophs, organisms that attain energy from light (sun) are termed phototrophs and those that use chemical energy (breaking bonds in chemicals) are categorised as chemotrophs.

   (*d*) Animals are also classified into five groups based on the nature of their excretory product: ammoniotelic (ammonia), ureotelic (urea), uricotelic (uric acid), aminotelic (amino acids) and guanotelic (guanine).

(*e*) Based on the habitat, living beings can be classified as aquatic and terrestrial. The aquatic biomes can be further subclassified as freshwater and marine; benthic and planktonic etc. Terrestrial inhabitants among plants can be variously classified: mesophytes, xerophytes, hydrophytes, halophytes, hygrophytes, psamnophytes etc.

6. The classification based on any one of the above-mentioned criteria is not discrete enough to differentiate between organisms, as organisms that fall into one group based on one criterion, can separate into very different groups when another criterion is considered. To group organisms based on close similarity in many features, molecular data was found useful.

7. Carl Woese classified all the living organisms based on the sequence of 16S (in prokaryotes) and 18S (in eukaryotes) rRNA gene. This classification is termed molecular taxonomy as it is based on the sequence of a macromolecule (DNA of the rRNA gene). It is the most accepted classification today. In this classification all living organisms fall into three major groups – the domains. The three domains are: Bacteria, Archaea and Eukaryota. The domain Bacteria has one kingdom – Eubacteria, domain Archaea has one kingdom – Archaebacteria and the domain Eukaryota has four kingdoms: Plantae, Animalia, Protista and Fungi.

8. As we understood the universality of life and overwhelming similarity at genetic level, it became clear that organisms with simpler structure, small genomes and those which can be bred and cultured in the laboratory easily, can be used to study process of development, mechanisms of genetic phenomena and progression of several human diseases like cancer, Parkinson's, Alzemier's, muscular dystrophy etc. Compounds which can serve as drugs and their efficiency can also be studied in such model organisms. Some of the popular model organisms that proved very useful are: *Escherichia coli* (the gut bacterium), *Saccharomyces cerevisiae* (baker's yeast), *Drosophila* (fruit fly), *Caenorhabditis elegans* (a minute nematode worm), *Mus musculus* (house rat) and *Arabidopsis thaliana* (a weed plant of the mustard and cabbage family).

2.11 SAMPLE QUESTIONS

2.11(a) Subjective Questions

Q.1. Describe the three-domain system of classification of living organisms.

Q.2. Why is classification necessary? Describe the hierarchal organization of complex multicellular organisms and taxonomic hierarchy.

Q.3. What is the need for model organisms? Why and how are they useful? What are the features based on which model organisms are chosen? Write about any two model organisms and how they were useful.

Q.4. Describe and differentiate prokaryotes and eukaryotes

2.11(b) Objective Questions

Q.1. A phylum in taxonomic hierarchy is:

    (*a*) At a higher level than an order.   (*b*) At a lower level than a class.

    (*c*) Is inclusive of domains.   (*d*) Has many kingdoms.

Q.2. A mitochondrion arose from

    (*a*) Protozoan.   (*b*) Archaebacteria.

    (*c*) Aerobic bacterium.   (*d*) Cyanobacterium.

Q.3. A prokaryote has

    (*a*) A nucleoid.   (*b*) Naked DNA.

    (*c*) 70S ribosome.   (*d*) Some of the above.

    (*e*) All of the above.

Q.4. Autotrophs utilize

    (*a*) Light or chemical energy.   (*b*) Only light energy.

    (*c*) Only inorganic carbon.   (*d*) Electron from organic source.

Q.5. Humans are

    (*a*) Ammotelic.   (*b*) Aminotelic.

    (*c*) Ureotelic.   (*d*) Uricotelic.

Q.6. Protists are

    (*a*) Multicellular.   (*b*) Photosynthetic organisms.

    (*c*) Eukaryotes that are not plants, animals or fungi.

    (*d*) Archaebacteria.

Q.7. Archaebacteria are

    (*a*) True bacteria.   (*b*) Exclusively extremophilic.

    (*c*) Are sensitive to antibiotics.

    (*d*) Have features different from eubacteria and eukaryotes.

Q.8. Chitin is present in the cell wall of

    (*a*) Bacteria.   (*b*) Protists.

    (*c*) Archaebacteria.   (*d*) Fungi.

ANSWERS

1. (*a*)   **2.** (*c*)   **3.** (*e*)   **4.** (*a*)   **5.** (*d*)   **6.** (*c*)   **7.** (*d*)   **8.** (*d*)

3

Genetics

The most profound change that genetics brings about might not be scientific at all. It might be mental and even spiritual enrichment: a more expansive sense of who we humans are, existentially, and where we came from, and how we fit with other life on earth.   **~ Sam Kean**

We have lived in a world where the discoveries of physics and genetics are far more awe-inspiring, as well as infinitely more liberating, than the claims of any religion.   **~ Christopher Hitchens**

3.1 INTRODUCTION

Genetics (a study of heredity determined by genes), is a science born in the beginning of 1900, from studies on inheritance in plants. It is a prologue to the exciting discoveries, that unfolded since mid 1950's in biological sciences and elevated the status of biology, to a plane equivalent to physical sciences. The appeal with physical sciences lies, in the immediate translation of knowledge obtained through discovery, to technology. Technology reduces drudgery, provides human comfort, and facilitates diagnosis and treatment in health care. Equivalent applications, are now being made from discoveries in biology, a hitherto descriptive science, due to the introduction of the discipline of genetic analysis.

An understanding of principles of heredity, led to planned breeding programs in plants and animals, which resulted in huge strides in agricultural production. The pest tolerant, high yielding varieties (HYVs) of crops, developed, based on sound knowledge of genetics, provided a solution to the demoralizing hunger problem in overpopulated and underdeveloped regions of the globe. The fact that a solution to food shortage, provided by the miracle HYVs developed by Norman Borlaug, awarded him a Nobel prize in 'peace', iterates the importance of genetics in the wellbeing of human society. Thousands of human diseases have been traced to genetic causes, and, a battery of prenatal diagnostic tests, have now become available. Genetic counselling for inherited diseases has come into vogue. The revelations from genetics, thus, had far-reaching consequences. Genetics is an interesting field of investigation, because, it is an inference-oriented science, where confident predictions are made, without actually being able to observe the underlying mechanisms.

From genetics, unfolded a new science – molecular biology, with the proposition of molecular structure of the gene. As a result of this molecular level dissection in genetics, the enlightenment dawned that, life happens due to the chemical behaviour of genes. It also became evident from molecular genetic analysis, that, the principles of life – storing, duplicating and expression of genes, are universal – from bacteria to humans. So, what is true for a bacterium, an insect, a plant or mice is true for humans as well! Results of experiments on a small insect, or even a plant (with no ethical issues involved in experimentation), can thus, be extrapolated to humans. At genetic level, a mouse was found to be very similar to humans – it is now considered, a 'pocket-sized man'. Therefore, it serves as a good model organism to carry out research, relevant to human disease and treatment. Why mice? Even a minute worm of ~1000 cells – *Caenorhabditis elegans* or the fruit fly (*Drosophila melanogaster*) can serve as a model organism to study genes responsible for such depressing diseases as Parkinson's and Alzheimer's.

Knowledge of genetic phenomena at molecular level, zeroed the gap between discovery and application, just as is the case in physical sciences. Some outcomes are: genetic engineering and transgenics, gene therapy, the myriad applications of biotechnology with strides in genomics, proteomics etc. and the near, real possibility of, personalized medicine.

Genetics started as a sub-discipline in biology in 1900, with the simultaneous rediscovery by three independent scientists, the principles of heredity, proposed earlier (1865), by an Augustinian monk – Gregor Johann Mendel. Between 1900 to 1950, termed the 'Mendelian era' in genetics, rapid strides in the understanding of heredity were made, due to investigations by many scientists, several of whom, were awarded a Nobel prize. By early 1950, the chemical nature and physical structure of genes came to be known. From then on, the science of genetics, branched out into several disciplines. With the post-Mendelian genetics since 1950's, came the revelation that understanding genes, is a way to comprehend the origin and process of life itself. Genetics tells us why we are as how we are – both physically, and, in terms of our behaviour.

3.2 MENDEL'S LAWS OF INHERITANCE

Mendel, called the father of genetics, was posthumously recognized for the laws of heredity that he proposed from the results of his experiments in garden pea during 1860's. Identical conclusions about heredity were arrived by three independently carried out investigations, in the beginning of 1900. With advances in genetic research, very soon, it was found that, there are more examples of exceptions to the laws proposed by Mendel, than those which conform to them. Then, why do Mendel's laws deserve (merit) mention in 21st century? Mendel's experiments are an example, par excellence, of the protocol of a meticulous scientific investigation. Though his discoveries are not indispensable, his experimental method is! The laws he proposed, provide a framework for understanding inheritance.

3.2.1 Mendel's Experiments

Mendel chose the pea plant (*Pisum sativum*) for his investigation of the principles of heredity. He conducted the experiments as described below:

1. Pea varieties were selected and paired based on their contrast for a particular character. Fourteen varieties, constituting seven pairs, were chosen for the experiments: tall vs. dwarf stature; terminal vs. axial flowers; purple vs. white flowers; green vs. yellow pod; inflated vs. constricted pod; green vs. yellow seed and round vs. wrinkled seed.

2. The selected varieties were raised for a few generations, each time with seed harvested from the previous generation, to examine if the character for which they were chosen, was consistently exhibited by all their progeny (true breeding).

3. Crosses were made, between varieties with contrasting character i.e., a tall was crossed with a dwarf, a purple flowered with a white flowered, etc. A cross, is a process of hybridization, that involves, transferring pollen (with male germ cells) from the flowers of one plant, to the stigma (female receptive organ) of the flowers on another plant.

4. Reciprocal crosses were made in each set i.e., alternatively using, a plant with a particular character, as the male in one cross, and, a female in another.

5. The hybrid progeny (first filial generation – F_1), was raised from the seed obtained through crossing, and, every plant in the progeny, was observed for the type of character it exhibited.

6. The hybrid plants were allowed to self-pollinate (the pollen of the flower falling on its own stigma), which is their usual mode of reproduction.

7. The seed obtained from hybrid plants was raised. This constituted the second generation(F_2) from the cross. All the plants in the second generation were examined to note: **a.** the character type each of them manifested; **b.** the number of plants showing a particular character type among the progeny.

In all the crosses made by Mendel, a consistent pattern was observed. In the first generation (F_1), all the progeny showed the character of one of the parents. In the second generation (F_2), approximately, three fourths of the plants scored, exhibited the character shown by the F_1 plants, and one fourth, – the character of one of the parents in the cross, which was not seen in the F_1 plants (**Figure 3.1, 3.2**).

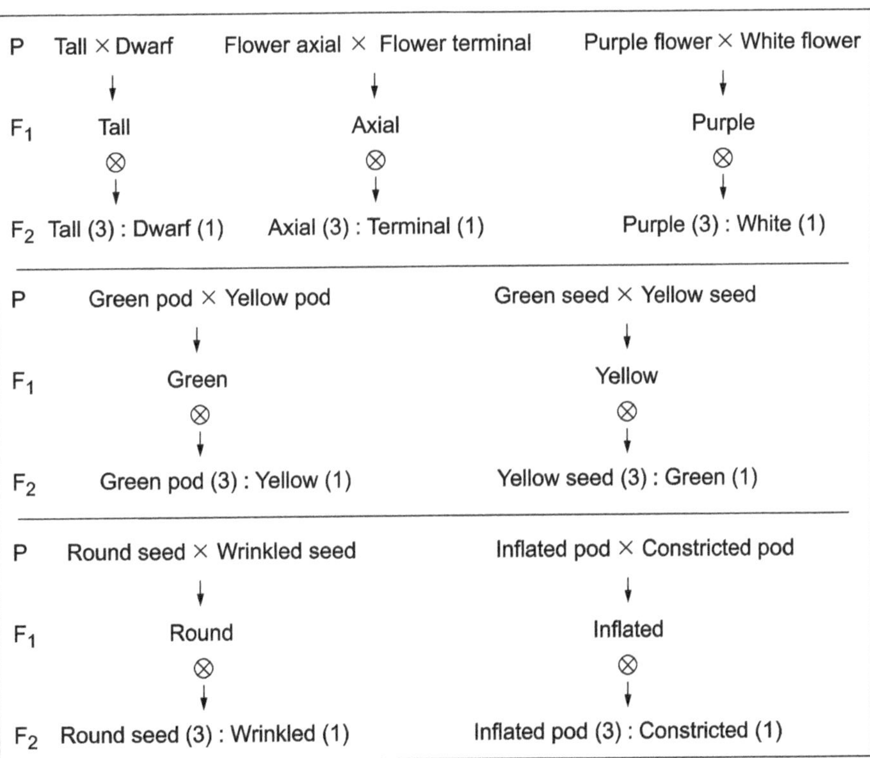

Figure 3.1: Monohybrid crosses in pea conducted by Mendel. The appearance of the first (F_1) and the second generation (F_2). *The progeny in the F_1 generation in all crosses was uniform – all showed the character type of one of the parents – the dominant character. In the F_2, 1/4th of the progeny showed the character of the other parent which was not expressed in the (F_1) – the recessive character and 3/4ths expressed the dominant character type. ⊗ symbolizes selfing – the pollen from the flower of a plant pollinates (and fertilizes) its own flowers.*

From these results, Mendel deduced as follows:

1. Each character expressed is due to a 'factor'.

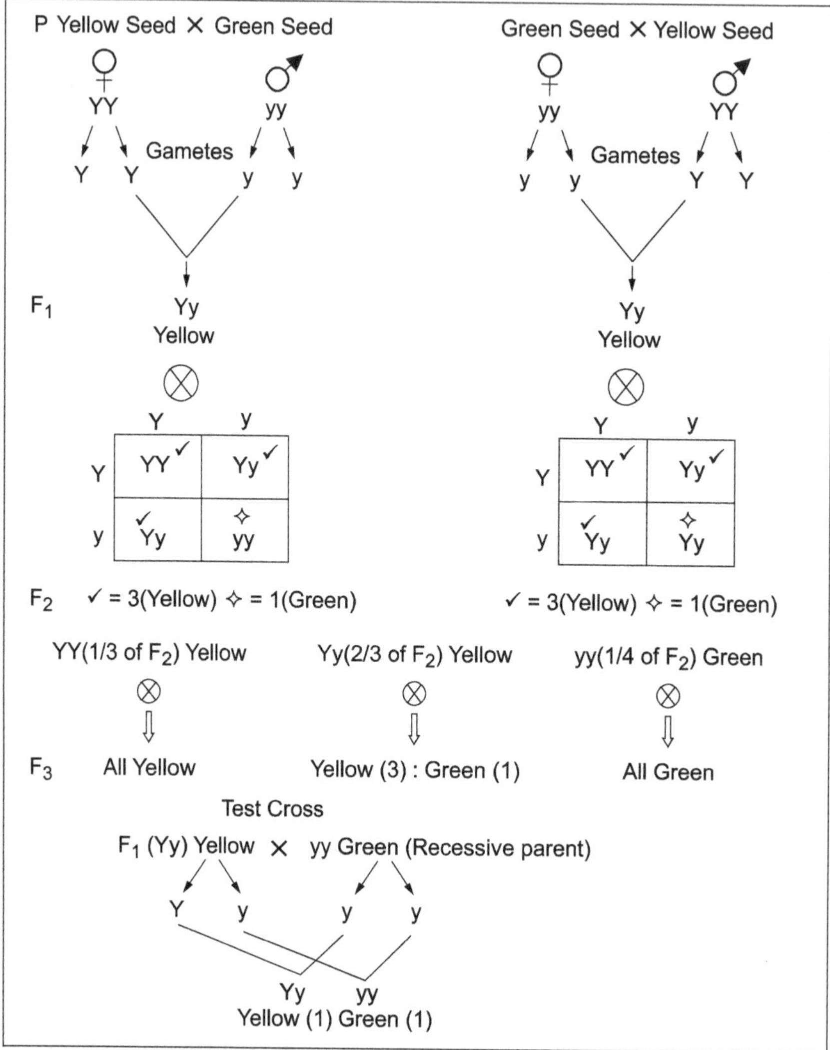

Figure 3.2: Results in the F₁, F₂, F₃ and test cross in reciprocal crosses between yellow and green seeded peas. *P stands for parent: ♀ is female, ♂ is male. ⊗ symbolizes selfing – the pollen from the flower of a plant pollinates its own flowers.*

2. An organism has a pair of factors for each expressed character. The factors in the pair may be identical (termed homozygous) or different (heterozygous).

3. In a heterozygous organism, one of the two dissimilar factors is expressed, and the other, not. The expressed factor is termed 'dominant' and the corresponding unexpressed factor – 'recessive'.

4. The two dissimilar factors for a character in a hybrid (heterozygote), do not affect, or, blend with each other, but, segregate one each into the germ cells (gametes). The gametes have one factor for each character.

This hypothesis of Mendel is popularly referred to as 'law of segregation'. The gametes of a hybrid fuse at random, producing zygotes which on development and maturation, constitute the progeny. Based on probability laws, such random fusions would result in a 3:1 ratio of plants with alternate expression of the character (3 dominant : 1 recessive) in the progeny of a hybrid (**Figure 3.2**).

Mendel then checked the third generation (F_3) progeny, raised from selfing, of all F_2 plants. He found that, the progeny of all the F_2 plants, with the recessive character, that constituted 1/4th of the F_2 progeny, bred true; i.e., they showed the recessive character. Among F_2 plants that showed dominant character (which constituted 3/4th of the F_2 population), 1/3rd bred true for the character; i.e., all their progeny showed the dominant character, while 2/3[rds] showed a 3:1 ratio of plants with dominant and recessive character respectively (**Figure 3.2**). The results observed in F_3 progeny, further lent proof to the law of segregation.

To further cross check his hypothesis on segregation of factors, Mendel performed a test cross i.e., crossing the F_1 hybrid, with a plant showing the recessive character (with a genotype similar to the recessive parent). The hypothesis of segregation of factors during gamete formation, and, random fusion of gametes with dissimilar factors stands, if the progeny contains equal proportion (1:1) of plants with contrasting character. This was actually realized in the test cross (**Figure 3.2**).

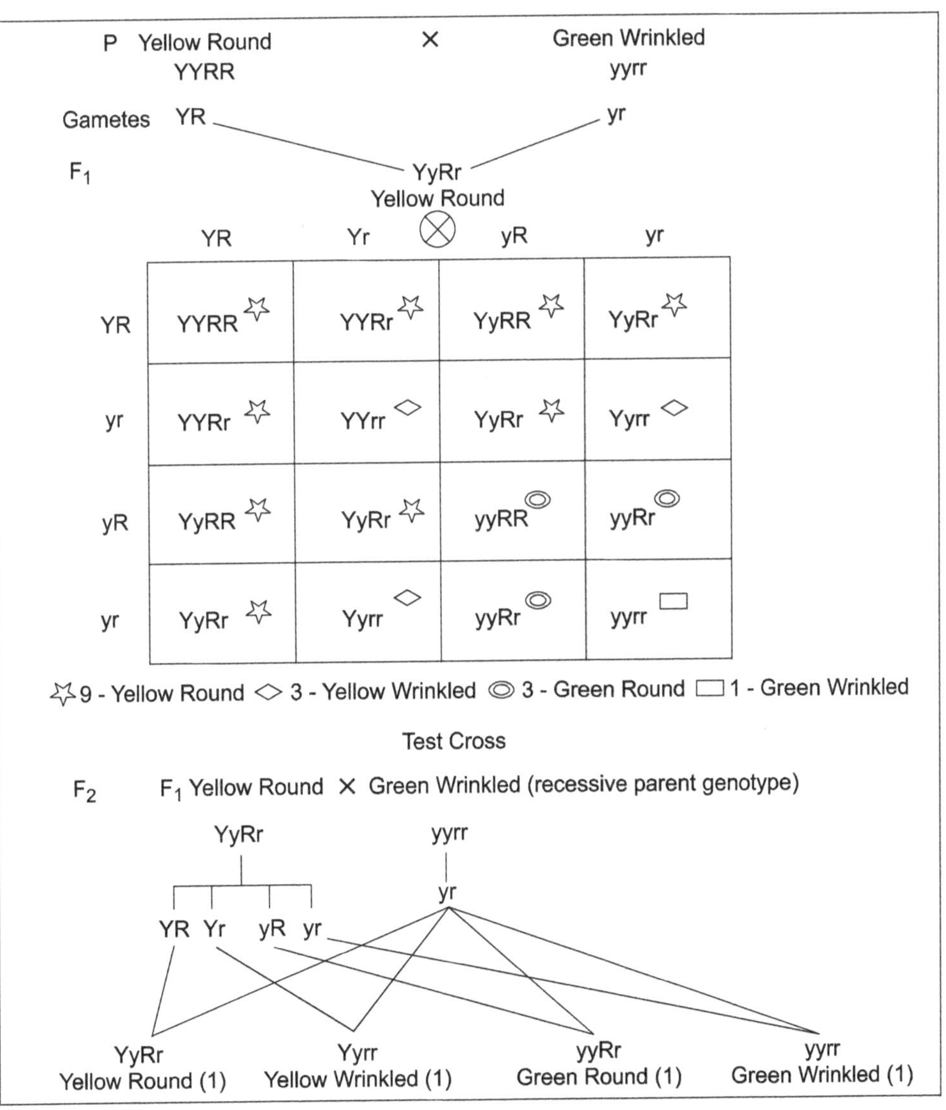

Figure 3.3: Dihybrid cross in pea resulting in a 9:3:3:1 ratio of four phenotypes in F_2 and 1:1:1:1 ratio in test cross. ⊗ *means selfing –the pollen from the flower of a plant, pollinates its own flowers.*

No difference in results was observed between the progeny from reciprocal crosses (**Figure 3.2**).

Mendel, next analysed the inheritance pattern of two characters at a time, in all pairwise combinations, of the seven pairs of contrasting characters. Crosses were made, between plants differing in two contrasting characters, say for example differing in seed colour and shape: plants that arose from yellow and round seed, with those, that grew from green and wrinkled seed (**Figure 3.3**). The progeny (F_1) of this cross, referred to as a dihybrid cross (between partners differing in two characters), was uniform. All F_1 plants had yellow and round seeds. The F_2 generation, raised from the F_1, had four seed type plants: yellow round, yellow wrinkled, green round and green wrinkled in a 9:3:3:1 ratio (**Figure 3.3**). A test cross of F_1 with a plant of the recessive parental type, with both recessive characters (green and wrinkled), resulted in a 1:1:1:1 ratio of yellow round, yellow wrinkled, green round and green wrinkled, in the progeny (**Figure 3.3**). The 9:3:3:1 ratio observed in F_2 of the dihybrid cross is actually a superimposition of the 3:1 segregation of each of the two-character types considered (**Figure 3.4**).

From these results, Mendel inferred that the segregation of one pair of factors is independent of segregation in another pair of factors. This observation is commonly referred to as Mendel's second law of inheritance – the 'law of independent assortment'.

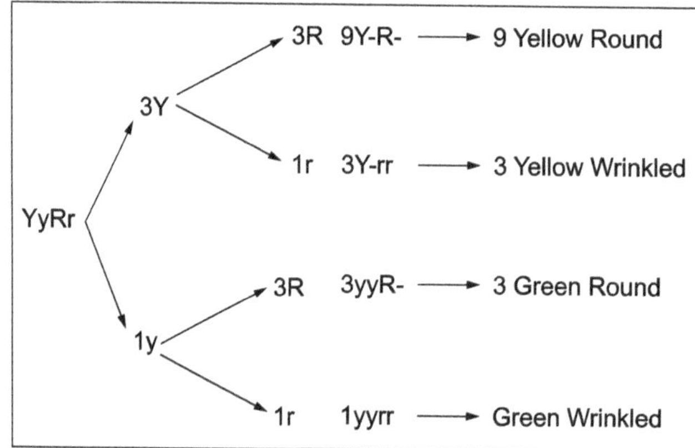

Figure 3.4: Dihybrid 9:3:3:1 ratio is the result of superimposition of two 3:1 monohybrid ratios. *- indicates the presence of upper- or lower-case alphabet of its corresponding partner gene.*

In genetics terms, the external appearance of an individual is termed as phenotype, and its genetic constitution, as the genotype. The dominant and recessive versions of a gene governing a character are called termed alleles.

3.2.1.(a) Highlights of Mendel's Experimental Method

3.2.1(a)i Choice of the Experimental Material

A material suitable for the problem of investigation was chosen. Mendel was assigned the task of developing a productive orchard in his monastery. To foster a highly productive orchard with high yielding fruit trees, knowledge of the underlying mechanisms of heredity is crucial. For this purpose, he did not work on the fruit trees which are his subject for improvement, but, instead, chose a pea plant. The pea plant is tailor fit for studying inheritance because: **a.** it is easy to make crosses manually (ease of handling), **b.** three generations of plants can be raised and observed within a year (facilitates speed of investigation) and **c.** bulky population of progeny can be analysed for inheritance from the large number of seeds borne by each plant (more data serves to draw a conclusion with higher level of confidence).

3.2.1(a)ii Splitting the Problem

The problem of investigation (study of heredity), which is highly complex, was subdivided into manageable modules. One module was handled at a time, and observations from all modules were put together, to arrive at the final, wholistic picture. Studying inheritance mechanism of all characters of the organism, as one whole unit, would be very confusing, with no chance of understanding inheritance. Therefore, Mendel set out studying the inheritance of one or two characters of the individual, at a time. This approach, made it possible to investigate the intricate subject of inheritance with success.

3.2.1(a)iii Precaution in Experimentation

Taking precaution is the hallmark of a good experiment. Mendel checked the pure breeding nature for the character under investigation, of all the pea varieties used in the experiments. This was an important precaution taken – if the plants that showed dominant character were heterozygous, and not pure breeding, it would not have been possible to observe consistent patterns in the several crosses performed, to lead to the proposition of generalized principles of heredity.

3.2.1(a)iv Applying Statistical Methods to Interpret Data

It is a famous maxim that 'all science is measurement'. Mendel in his experiments, noted the number of plants with each character in the F_2 and test cross progeny. From the analysis of the numerical data, he could interpret that, the pattern of inheritance, conformed to mathematical laws of probability. He could thereby identify factors (genes), as the basis of heredity, and, propose the general principles of inheritance: **1.** segregation of alleles of a gene pair during gamete formation, **2.** independent assortment of different gene pairs.

3.2.1(a)v Proposing a Hypothesis Based on Preliminary Results and Further Testing its Validity

After Mendel proposed a hypothesis, in both his monohybrid and dihybrid crosses, he tested its validity, through follow up observation of F_3 progeny, and, conducting a test cross. What was even more ingenious, was that, for examining F_3 progeny, first, he chose the hybrids of crosses between seed morphology variants: green/yellow and round/ wrinkled. The seeds of the F_3 are borne on the F_2 plants. Their morphology can be examined without raising plants from them. Therefore, one generation time (~ three months) is saved in the experimentation time of validating his hypothesis.

3.2.1(a)vi Stating the Results with Humility

The famous axiom of Karl popper goes thus: "Science stands on the frail foundations of falsifiability", i.e., a hypothesis stands, only, as long as it is not falsified. Therefore, hypotheses even when backed with sound experimental evidence, are to be proposed humbly, lest they be disproved, with further evidence. Mendel just described what he observed, and, mentioned his inferences. He did not state them as laws. It is in later reference to Mendel's work, that scientists called them laws! Also, he left the obvious observations of great importance, but controversial in his time, without mention, leaving it to the audience, for the conclusion to be drawn from his results. At a time, when a female parent was believed to have no contributary role in heredity, he demonstrated through reciprocal crosses, the equal genetic contribution, of both, the maternal and paternal parents, to the progeny. But he did not make a specific statement about it!

3.3 CONCEPT OF DOMINANCE AND RECESSIVENESS

A gene is said to be dominant, when it is expressed, even in the presence of its mutant allele. If an individual with a genotype (genetic constitution) say *Aa*, shows the phenotype (external appearance) expressed by *A*, though *a* gene is also present, then *A* is said to show dominance over *a*. In a heterozygote *Aa*, only *A* is expressed and therefore called dominant, because, *a*, has no expression and hence remains recessive. The allele *A* is expressed because it is functional, while, its mutation to *a*, results in loss of function and therefore, has no expression. Dominance is of three types:

(*i*) **Complete dominance**: The expression of the dominant allele is to such an extent, that, there is no distinction in the phenotype of homozygous (*AA*), and heterozygous (*Aa*) individuals (**Figure 3.5**). All the characters studied in the pea plant by Mendel showed complete dominance.

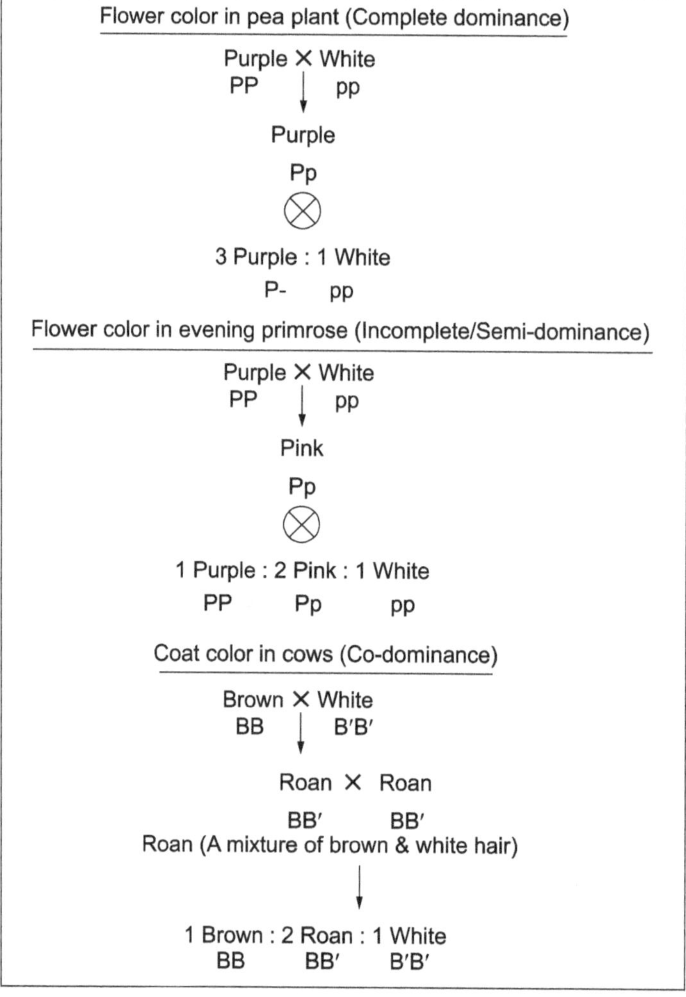

Figure 3.5: Complete, incomplete and co-dominance of alleles. *Complete dominance results in an F_1 with the appearance of one of the parents and a 3:1 ratio in F_2 (- can be either P or p). Incomplete dominance results in an F_1 with an appearance, intermediate between the two parents, and, a 1:2:1 ratio in F_2. Co-dominance results in an F_1 showing a mix of appearance of both parents, and, a 1:2:1 ratio in F_2. Roan is a colour resulting from the presence of both brown and white hairs on the coat.*

(*ii*) **Incomplete/Semi-dominance**: Here, the level of expression of the dominant allele depends on the number of its copies. Homozygous dominant, with two dominant alleles (*AA*), has a higher (usually double) level of intensity of phenotypic expression of *A*, than, when there is one dominant allele (*Aa*). In such cases, the heterozygote (*Aa*) shows a phenotype, intermediate between the homozygous dominant (*AA*) and homozygous recessive (*aa*) (**Figure 3.5**).

(*iii*) **Co-dominance**: Both alleles in the gene pair are expressed. Because, both alleles are expressed, they are denoted in capital letters as *A* and *A'*. Individuals of *AA*, *AA'* and *A'A'* genotypes, have different phenotypes: *AA* shows the phenotype of *A*; *A'A'* – that of *A'*, and *AA'* – the phenotype of both *A* and *A'* (**Figure 3.5**).

Perceiving the expression level of dominance of a gene, however, depends on, which aspect of the phenotype governed by it, is examined. To illustrate the point, the example of the seed surface variants in pea studied by Mendel, is taken.

Externally examined for seed morphology, an even surfaced round seed is found completely dominant over wrinkled (**Figure 3.6**).

Genotype Phenotypic level	RR	Rr	rr	Remark
Seed external appearance	Round	Round	Wrinkled	Complete dominance
Starch grains under microscope	Entire	Entire & Fisssured	Fisssured	Co-dominance
Test for free sugars	No	Medium	High	Incomplete Semi-dominance

Figure 3.6: *Dominance relationship between alleles of the same gene perceived differently when the phenotype is analysed at different levels.*

When the seed is sectioned, and observed under a microscope, all the starch grains in the cells of homozygous round (*RR*) seed appear oval shaped; the wrinkled seeds (*rr*) have all fissured starch grains, and the heterozygote (*Rr*), shows both types of starch grains (**Figure 3.6**). Thus, when observed at microscopic (cytological) level, the relation between *R* and *r* appears as co-dominant, since both starch grain phenotypes are observed.

In a biochemical analysis of estimation of free sugars in the seed, the round seeded homozygote (*RR*), has no trace of soluble sugars in the seed; the wrinkled seeded (*rr*), has very high levels of free sugars; while, the round seed with a heterozygous genotype (*Rr*), has approximately half the amount of free sugars compared to homozygous wrinkled (*rr*). Thus, at biochemical level, the relation between *R* and *r* can be termed as incomplete or semi-dominance (**Figure 3.6**).

The gene for round and wrinkled nature of seed in peas, was later, found to be coding for a starch branching enzyme (*Sbe1*). The enzyme catalyses polymerization of sugars to starch. When mutated to *sbe1*, the synthesis of starch from sugars, does not take place. The *R* and *r* alleles in Mendel's crosses thus technically represent *Sbe1* and *sbe1* genes respectively.

From this knowledge of the function of the seed surface gene *R*, that it polymerizes sugars to starch, the results of both, biochemical (free sugar estimation) and cytological (morphology of the starch grain in the cells) analysis of wrinkled, homo and heterozygous round seeds, can now be interpreted as, due to incomplete dominance of *R* (*Sbe1*) over *r* (*sbe1*).

Due to incomplete dominance, the amount of starch branching enzyme produced in a heterozygote (*Rr/ Sbe1sbe1*), by a single dominant allele, is not sufficient to polymerise all the sugar molecules. Nearly, half the amount of sugars, remain unpolymerized and free. Therefore, it has about half the amount of free sugars as that of a homozygous recessive – *rr /sbe1sbe1*, in which, there is no starch polymerization at all, and all sugars molecules remain free. This explains the results of biochemical analysis of the different seed shape genotypes (*RR*, *Rr* and *rr*) of pea.

The phenotype of the starch grains in the seed, as seen under the microscope, can be rationalised thus: in the homozygous dominant (*RR/Sbe1 Sbe1*) genotype, all sugars are polymerised to starch. Hence, all grains are filled with starch, with the starch molecules being deposited concentrically around a central nucleus, resulting in an oval shape. In seeds of homozygous recessive genotype (*rr /sbe1sbe1*), the starch is not formed, due to the absence of the starch branching enzyme. Therefore, there is no deposition of starch in the grains. Being empty, the membrane of the grains becomes folded, and therefore, they appear fissured. In the heterozygotes (*Rr /Sbe1 sbe1*), the amount of starch branching enzyme, produced by a single allele of *R* (*Sbe1*), is not sufficient to polymerize, all the sugars to starch. Therefore, some grains are filled with starch, which appear oval, while, some grains, remain empty, and are fissured. Thus, when interpreted from the functional aspect of the gene, the expression of starch grain morphology – the presence of both oval and fissured types in the heterozygote, is not due to co-dominance between alleles of the gene pair, but, a result of incomplete dominance.

3.4 CONCEPT OF ALLELE

In Mendel's experiments, each character examined has two alternate expressions, determined by two different versions of what he termed, a 'factor' (later called a gene). The factor determining each alternate expression of a character, say for e.g. tall and dwarf, represented by *T* and *t* respectively, are termed alleles. Thus, alternate forms of a gene are called alleles. Alternate forms occur because of a mutation, in a gene. The allele which is more prevalent in the population, is termed the wild type allele and is usually, dominant in expression.

A gene, due to mutation, can have not just one alternate form, but due to several independent and different mutational events, multiple forms, can exist. They are termed multiple alleles. The multiple alleles of a gene constitute a series. The number of multiple alleles in a series, vary for different genes. For example, the gene coding for blood group type in humans has three alleles; that which determines the coat type in rabbits has four alleles. In a diploid organism, any two alleles in the multiple allelic series are present.

The blood group types in humans: A, B, AB and O are governed by multiple alleles: L^A, L^B, and *l*. The L^A and L^B alleles are codominant, while, *l* allele is recessive to both L^A and L^B (**Table 3.1**). The alleles governing blood group types are named *L* (with a superscript of the blood group type), in honour of Landsteiner, who discovered blood groups in humans.

Another example of multiple allelic series is the gene that determines coat colour in rabbits: C, C^{ch}, C^h, c for agouti, chinchilla, Himalayan and albino types. The dominance relationship of these alleles is in the order of $C > C^{ch} > C^h > c$ (**Table 3.2**).

Table 3.1: Multiple alleles governing blood type in humans

Blood Group	Genotype	Blood serum contains	Can		Remarks
			Receive	Donate	
A	$L^A L^A$, $L^A l$	Protein A	A and O	A	
B	$L^B L^B$, $L^B l$	Protein B	B and O	B	
AB	$L^A L^B$	Proteins A & B	A, B, AB & O	AB	Universal receiver
O	ll	No Protein	O	A, B, AB & O	Universal donor

Table 3.2: Multiple alleles governing coat colour in rabbits

Coat Type	Genotypes
Agouti	CC, CC^{ch}, CC^h, Cc
Chinchilla	$C^{ch}C^{ch}$, $C^{ch}C^h$, $C^{ch}c$
Himalayan	C^hC^h, C^hc
Albino	Cc

Alleles are of five types: **1.** amorph, **2.** hypomorph, **3.** hypermorph, **4.** neomorph and **5.** antimorph.

An **amorph** allele, has complete loss-of-function, and therefore, known as a "Null" mutation.

Hypomorphic alleles, have only a partial loss-of-function, and are called "leaky" mutants, because, they do not provide complete, normal function, but, result in, 'some' function. Both amorphs and hypomorphs, are usually, recessive to their wild type allele.

Hypermorphic alleles, produce quantitatively more, of the same active product, as the wild type allele. A mutant allele, that produces more proteins, or, proteins that have increased effectivity than wild-type, is considered a hypermorph. Hypermorphic mutations of most genes, usually, act as dominant to wild type, since they are a gain-of-function.

Neomorphic alleles, produce a product, with a new and different function, than the wild type allele. Neomorphic mutations of most genes, usually, act as dominant to wild type, since, they are a gain-of-function.

Antimorphic alleles, arise due to mutations, that lead to the gene having a new activity, that is dominant, and, opposes the wildtype function. In an antimorph, the protein from the mutant allele, binds to the wild-type protein, thereby inactivating it. Antimorphs are relatively rare.

All mutant versions of a gene that constitute a multiple allelic series, affect the same character of an organism. But there are several instances, where the same character, can be affected by different genes; these are termed, multigene traits. For example, in humans, the ability to hear, is affected by approximately, 57 genes.

3.4.1 Test of Allelism

To test, if the different versions of a mutant phenotype for a character, are due to allelic genes (different mutant versions of the same gene e.g. A to: a, a' a'', A', A'' etc.), or non-allelic genes (mutants of two different genes A and B: a and b), a complementation test is done.

In a complementation test, two different mutant types for a character both of which are recessive to the wild type (dominant) phenotype, are crossed, and the progeny is examined. If all the progeny shows a wild type character, then, the genes governing the character, are concluded to be different, i.e., non-allelic (e.g. *A* and *B*); if all the progeny, have a mutant phenotype (of either of the parent), then, the genes governing the mutant phenotype in the two individuals, are considered allelic, i.e., they arose, due to different mutations in the same gene (e.g. *A* to *a*, *a'*, *a''*, *A'*, *A''* etc.) (**Figure 3.7**).

For example, in a cross between pure bred rabbits with chinchilla and Himalayan coat types, the progeny is all chinchilla; none of the progeny, have the wild type agouti coat type. Therefore, the genes governing the coat colour in chinchilla and Himalayan types are considered allelic.

When two genes *A* and *B*, govern the same character, they control two essential steps leading to that character. In such an instance, mutant phenotype for that character can arise from a genotype of either *AAbb* or *aaBB*, because, one of the steps leading to expression of the normal wild phenotype for that character, cannot be carried out in them, due to double recessive condition in one of the genes. In a cross between these two individuals, with mutant phenotype for that character (a cross for test of allelism), the progeny will all be *AaBb*, and, have wild type phenotype, because, they have dominant genes, for both *A* and *B*. This is called complementation (**Figure 3.7**). Test for allelism, is therefore, called complementation test.

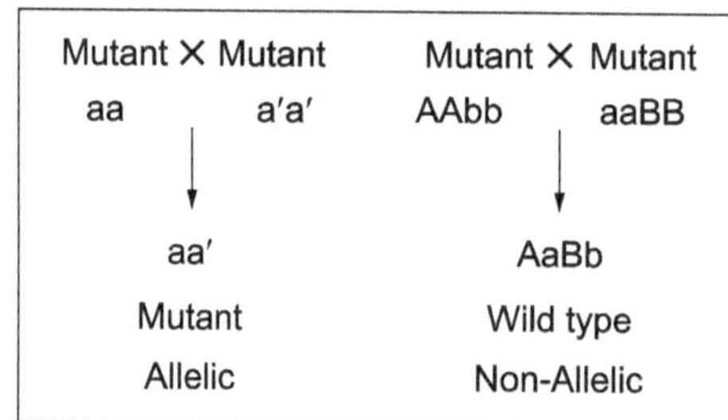

Figure 3.7: Allelism test. *If a cross between two different mutants affecting the same character result in a progeny with a mutant phenotype, the genes causing the two mutant phenotypes are considered allelic i.e., they have resulted from mutation of the same gene, but, in two different ways. If a cross between two different mutants affecting the same character result in progeny with wild phenotype, the genes causing the two mutant phenotypes for that character are concluded to be non-allelic i.e., they have resulted from mutation in two different genes affecting the same character.*

Different recessive mutant alleles of a single gene, cannot complement each other, and, show mutant phenotype, because, they all result from mutation in the same gene and the function of the gene is lost due to mutation. If the two individuals with altered expression were due to allelic mutants, say *a* and *a'*, the progeny of the cross between two mutants *aa* and *a'a'* would all be *aa'* and being double recessive, will have a mutant phenotype (**Figure 3.7**).

A very clear example for complementation is observed in human pedigrees, with profound hearing loss. Deafness, is caused due to a mutation in any one of the 57 genes governing the ability to hear. Therefore, in many instances, progeny of two deaf mates, have normal hearing – the parents being double recessive for two different genes among the 57 genes causing deafness. Since the mutant genes causing deafness are non-allelic, they complement each other in the progeny, who therefore have, normal hearing ability. It must also be appreciated by an individual with normal hearing, that, it must be due to providence and immense luck, that, he has a dominant version of all the 57 genes! A double recessive condition, for any one of the 57 genes, would, cause deafness.

3.5 SINGLE GENE DISORDERS IN HUMANS

When a mutated version of a wild type gene causes a disease or an abnormal condition, it is termed a single gene disorder. The phenotype in single gene disorders is overwhelmingly due to the effect of mutation of a single gene, with little contribution, from other genes. Compared to polygenic diseases, environmental influences tend to be less, in monogenic (single gene) disorders. Since they are caused by defect in a single gene, they show a simple pattern of Mendelian inheritance. Therefore, they are also referred to as Mendelian disorders. A catalogue of such gene mutations in humans and the diseases/disorders they result in, are listed in the website: 'Online Mendelian Inheritance in Man (OMIM, http://omim.org/).

There are over 10,000 known single gene disorders in humans. Some examples of single gene disorders are: cystic fibrosis, sickle cell anaemia, thalassemia, Huntington's muscular dystrophy, polycystic kidney disease, haemophilia, red-green colour blindness, Duchenne's muscular dystrophy, Tay-Sachs disease, severe combined immunodeficiency disease (SCID), popularly called the 'bubble baby' disease, alkaptonuria and achondroplasia. About 1% of human population, suffer from some type of single gene disorder. Thalassemia – a blood disorder, is the most common inherited single-gene disorder in the world, specially, in regions, where malaria was or is still, an endemic. It is a major health concern in Iran, where an estimated 8,000 pregnancies annually are at risk.

Many single-gene disorders run in families, because, they are inherited. These disorders can be easily tracked in human families, from the family tree. In humans, there are 46 (23 pairs) chromosomes, two (a pair) of which, are sex chromosomes (determine sex of the individual – whether male or female); the rest, are called autosomes.

Depending on the location (sex chromosomes or autosomes) of the affected gene, the inheritance pattern varies. In humans, the sex chromosomes are of two types: X and Y. A female has two X chromosomes, while, a male has an X, and a Y, chromosome. The Y chromosome does not have many genes. Almost all single gene disorders due to mutant genes on sex chromosomes, are present on the X chromosome.

Based on the location (autosome/sex chromosome), and expression (whether the affected gene is dominant or recessive), single gene disorders are classified into four types: **1.** autosomal dominant, **2.** autosomal recessive, **3.** sex(X)-linked dominant and **4.** sex(X)-linked recessive.

3.5.1 Autosomal Dominant Disorders

A person heterozygous for the mutation (i.e., *Aa*), manifests clinical symptoms of the disease. Examples of diseases due to dominant genes are: Huntington's disease, Marfan syndrome, brachydactyly and congenital cataracts. For disorders due to dominant genes, in a family situation of an affected person, one of whose parents was normal (heterozygous for the gene), and a normal spouse, the chance of a child inheriting the disease, is 50 percent, i.e., if they have four children, it is possible that two of those four children, inherit the disease (**Figure 3.8A**).

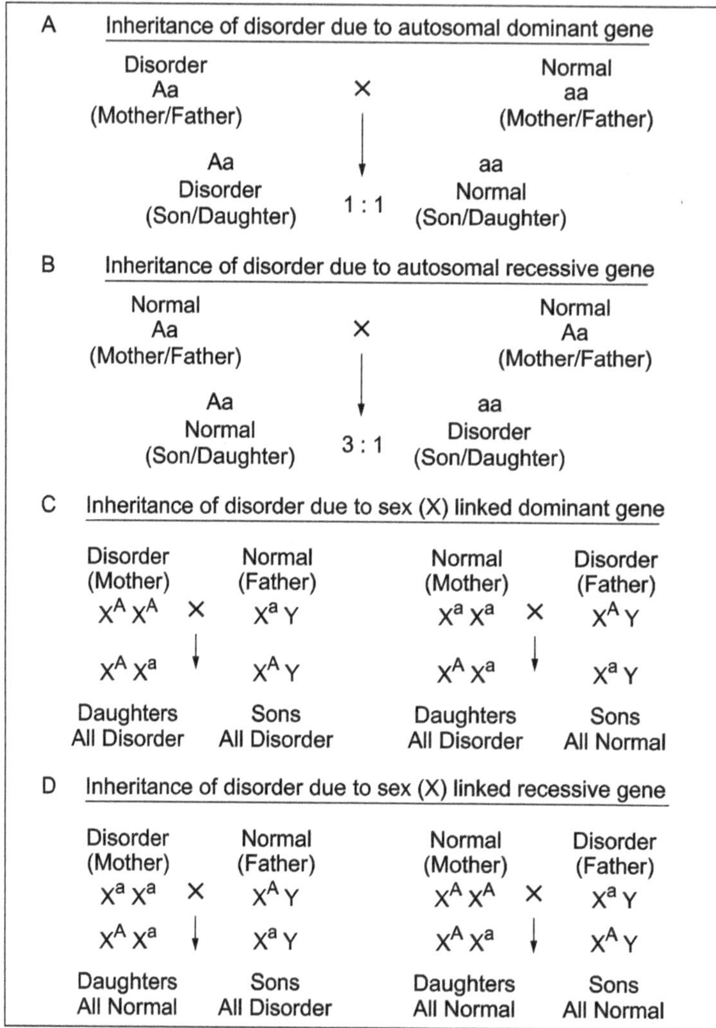

Figure 3.8: Inheritance pattern of genes causing disorders in humans when present on autosomes and sex(X) chromosomes. *In autosomal gene disorders, inheritance of the disorder is not affected depending on which parent (mother/ father) has the disorder and there is no bias in the inheritance between boys and girls in the progeny. In sex(X) linked gene disorders however, the inheritance pattern depends on which parent (mother/father) has the disorder and if the affected gene is dominant or recessive with a bias on whether it is passed down to boys or girls in the progeny.* **A.** *When disorder is due to an autosomal dominant gene, 50% of the offspring might be affected, with no bias seen, between the boys and girls.* **B.** *When the disorder is due to an autosomal recessive gene, 25% of the offspring might be affected, with no bias seen, between the boys and girls.* **C.** *In disorders due to a X (sex) linked dominant gene, when mother has disorder and is homozygous, all the progeny has the disorder; when father has the disorder, all daughters have the disorder and all sons are normal.* **D.** *In disorders due to a recessive gene on X chromosome (sex chromosome): when mother has the disorder, all sons have the disorder and all daughters are normal; when father has the disorder, both daughters and sons are normal.*

3.5.2 Autosomal Recessive Disorders

Many of the single gene disorders are autosomal recessive. Autosomal recessive disorders, are clinically manifested, only when the individual is homozygous for the disease-causing gene (i.e., both copies of the gene are mutant – *aa*). Some common examples of autosomal recessive disorders are: sickle cell anaemia, thalassemia, cystic fibrosis, Tay-Sachs disease, albinism and alkaptonuria. In a situation, when both parents are normal, but, have a family history of disorder and carry the recessive disease gene (heterozygous condition –*Aa*), the chance of a child inheriting the disorder is 25 percent i.e., if they have four children, it may be more likely, that, only one child among them, might develop the disease (**Figure 3.8B**). Autosomal recessive disorders are more frequently seen in consanguineous marriages, when there is a family history of the disease, as, the likelihood of both the partner cousins being carriers (heterozygous), is high.

3.5.3 Sex (X) Linked Dominant Disorders

Relatively few X-linked dominant disorders have been described; one example is vitamin D resistant rickets. Disorders due to X-linked dominant genes, are approximately, twice as frequent in females, as in males. They are characterized by transmission of the disorder, from affected men, to all daughters, but to none of the sons when their woman partner is normal (**Figure 3.8C**). When an affected female both of whose parents have the disorder and has a homozygous genotype for the affected dominant gene, has a normal male partner, all their offspring, both sons and daughters, are affected (**Figure 3.8C**).

3.5.4 Sex (X) Linked Recessive Disorders

X-linked recessive disorders are observed, primarily in males, and, in a lower frequency, in females. They are never transmitted from affected fathers to sons, but, always from affected mothers, to sons (**Figure 3.8D**). When the mother is normal with no family history of the disorder (homozygous for the dominant allele of the gene for disorder) and the father is affected, none of the offspring either boys or girls have the disorder (**Figure 3.8D**). Examples of X-linked recessive disorders are: red-green colour blindness, haemophilia, Duchenne's muscular dystrophy.

The inheritance of sex-linked gene disorders follows the inheritance pattern of the sex chromosomes, X and Y (**Figure 3.8C and D**). There is a bias in expression, being more, or less, frequent, in a particular sex, in the offspring, depending on, which parent carries the affected gene, and, whether it is dominant or recessive (**Figure 3.8C and D**). In autosomal single gene disorders, no such difference in expression of the disorder between the different sex (male, female) offspring is observed (**Figure 3.8A and B**).

Many genetic disorders are so serious, that, children who have them are extremely sick (with hospitalisation required), or, cannot survive after birth. Some, are relatively easy to manage, and with proper care, people who have them, have normal fulfilling lives. The chances of a good treatment outcome, are much higher, if the condition is identified soon after birth, or even before. Pre-natal diagnostic tests are available for some single gene disorders. Prospective parents, who have a risk, of offspring with a particular genetic disorder, can have the pre-natal tests done, to take appropriate decisions. Genetic testing cannot however, tell everything about an inherited disease. For example, for some disorders, a positive result does not always mean, the offspring with the affected gene will develop the disorder for sure, and it is hard to predict, at what age, and how severe, the symptoms, may be.

Correction of single gene disorders by gene therapy – compensating for the faulty copy of the gene by providing a fully working copy, remains the ultimate goal. Gene therapy, however, is technically challenging and expensive. The first condition for which gene therapy was used clinically was adenosine deaminase (ADA) deficiency, one of the causes of severe combined immunodeficiency disease (SCID) popularly known as the bubble baby disease (because the child is kept in an oxygen bubble to protect him from infections since he has no immunity to fight an infection).

As we unlock the secrets of the human genome (the complete set of human genes), we are learning that nearly all diseases have a genetic determinant.

3.6 Gene Interaction

Mendel's laws of inheritance, put forward, based on such meticulously planned experiments with sound evidence, as described earlier, now, stand as an exception, rather than a rule. Even, structurally simple single celled organisms, have several thousands of genes, and, it is but logical, that the expression of a gene, is influenced by, more than one gene. Characters governed by a single gene, though prevalent, are rare in occurrence. Thus, a phenotypic character, can be a consequence of the interaction of several genes. Tracking the inheritance pattern of characters resulting from interaction of several genes, through designed crosses, or, following the pedigrees (ancestry in family tree) is difficult. At best, the cases of two gene interactions can be studied, in planned hybridization experiments. Some of the classical examples of phenotypes resulting from two gene interaction are described below.

3.6.1 Comb Types in Fowl – 9:3:3:1 F_2 Ratio

Distinctive comb types: single, pea, rose and walnut are seen in different breeds of fowls (**Figure 3.9**).

The comb type breeds true within each fowl breed, i.e., when interbred within a breed, all the individuals in the progeny have the same type of comb as their parents. In the early 1900's, the genetic basis of these comb types, was studied by mating between different breeds.

Mating between rose and single, yielded F_1 hybrid progeny, all with a rose comb. When the F_1's were inter-mated, $3/4^{th}$ of the progeny had rose comb, and, $1/4^{th}$, a single comb (**Figure 3.10**). Thus, the gene for rose comb is dominant over single. Similarly, matings between pea and single, resulted in progeny, all with

Figure 3.9: Types of combs in fowls.

a pea comb. The progeny from inter-mating the F_1's, resulted in $3/4^{th}$ of the progeny with pea, and, $1/4^{th}$, with single comb (**Figure 3.10**). Thus, pea is dominant over single.

In the mating between rose and pea, the F_1 hybrid had a new comb phenotype – the walnut; mating between the F_1 walnuts, resulted in a progeny with a 9:3:3:1 ratio of walnut: rose: pea: single. In the F_2, an unexpected dihybrid ratio was observed, and, a phenotype, not seen in the parents or F_1 – the single, was observed (**Figure 3.10**). From these results, it was inferred that comb type is governed by two genes – *R* and *P*. The rose is due to *R-pp* genotype, pea: *rr P-*, walnut: *R-P-* and single is *rrpp* (**Figure 3.10**). (- indicates the presence of either a dominant or recessive allele of that gene). In this case, the dominant allele of one gene (*R* or *P*), is dominant over not only its own recessive allele, but, masks the expression of the recessive allele of another gene. In rose (*R-pp*), R masks the expression of *pp*, and, in pea (*rrP-*), *P* is masking the expression of *rr*. This phenomenon, of a dominant gene masking the expression of the double recessive condition of another non-allelic gene, is termed epistasis. Here, the interaction of dominant alleles of the two genes *R* and *P* produces a new phenotype – walnut, and, the interaction between the recessive alleles *r and p*, in double recessive condition (*rrpp*), results in a single type comb (**Figure 3.10**).

The characteristic feature of phenotypes resulting from two gene interactions, is a dihybrid ratio in F_2 in a cross between individuals differing in a single character.

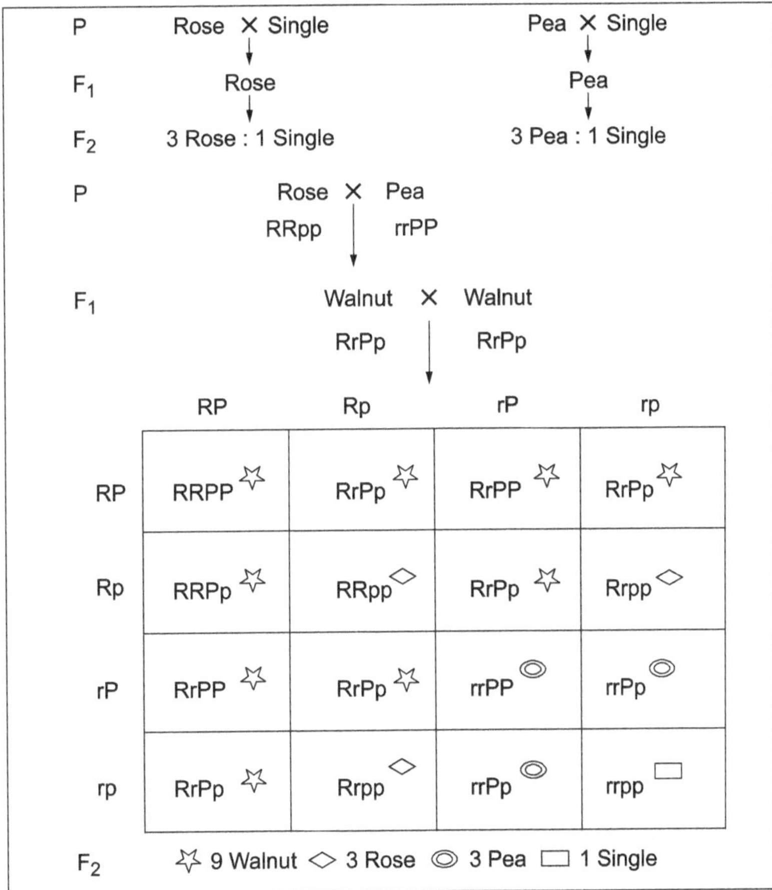

Figure 3.10: Inheritance pattern of comb type in fowls due to interaction between two genes. *Interaction between two genes to produce a character results in a dihybrid ratio of 9: 3: 3:1 in the F_2 with the production of four kinds of phenotypes. Presence of both P and R results in walnut, R-pp is rose, rrP- is pea and rrpp is single. (- indicates the presence of either a dominant or recessive allele of that gene).*

3.6.2 Fruit Colour in Capsicum due to Two Gene Interaction – 9:3:3:1 F_2 Ratio

Yet another example of epistasis, is the raw fruit colour in capsicum. A cross between green fruited with cream fruited, resulted in an F_1 that was green. Selfing of F_1 resulted in a F_2 with a 9:3:3:1 of green, peach, orange and cream (**Figure 3.11**).

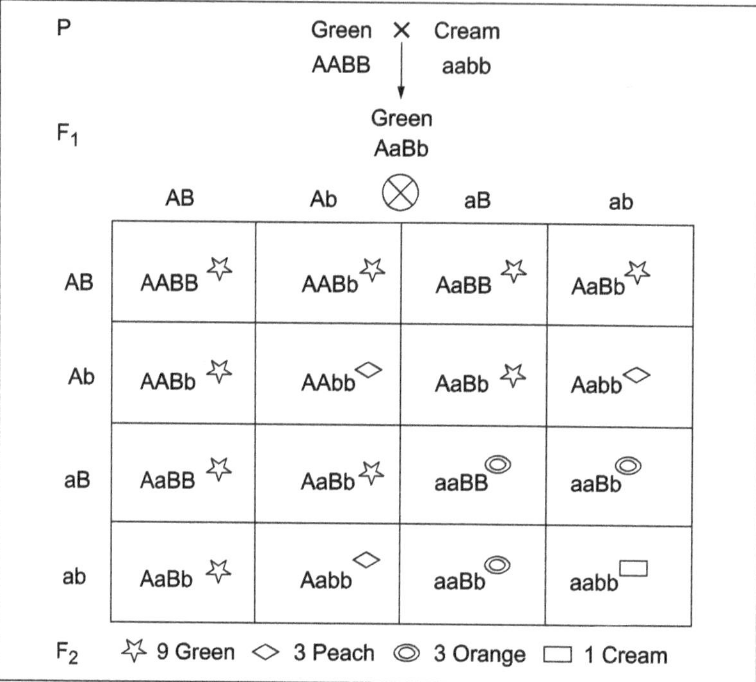

Figure 3.11: Inheritance pattern of fruit colour in capsicum due to interaction between two genes. *A 9:3:3:1 ratio of four different colour phenotypes of the fruit result in F_2. The F_2 has two new phenotypes (peach and orange) not seen in either of the parents nor the F_1. Presence of both dominant genes A-B- results in green, A-bb are peach, aaB- are orange and aabb are cream. (- indicates the presence of either a dominant or recessive allele of that gene). $\otimes$ indicates selfing – the pollen from the flower of a plant pollinates its own flowers.*

In most two gene interactions, the typical 9:3:3:1 dihybrid ratio is not observed; instead, modified dihybrid ratios like 9:7, 9:3:4, 13:3, 15:1 are observed.

3.6.3 Complementary Gene Interaction – 9:7 F_2 Ratio

A two gene interaction, affecting flower colour, was observed in pea. Two varieties of pea with white flowers, when crossed, produced a hybrid, which was purple flowered. Selfing the F_1 resulted in a progeny, showing a ratio of 9 purple:7 white flowered plants. In this case, the purple (presence of anthocyanin) flower colour, is produced due to the function of two genes *C* and *P*. When both dominant genes *C* and *P* are present, either in homo, or, heterozygous condition i.e., *C-P-* (- indicates an upper- or lower- case alphabet of the corresponding partner), the flowers are purple. When either, or both of them, are in double recessive condition (*ccP-, C-pp, ccpp*), no colour is produced, and, the flowers are therefore, white. This type of two gene interaction, is referred to as 'complementary gene interaction', because, neither of the two genes in dominant condition, can complement the double recessive condition in the other, to produce colour phenotype; colour is produced, only when, dominant alleles of both genes, are present (**Figure 3.12**).

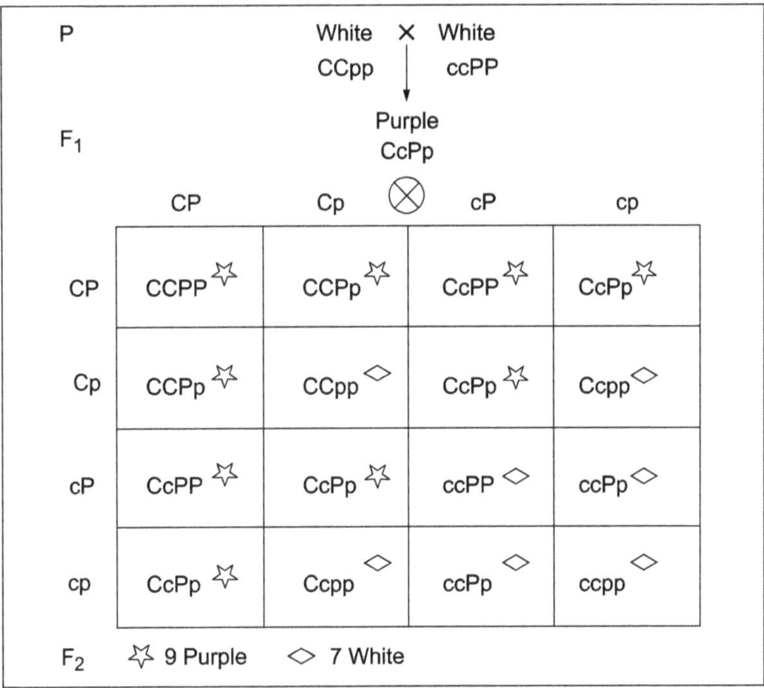

Figure 3.12: Complementary gene interaction with a 9:7 ratio in F$_2$. *Presence of both dominant genes: C-P-results in development of colour in flowers. Double recessiveness at either or both: ccP-, C-pp, ccpp results in failure to produce colour and the flowers are therefore white. (- indicates the presence of either a dominant or recessive allele of that gene). ⊗ indicates selfing – the pollen from the flower of a plant pollinates its own flowers.*

3.6.4 Flower Colour in Blue Eyed Mary Plants – 9:3:4 F$_2$ Ratio

The flower colour in blue eyed Mary plants was found to be due to interaction of two genes of yet another type of epistasis. A hybrid between pure breeding blue flowered variety with a white flowered one, had blue flowers. The selfed progeny of the hybrid consisted of blue, magenta and white flowered in 9:3:4 ratio (**Figure 3.13**). The dominant gene of *B* is required for pigment formation. When its interacting gene *W* is also dominant, i.e., the genotypes *B-W-*, flowers develop blue colour. When *W* is double recessive, in the genotypes with dominant *B*, i.e., *B-ww*, the flowers develop magenta colour. If B is also double recessive, along with *W*, i.e., *bbww*, no pigment is produced, and, the flowers are white. (**Figure 3.13**).

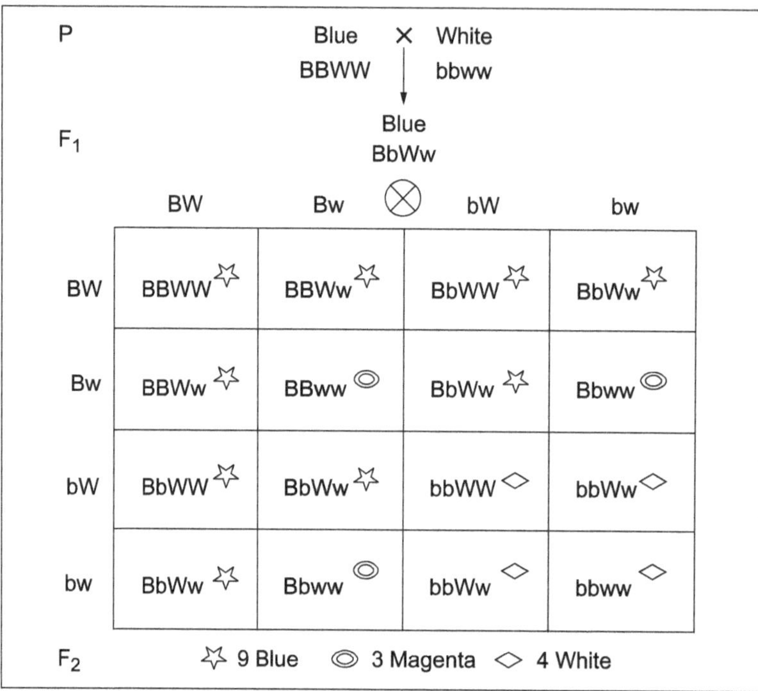

Figure 3.13: Two gene interaction resulting in a 9:3:4 ratio in F$_2$. *The dominant gene of B is required for pigment formation. When its interacting gene W is also dominant: the genotypes B-W-, flowers develop blue colour. When W is double recessive i.e. in the genotypes with dominant B, i.e., B-ww, the flowers develop magenta colour. If B is double recessive i.e., bbW or bbww, no pigment is produced, and, the flowers are white. (- indicates the presence of either a dominant or recessive allele of that gene). ⊗ indicates selfing – the pollen from the flower of a plant pollinates its own flowers.*

3.6.5 Inhibitory Gene Interaction – 13:3 F$_2$ Ratio

In chickens, a cross between two white feathered breeds resulted in an F$_1$ which were white feathered. When the F$_1$'s were inter-mated, the progeny contained white and coloured birds in a 13:3 ratio. This is a case of 'two gene inhibitory gene interaction'. Two genes: *I* (inhibitory gene) and *C* (colour gene) are involved in feather colour. All individuals with a dominant *I* (*I-C-* and *I-cc* genotype) are white, irrespective of presence of the colour gene *C* in dominant condition. This is because the inhibitory gene prohibits colour formation. Individuals with *ii C-* are coloured, since, the inhibitory gene is recessive, while, the *iicc* are also white, because the colour gene is recessive (**Figure 3.14**).

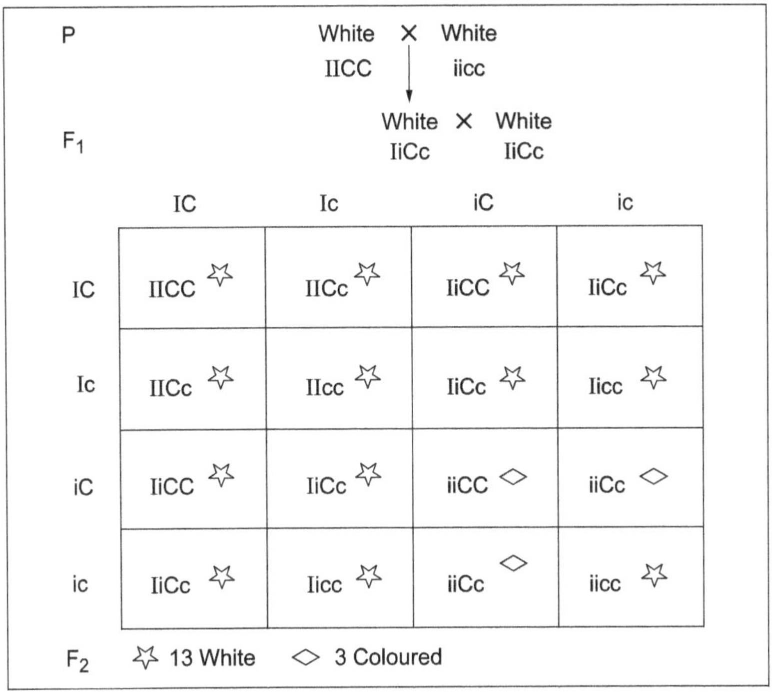

P		White × White
		IICC iicc
		↓
		White × White
F_1		IiCc IiCc

	IC	Ic	iC	ic
IC	IICC ☆	IICc ☆	IiCC ☆	IiCc ☆
Ic	IICc ☆	IIcc ☆	IiCc ☆	Iicc ☆
iC	IiCC ☆	IiCc ☆	iiCC ◇	iiCc ◇
ic	IiCc ☆	Iicc ☆	iiCc ◇	iicc ☆

F_2 ☆ 13 White   ◇ 3 Coloured

Figure 3.14: Two gene inhibitory interaction with a 13:3 dihybrid ratio in F_2. *I is the inhibitory gene and C is the colour gene. All genotypes with a dominant I, i.e., I-C- are white because the inhibitory gene prevents production of pigment. The double recessive iicc is also white because the colour gene is also recessive. Genotypes double recessive for I with a dominant C: iiC- are coloured. (- indicates the presence of either a dominant or recessive allele of that gene).*

3.6.6 Duplicate Gene Interaction – 15:1 F_2 Ratio

The wheat grains are either dark brown or cream coloured. When a pure breeding dark brown seeded wheat plant was crossed with a cream seeded one, the F_1 progeny had dark brown seed. In the F_2 progeny, dark brown seeded plants and cream seeded ones occurred in a 15:1 ratio. This modified dihybrid ratio, is due to interaction of two genes *A* and *B*. A dominant condition at either of the two gene loci gives rise to dark brown seed; double recessive condition at both, results in cream colour seeds. Since either of the two genes results in dark brown seed, this type of gene interaction is termed 'duplicate gene interaction' (just as either of the duplicate keys open a lock) (**Figure 3.15**).

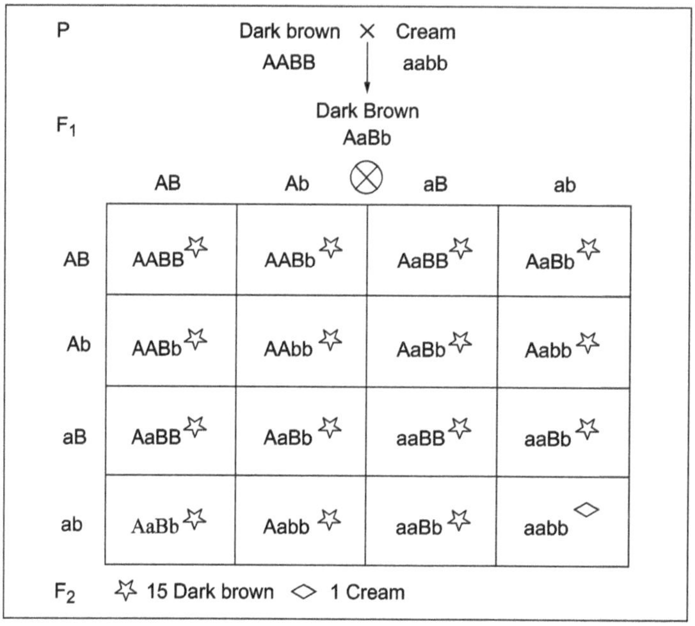

Figure 3.15: Two gene duplicate factor interaction with a 15:1 ratio in F$_2$. *Dominant allele of either of the interacting genes (genotypes A-B-, aaB-, A-bb) is sufficient to produce pigment in the seed. Only when both genes are double recessive, i.e., aabb are the seeds with no pigment and appear cream. (- indicates the presence of either a dominant or recessive allele of that gene). ⊗ indicates selfing – the pollen from the flower of a plant pollinates its own flowers.*

3.7 CHROMOSOMAL THEORY OF INHERITANCE

Named after the proposers of the hypothesis as Sutton-Boveri chromosome theory, it states that, chromosomes that are seen in all dividing cells, and, get distributed to the daughter cells, are the basis, for genetic inheritance. It is a fundamental unifying theory of genetics, which identifies chromosomes, as the carriers of genes. It states that, inheritance patterns may be generally explained, by assuming that, genes are located in specific sites (loci), on chromosomes. The behaviour of chromosomes during cell division is an impressive, visual enactment of segregation and independent assortment of alleles, stated by Mendel. Apart from the difference in gene content, chromosomes physically differ from each other in their length and position of the centromere (a visible constriction along their length).

Most higher organisms are diploid, i.e., they have each type of chromosome, represented twice. The two similar chromosomes of each set of a chromosome pair, are referred to as homologous chromosomes. The two members of the homologous pair, are inherited, one from the maternal(female), and, the other from paternal (male) parent. While the number of genes in an organism totals to thousands and tens of thousands (humans have approximately 21,000), the number of chromosomes is in units or tens (humans have 46 chromosomes – 23 pairs of homologues). Thus, the 21,000 genes in humans are distributed on 23 chromosomes; each gene has its allele partner on its homologous chromosome. Due to grouping of several (hundreds and thousands) genes into chromosomal units of much fewer number (units or tens), the distribution of the genes equally to both daughter cells becomes possible.

The genes present in the same chromosome are said to be linked genes. Linked genes on a chromosome form one linkage group. An organism has as many linkage groups as the number of homologous chromosome pairs. The position of the genes on chromosomes is conserved in a species with the members within a species having identical linkage groups.

Very soon after the rediscovery of Mendel's principles of heredity, exception to independent assortment of genes was observed in sweet pea. The genes governing flower colour and pollen shape did not assort independently. Often, they tended to pass together. As a result, the typical dihybrid ratio of 9:3:3:1 ratio was not observed in crosses between plants differing in these two characters. Instead, the F_2 progeny consisted of, a very large proportion of individuals with parental combination of the two traits, and, a few recombinants. The passing together of the two traits was inferred to be due to linkage between the two genes governing the traits i.e., the two genes reside close together on the same chromosome. The closely linked genes are passed on together to the gametes because independent assortment is not possible when they are present on the same chromosome. Recombinants are a result of, an occasional crossing over in the region between the linked genes, resulting in exchange of parts between homologous chromosomes.

3.8 MITOSIS AND MEIOSIS AS DEMONSTRATION OF CHROMOSOMAL THEORY OF INHERITANCE AND RECOMBINATION

Multicellular organisms have two types of cells: the vegetative, or somatic cells, that form the various body parts, and, the reproductive, or germ cells. Cell division that occurs in the somatic cells, is termed mitosis and that which happens in germ cells, is called meiosis.

3.8.1 Mitosis for Faithful Inheritance of Genetic Material

The cell accomplishes to duplicate itself, and increase cell number, through mitotic cell division. In order that the genetic constitution of an organism remains conserved, the daughter cells should be identical

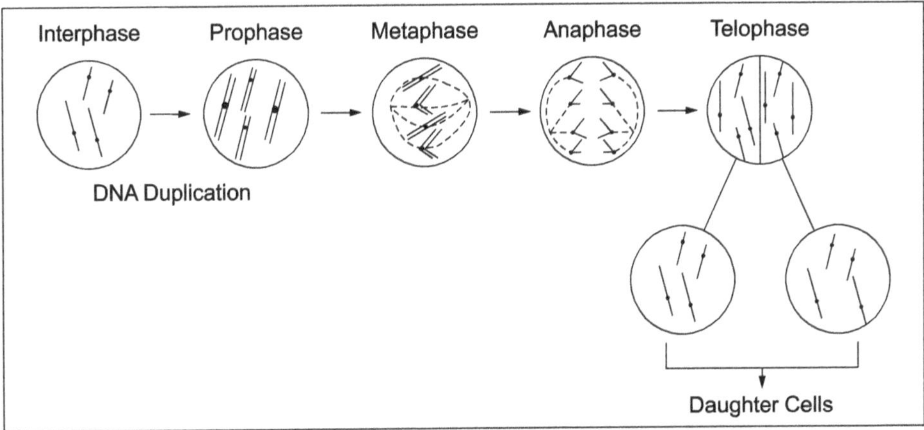

Figure 3.16: Chromosomes as vehicles of inheritance: A diagrammatic representation of mitosis as a visual evidence of the chromosomal theory of inheritance: *The chromosomes in the daughter cells, are identical in genetic content, to each other, and also to their mother cell, because of the faithful duplication of DNA during interphase (cell growth phase). The daughter cells receive equal number of chromosomes due to the organized events during mitosis, facilitating their equitable distribution. At the metaphase stage, the chromosomes orient themselves on a metaphase plate, and get attached to spindle fibres, at their centromere region. During the anaphase stage, the two strands (chromatids), of each chromosome, are pulled to opposite sides, by the spindle fibres. During telophase, a partition is formed in the cell, with the two regions enclosing the chromatids that moved to their side. Each half, then develops into a new daughter cell, with both daughter cells having an identical set of chromosomes to each other, and to the mother cell.*

to their parent. The cell accomplishes it through faithful duplication of the chromosomes (DNA in the chromosomes) during the growth phase of the cell cycle (interphase) and equal distribution of the chromosomes to the daughter cells during the division phase of cell cycle (mitosis) (**Figure 3.16**). The inheritance of genetic material can be visualised as the passing down of chromosomes in equal numbers to daughter cells during mitosis. The chromosomes therefore are inferred to carry the genes – the tenet of chromosomal theory of inheritance.

3.8.2 Meiosis for Recombination of Genes Resulting in Shuffling of Gene Combinations Leading to Variation Among Progeny

If the sex cells (sperm and ovum) have the same number of chromosomes as the vegetative cells, the zygote which is formed by fusion of the sperm and ovum, will have double the number of chromosomes. The embryo, that develops from the zygote, and grows into an adult individual, would then, have double the number of chromosomes as the parents. In this way, the number of chromosomes would geometrically increase in each progressive generation. To offset this contingency, the number of chromosomes in the germ (sex) cells is halved through a division process called meiosis.

An equally essential function to that of halving the chromosomes, happens due to meiosis – different gene combinations arise in the progeny due to recombination. Recombination results in variation. Variation is fundamental for the success of a species.

The mechanism of segregation and independent assortment of genes, observed in the progeny of monohybrid and dihybrid crosses, which form the basis of heredity, can be visualised under the microscope by observing the events occurring during meiosis.

Meiosis is actually two cell divisions: meiosis I and meiosis II. Due to the events in meiosis I, the chromosome number is halved with the two daughter cells having half the number of chromosomes as the mother cell. Meiosis II is similar to mitosis (**Figure 3.17**).

During meiosis I, the homologous chromosomes pair. The paired homologous chromosomes orient themselves on the metaphase plate and one member of each pair, is distributed to the daughter cell through the anaphase I and telophase I stages. As a result, the number of chromosomes is reduced to half in the resulting daughter cells, at the end of meiosis 1 (**Figure 3.17**). This results in segregation of alleles on homologous chromosomes which explains the law of segregation (**Figure 3.17**).

Two types of recombinants result due to meiosis – recombinants of gene combinations for genes present on different chromosomes and recombinants of linked genes present in the same chromosome.

3.8.2(a) Recombinants of Gene Combinations of Genes Present on Different Chromosomes in a Dihybrid (YyRr) Due to Difference in Orientation of Homologous Pairs of Chromosomes at Metaphase I

The distribution of the two members (maternal and paternal) of each pair of homologous chromosomes, to the two descendant cells, is random. If for example a dihybrid *YyRr* from *YYRR* and *yyrr* parents is considered, it has a set of two homologous chromosomes, with one pair having the *Y* and *y* gene alleles, and, the other pair, with *R* and *r* alleles. The distribution of the members of these two homologous chromosome pairs to the gametes, during meiosis can be visualised as shown in **Figure 3.17**. The distribution of the members of the two pair of homologous chromosomes, depends on the orientation of the chromosome pairs, at metaphase 1 of meiosis. For two pairs of chromosomes both heterozygous for a gene (*Y* and *R*), two orientations are possible (**Figure 3.17**). As a result, the segregation of one pair of genes, becomes independent of the other, resulting in gametes with four different genotypes: *YR*, *yr*, Yr and yR. This explains independent assortment of genes present on different chromosomes rationalizing the law of independent assortment. The *Yr* and *yR* gametes have recombinant genotype compared to the *YR* and *yr* gametes of the parents of the dihybrid (*YYRR* and *yyrr*). Thus, recombinant gene combinations of genes present on different chromosomes arise, due to different orientations of the homologous pairs of chromosomes at metaphase I stage of meiosis (**Figure 3.17**).

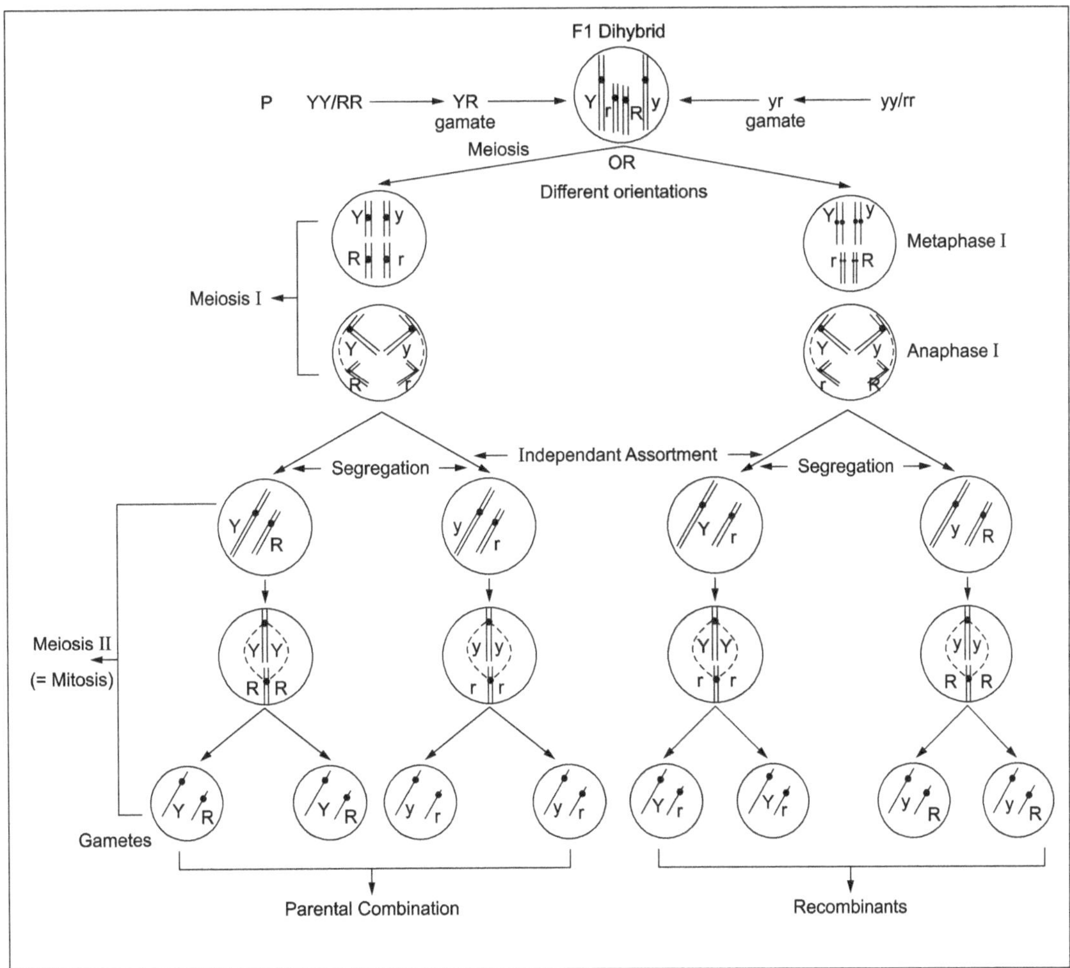

Figure 3.17: Events during meiosis showing segregation of alleles of a gene pair, independent assortment of two pairs of genes on different chromosomes and the origin of recombinant gene combinations in a dihybrid *YyRr*. *Segregation happens due the distribution of one member each, of a homologous chromosome pair, to each gamete in anaphase I. Independent assortment of different genes takes place due to different orientation of different homologous chromosome pairs during metaphase I. Different orientation of homologous chromosome pairs in metaphase I also results in recombinant gene combinations: the parents (YYRR and yyrr) of the dihybrid (YyRr) have gamete types YR and yr, while, the products of meiosis of the dihybrid have, in addition to the two parental gamete types: YR and yr, the recombinant gamete types: yR and yr.*

3.8.2(b) Recombinants of Gene Combinations for Genes Present on the Same Chromosome (linked) in a Dihybrid (YR/yr) Due to Crossing Over During Meiosis 1

When the homologous chromosomes are paired during meiosis I, exchange of parts between them, technically described as crossing over, happens, resulting in recombination between linked genes. (**Figure 3.18**). At a given region of recombination, only one of the two chromatids of each member of the homologous pair of chromosomes, exchanges parts. Therefore, the maximum recombination between linked genes does not exceed 50% (**Figure 3.18**). Thus, for linked genes, recombinants arise, due to crossing over between paired homologous chromosomes, in the region between them. The daughter cells resulting from meiosis I, have either the parental or recombinant gene combination of the linked genes (**Figure 3.18**). The chromosomes with recombinant gene combination are separated out into different gametes during meiosis II (**Figure 3.18**). Thus meiosis II serves this important function of separating the parental and recombinant gene combinations in addition to increasing the number of gametes.

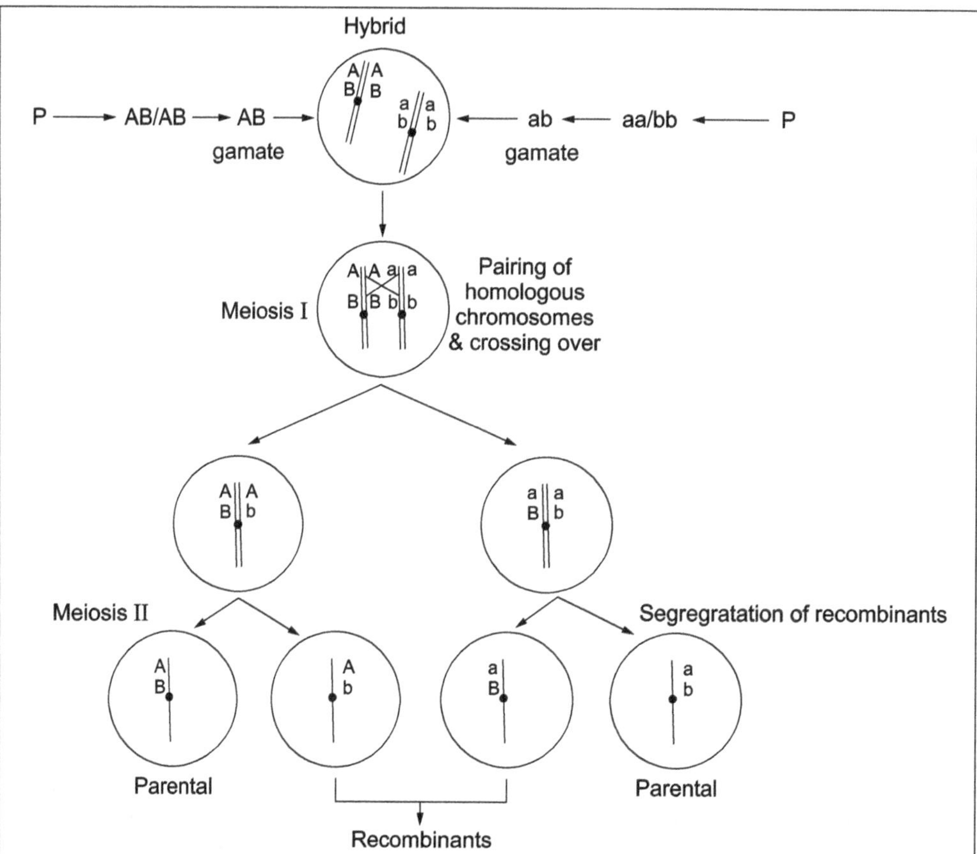

Figure 3.18: Recombination between linked genes during meiosis resulting in recombinants. *The recombinant gene combinations (Ab and aB) segregate from the parental combinations AB and ab during Meiosis II.*

The frequency of recombination between linked genes depends on the distance between them. Recombination or crossing over, is more frequent when the genes are further from each other than when they are closer together. The possibility of crossing over between two genes increases when the two genes are further apart, because, exchange at any region between them would result in a recombinant; when they are closer, the space in which crossing over should occur becomes restricted and the likelihood of its occurrence is thus reduced (**Figure 3.19**).

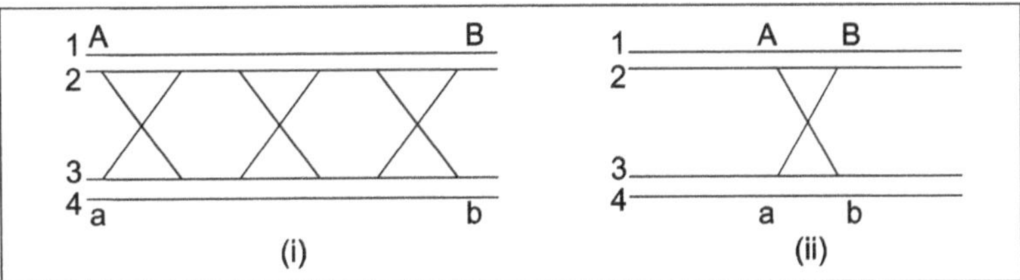

Figure 3.19: Frequency of recombination between linked genes is dependent on the distance between them. i. *Crossing over can occur at multiple sites (marked as X) when the linked genes are far apart. Therefore, probability of recombination increases when the linked genes are far apart.* **ii.** *Crossover region between two closely linked genes being short has space restriction decreasing the chance of a recombination to take place. 1, 2, 3 and 4 are strands (chromatids) of the two homologous chromosomes.*

3.9 GENE MAPPING

Gene mapping is a process in which, through planned crosses between individuals of specific genotypes, the order of linked genes on a chromosome, and, the distance between them, is determined. The cross that is done for this purpose, is termed a three-point test cross. Through this method, three genes can be ordered at a time. For example, if a gene *A* is known to be linked to *B* and if *B* is also known to be linked to gene *S*, then through a three-point test cross (*ASB/asb* X *asb/asb*) it is possible to decide if the gene order is *ASB* or *SAB* or *ABS*, and, the distance between the genes, can be computed from recombination frequency. Gene maps were constructed through this laborious process in *Drosophila* and maize in 1930's. In today's genomics world, it is possible to know the order of all genes present on a chromosome and the exact distance between them through sequencing DNA. Chromosomes are packed DNA molecules and genes are stretches of DNA.

3.10 CONCEPT OF MAPPING: PHENOTYPE TO GENES

In Mendelian sense, genotype is the genetic constitution of an organism on which its external appearance (physical, biochemical, developmental, behavioural) – the phenotype is dependant. The genotype, is thus, considered as the genetic blue print for the type of phenotype exhibited by an individual. As understanding of heredity in terms of Mendel's stated principles dawned, it became slowly, but, increasingly evident, that, the relationship between a gene to phenotype, is most often, not direct. Even monozygotic twins with almost identical genotypes, do not have exactly identical phenotypes – their fingerprints are different.

The expression of genes is influenced by several factors, chief among them, being the surrounding environment, and the effect of the other genes. The degree to which an organism's phenotype, is determined by its genotype, is described as phenotypic plasticity. Phenotypic plasticity is high when environmental factors have a strong influence on the gene, as to the particular phenotype that it produces. In such instances, the phenotype of the individual, cannot be predicted from its genotype. The phenotype of an organism, can be reliably envisaged from knowledge of the genotype, when there is little phenotypic plasticity.

In contrast to phenotypic plasticity, the concept of genetic canalization, addresses the extent to which, an organism's phenotype allows conclusions about its genotype. A phenotype is said to be canalized, if the effect of a mutation is not phenotypically (physically) apparent. A canalized phenotype, may thus, result from a wide variety of different genotypes. Therefore, it is not possible to exactly predict the genotype by observing the phenotype. Thus, the genotype-phenotype map is not invertible. When there is no canalization, even small changes in the genome have an immediate effect on the phenotype.

The mapping of a set of genotypes, to a set of phenotypes, is referred to as the genotype-phenotype map.

The predicting of the entire phenotypic outcome, of a total genotype, taking into account all the factors affecting the expression of the genes, is a challenging task. This is termed as mapping of a phenotype to genes. The number of genes of an organism, even in a very simple organism, runs into thousands. The interactions between them, and with the surrounding environment is very complex. Therefore, it is very difficult to fathom and predict the resulting phenotype. However, it is being attempted.

Understanding how metabolic reactions, cell signalling, and developmental pathways translate the genome of an organism into its phenotype is a grand challenge in biology. Genome-wide association studies (GWAS), statistically connect genotypes to phenotypes, without any recourse to known molecular interactions, whereas a molecular biology approach directly ties gene function to phenotype through gene regulatory networks (GRNs). Using natural variation in allele-specific expression, the GWAS and GRN approaches can be merged into a single framework via structural equation modelling (SEM). Such an approach facilitates genotype to phenotype mapping.

3.11 SUMMARY

1. Genetics is the study of heredity. Advances in understanding of the mechanism of inheritance has greatly improved the quality of human life. Application of the knowledge in genetics resulted in increased agricultural production, superior breeds in animal husbandry and better diagnosis and treatment options in human health and medicine.

2. The discipline of genetics originated in 1900, with the rediscovery of principles of heredity separately, by three individual scientists. Thirty-five years earlier (1865), Mendel, now called as the father of genetics, explained the mechanism of heredity through his investigations on pea plant. Mendel's experimental methodology, is an example par excellence, of a good scientific investigation. It is Mendel's method of investigation, rather than the discoveries he made, that merit more applause and recognition.

3. Mendel chose 14 pea varieties, constituting seven pairs, each pair, having contrasting feature, for a particular trait (character). Through crossing individuals with contrasting characters and statistically analysing the data of three generations of the monohybrids (differing in one pair of contrasting characters) and dihybrids (differing in two pairs of contrasting characters) he inferred as follows:

   (a) Every character exhibited is due to the presence of a factor (later named gene). An individual carries two factors for each character. The factors for the contrasting characters of a trait are called alleles. Between the two alleles governing a particular trait, one is expressed and called dominant; the other termed recessive is not expressed. Individuals who have similar alleles for a character, are termed homozygotes, and those, with different alleles – heterozygotes.

   (b) The different alleles present in a heterozygote, do not blend with each other, but, segregate during gamete formation. This is known as the law of segregation. The segregation of one pair of genes, is not influenced by the segregation of another pair of genes. This is called the law of independent assortment.

4. The observable trait of an organism is called its phenotype, the genetic constitution for the phenotype is termed the genotype.

5. Dominance relationship between alleles can be: **a.** complete – only dominant allele is expressed; **b.** incomplete – only dominant gene is expressed, but, its level of expression depends on the number of copies – the homozygous dominant (AA), having a higher level of expression than the heterozygotes (Aa) or; **c.** co-dominant, where both alleles are expressed (AA'). The expression of dominance can be interpreted differently, based on the level at which the phenotype is examined: external, biochemical or cellular.

6. The alternate form of a gene is called an allele. Alleles arise due to mutation. Due to several independent and different mutational events in a gene, it can exist in multiple forms, called the multiple alleles. Different blood groups in humans and coat colour in rabbits are due to multiple alleles.

7. Two mutant phenotypes of the same character may be due to mutation in two different genes or two mutant alleles of the same gene (alleles). To determine if it is an instance of two genes or two alleles, test for allelism, called the complementation test, is conducted. A cross between two mutant phenotypes is made, and the progeny examined. If the progeny are mutant, the two mutant types are concluded to be due to allelic genes. If the progeny are wild type, the two mutant types, are due to mutation in two different (non-allelic) genes.

8. Mutations in a gene, resulting in loss of function, or, gain of function, or, altered function, can cause an aberrant condition or disorder. More than 10,000 single gene disorders are known in humans. Their pattern of inheritance, depends on ,whether the mutant gene for the disorder causing disease, is dominant or recessive, and if it present on the sex (X) chromosome, or normal chromosome (autosome).

9. A trait is usually the result of interaction between several genes. The inheritance pattern of traits, resulting from interaction of several genes, is very complex, and it not possible to follow, through genetic crosses. Two gene interactions can be studied through planned genetic crosses. Two gene interactions result in dihybrid ratios in F_2. Normal dihybrid (9:3:3:1), and modified dihybrid ratios such as 9:7, 13:3, 9:3:4, 15:1 are observed in two gene interactions, depending on their type of interaction.

10. The microscopic observation of the events during mitotic cell division, lend proof to the chromosomal theory of inheritance. The chromosomes are duplicated during the interphase, and the mechanism of distribution exactly equally, in number and content, is evident in the different stages of the mitotic cell division.

11. Segregation of alleles during gamete formation, independent assortment of genes present on different chromosomes, recombination between linked gene combinations can be visualised in the events occurring during meiosis, a special type of cell division of germ (sex) cell mother cells.

    (*a*) Segregation of alleles of a gene pair is evident from the segregation of the members of homologous chromosomes at anaphase 1 stage of meiosis.

    (*b*) Independent assortment of genes on different chromosomes happens due to different orientation of the homologous chromosome pairs on the metaphase 1 plate during meiosis.

    (*c*) Independent assortment of genes on different chromosomes results in recombinant gene combinations. Recombination of genes present on the same chromosome, is due to exchange of parts, between members of a homologous chromosome pair, in the region between linked genes. This exchange, happens during meiosis 1, and visible as, cross overs between the homologous chromosome pairs. The frequency with which recombination takes place between genes on the same chromosome, depends on the distance between them: the farther they are, the more frequent is recombination.

12. Gene number, even in simple organisms is in thousands, while, the number of chromosomes, is in units or tens. Chromosomes have thousands of genes. Genes present on the same chromosome are termed linked genes and they form one linkage group. The sequence of genes present on a chromosome and the distance between them (gene mapping) was found out during 1900s by conducting three-point test crosses; now DNA sequencing data gives detailed gene maps.

13. Predicting phenotype of an individual, from the knowledge of its genotype, called mapping phenotype to genotype, is not direct, and therefore, not simple. Gene expression is influenced by, the effect of other genes in the genome, and, by environmental conditions. The degree to which, a phenotype is the result of its genotype, is termed phenotypic plasticity. The extent to which an organism's phenotype helps in predicting the genotype, is termed genetic canalization. Mapping the phenotype to the genotype is now attempted using computer-based programmes: genome-wide association studies (GWAS) and genome regulatory networks (GRN) merged into a single structural frame work through structural equation modelling (SEM). These programmes are built from the knowledge of the genomes (entire genetic content) of the organism.

3.12 SAMPLE QUESTIONS

3.12(a) Subjective Questions

Q.1. Describe the experiments that led to the discovery of the principles of heredity. State the principles.

Q.2. What is dominance? Describe the various levels of dominance. How does the judgement of the level of dominance of a gene vary depending on the type of analysis?

Q.3. What is gene interaction? Describe two gene interactions with five examples.

Q.4. What are single gene disorders? How are they classified? Describe the inheritance pattern in different types of single gene disorders.

Q.5. What is an allele? What are multiple alleles? Describe with examples.

Q.6. Describe how mitosis helps in inheritance.

Q.7. Detail with diagrams the steps in meiosis which lead to segregation and independent assortment of genes on different chromosomes. Add a note on how recombinant gene combinations of genes present on different chromosomes arise during meiosis.

Q.8. A couple are both of O blood type. Predict the type of blood group of their children. Explain it with a diagrammatic representation of the mating results showing the genotypes and phenotypes of the parents and offspring.

Q.9. A red fruit in tomato is dominant to orange, entire leaf morphology is recessive to dissected leaf. The genes for fruit and leaf morphology are on different chromosomes. Describe and detail the results of a cross between pure breeding red fruit and dissected leaf with an orange fruit and entire leaf plants up to two generations of offspring. What do the results indicate?

3.12(b) Objective Questions

Q.1. Mendel's stated principles of heredity are based on

(*a*) Mathematical principles.   (*b*) Phenotypes.

(*c*) Genotypes.   (*d*) Alleles.

Q.2. Two gene interactions

(*a*) Affect the same character.   (*b*) Result in dihybrid ratios.

(*c*) Sometimes produce progeny with new phenotypes

(*d*) All statements above are true.   (*e*) Some statements among (*a*), (*b*), (*c*) are true.

Q.3. Alleles of a wild type gene

(*a*) Can be multiple.   (*b*) Result from mutation.

(*c*) May or may not have expression.   (*d*) All statements above are true.

(*e*) Some statements among (*a*), (*b*), (*c*) are true.

Q.4. Sex linked single gene disorders

(*a*) Are recessive always.

(*b*) Show different inheritance pattern depending on which parent has the disorder.

(*c*) Cause death (lethal).

(*d*) Are always more frequent in males than females.

Q.5. Incomplete dominance results in a hybrid genotype to have a phenotype

(*a*) Intermediate between the two parents.

(*b*) Of both parents.   (*c*) Of either parents.

(*d*) Of one of the parents.

Q.6. Linked genes show recombination because of

(*a*) Crossing over between chromosomes.

(*b*) Pairing of chromosomes.   (*c*) Segregation of chromosomes.

(*d*) Segregation of chromatids.

Q.7. Meiosis II is essential to

(*a*) Increase the number of sex cells.

(*b*) Reduce the chromosome number to half.

(*c*) Facilitate recombinants of linked genes to segregate into different gametes.

(*d*) All the above.

(*e*) Some of (*a*), (*b*), (*c*) are true.

Q.8. Independent assortment of genes happens

(*a*) When genes are present on different chromosomes.

(*b*) When genes are present on the same chromosome.

(*c*) When genes are present very close to each other.

(*d*) During mitosis.

Q.9. Single gene disorders in humans can be

(*a*) Corrected with one course of medication.

(*b*) Require life time of continuous treatment.

(*c*) Can be prevented by vaccination.

(*d*) Have no treatment options.

ANSWERS

1. (*a*)   **2.** (*d*)   **3.** (*d*)   **4.** (*b*)   **5.** (*a*)   **6.** (*a*)   **7.** (*e*)   **8.** (*a*)   **9.** (*b*)

4 Biomolecules

Life is a relationship between molecules.

Biology today is moving in the direction of Chemistry. Much of what is understood in the field is based on the structure of molecules and the properties of molecules in relation to their structure. If you have that basis, then biology isn't just a collection of disconnected facts.

It is possible with ... carbon ... to form very large molecules that are stable. This results from the stability of the carbon-to-carbon bond. You must have complexity in order to achieve the versatility characteristic of living organisms. You can achieve this complexity with carbon forming the molecular backbone.
~ Linus Pauling

The elements of the living body have the chemical peculiarity of forming with each other most numerous combinations and very large molecules, consisting of five, six or even seven different elements.
~ William Thierry Pre

4.1 INTRODUCTION

All living organisms from microbes to plants to animals, though extremely diverse, have the same chemical constituents, consisting mainly of carbon, hydrogen, nitrogen and oxygen. All cells evolved from a common ancestor and use the same kinds of carbon-based molecules. The composition of carbon in the cell depends on the surroundings and function of the cells. Non-living matter, like the earth's crust and rocks, have similar chemical compounds as the living beings, but, the relative proportion of carbon and hydrogen, is lower. All carbon compounds found in living organisms are termed biomolecules. The carbon skeleton in biomolecules can be linear, branched, cyclic, or aromatic. Biomolecules are thus organic molecules made by living organisms. They are the building blocks of cells and form the basis of life – growth and maintenance of living beings. Other than in living organisms, organic molecules including basic biomolecules like amino acids and carbohydrates are found only in trace amounts in the physical world like earth's crust, sea and the atmosphere.

Biomolecules are larger than typical organic molecules. Small biomolecules have molecular weights over 100 Da, while most biomolecules have molecular weights in thousands, millions, or even billion Da. The small biomolecules – monosaccharides, amino acids, nucleotides, glycerol and fatty acids, are the building blocks of larger biomolecules. Large biomolecules constitute polysaccharides (carbohydrates), polynucleotides – DNA and RNA (nucleic acids), polypeptides formed from linking of amino acids (proteins), and lipids (glycerol + fatty acid). Macromolecules are non-covalently held together into supramolecular complexes like, cell wall (polysaccharides), cellular membranes (lipids and proteins), ribosomes (RNA and proteins) and chromosomes (nucleic acid (DNA) and proteins).

The biomolecules in living beings are analysed by grinding their mass or tissue in trichloroacetic acid and filtering. The filtrate consists of acid soluble small biomolecules like vitamins, pigments, alkaloids, amino acids, nucleotides and lipids (with a molecular weight less than 1000 KD), while the retentate contains acid insoluble large biomolecules like proteins, polysaccharides and nucleic acids (more than 1000 KD). Lipids are non-polymeric acid insoluble biomolecules, whose molecular weight does not usually exceed 800 Da; non polar lipids with a molecular weight of 1500 Da have been reported. Lipids are actually small biomolecules. However, large number of lipid molecules, may bind non covalently, to form very large structures like vesicles. Lipids, in addition to being present in free form in the cytoplasm, are major components of the membrane system of a cell. When the living tissue/mass is ground, the membranes of the cell are broken into pieces. The free lipids being water insoluble, form vesicles, and, along with the broken bits of the membranes, get separated with the acid insoluble pool of macromolecular fraction. Lipids are thus, not strictly macromolecules.

Biomolecules, besides being differentiated as micro and macro, are also, classified as primary and secondary metabolites. Biomolecules with known physiological functions like carbohydrates, proteins, nucleic acids and lipids are termed primary metabolites. The secondary metabolites have no known physiological functions in the producer plants – they are the pigments, scents, latex, toxins, alkaloids, tannins etc. Some of the secondary metabolites have accessory functions in the producer organisms like conferring immunity. Some secondary metabolites are valued for their physiological effects in humans e.g. morphine, caffeine, nicotine, quinine etc.

Living organisms also contain several inorganic compounds. These inorganic compounds are found in the ash, when the living tissue is burnt, allowing the carbon compounds to be burnt.

The components of living organisms in terms of percentage of the total cellular mass are: water — 70–90%, proteins —10–15%, nucleic acids — 5–7%, carbohydrates — 3%, lipids — 2%, and, inorganic substances — 1%.

All living organisms, despite their uniqueness, are made of the same kind of biomolecules which perform similar functions in all of them. The vast diversity in living beings is due to the differences in the hereditary (genetic) material constituted by DNA/RNA – the macro biomolecules. The differences in the nucleic acid content are directly reflected in the diversity of the proteins, since they are the products of expression of the genetic material. Proteins constitute the major component of the macro biomolecules in a cell and are highly diverse in the living world. Each protein and nucleic acid have a characteristic information in the sequence of its sub units. Carbohydrates and lipids are of fewer types and less diverse among the living beings.

Large biomolecules have a typical 3D structure (shape). All biomolecules share in common, a fundamental relationship between structure and function. Nucleic acids among biomolecules are self-replicating.

The important biomolecules – carbohydrates, proteins, lipids and nucleic acids and their constituents are described in this module. They are listed in **Table 4.1**.

Table 4.1: Biomolecules

Biomolecule	Name	Constituents
Carbohydrate	Monosaccharide	Three to seven carbon compound.
	Disaccharide	Two monosaccharides.

Biomolecule	Name	Constituents
	Oligosaccharides	Three to ten monosaccharides.
	Polysaccharides	More than ten to several hundred monosaccharides.
Proteins	Peptide	A chain of up to fifty amino acids.
		One or more peptides.
Nucleic acids: DNA & RNA – Deoxyribo/Ribo Nucleic Acid	Nucleotide	A unit with a pentose sugar (ribose/deoxyribose), a nitrogen base and a phosphate group.
		Polymers of nucleotides.
Lipids	Fatty acids	Hydrocarbon chain with an acid group.
	Fats and oils	Glycerol + three fatty acids.

4.2 CARBOHYDRATES

Carbohydrates are primarily synthesized by plants, algae and photosynthesizing bacteria through photosynthesis. Annually, more than 100 billion metric tons of CO_2 and H_2O are fixed by plants as carbohydrates. They are the most abundant organic compounds in the plant world. Carbohydrates account for approximately three-fourths of the dry weight of plants. Animals (including humans) get their carbohydrates by eating plants. They do not store much of what they consume, but utilize them for life activities. In fact, less than 1% of the body weight of animals, is made up of carbohydrates.

Living organisms mainly use carbohydrates to generate energy for their activities. The energy released due to breakdown of carbohydrate during respiration is the chief source of energy for living beings. Carbohydrates are stored in the form of starch in plants and glycogen in animals. They also serve as structural components. The cell wall of plant cells is made of cellulose and those of fungi and exoskeleton of crustaceans is made of chitin.

Carbohydrates are also called saccharides (meaning sugar in Latin) since most of them are sweet to taste. Carbohydrates are aldehyde (RCHO) and ketone (RCOR) derivatives of polyhydroxy alcohol. They are expressed by the molecular formula $C_x (H_2O)_y$. Carbohydrates can be classified as monosaccharides, derived monosaccharides, disaccharides, oligosaccharides and polysaccharides.

The simple micro carbohydrates (with molecular weights of around 100–200 Da) are termed mono and disaccharides. They are the major source of energy for living organisms being the substrates for respiration. Polymers of monosaccharides with molecular weights often in the hundreds of kilodaltons are called polysaccharides. Polysaccharides have definite shapes and are either constituents of structural elements or serve as source of stored metabolic energy. Monosaccharides and disaccharides are sweet, crystalline and water-soluble. Polysaccharides are not sweet nor in crystalline form and are insoluble in water.

Carbohydrates are associated with all the other three polymer biomolecules – proteins, lipids and nucleic acids. Carbohydrates linked to proteins are called glycoproteins. Carbohydrates attached to lipids are termed glycolipids. Carbohydrates – ribose or deoxyribose are present in DNA and RNA, constituting the backbone of their helical strands.

Monosaccharides contain three to six carbon atoms. The most common monosaccharides are glucose, fructose and galactose. These are hexose sugars with the carbon atoms in a linear chain, or, alternatively as a ring. The pentose sugar ribose and its derived form, deoxyribose (removal of oxygen from ribose), are a constituent of nucleic acids and coenzymes (ATP, NAD and NADP).

Monosaccharides that are chemically modified are termed derived monosaccharides; e.g. glucosamine (with an amino group), glycosides, sugar phosphates, amino sugars and vitamin C. Glucuronic acid, an oxidised form of glucose, is an important coupling agent. Many drugs, pesticides and hormones are coupled with glucuronic acid and excreted in urine or bile, as glucuronides.

Disaccharides are formed from two monosaccharides; e.g. sucrose (fructose + glucose), maltose (two glucose molecules), lactose (glucose + galactose). The bond that holds the monosaccharides together is called a glycoside bond.

Micro carbohydrates, that are formed by condensation of three or more monosaccharides, are termed oligosaccharides; e.g. raffinose (trisaccharide), trehalose (tetrasaccharide). Short chains of fructose – fructo oligosaccharide, and galactoside – galacto oligosaccharide are present in fruits and vegetables and constitute dietary fibre. Oligosaccharides give specificity to the blood as O, A, B and AB types (**Figure 4.1**). These surface-bound oligosaccharides on the red blood cell, designated as A, B, AB and O, act as antigens in the case of A, B and AB blood group types. The oligosaccharide attached to the erythrocytes in O blood group individuals does not have antigenic property. Oligosaccharides play an important role in cell recognition and cell adhesion and signalling pathways. They are extensively used as pharmacological supplements and in cosmetics.

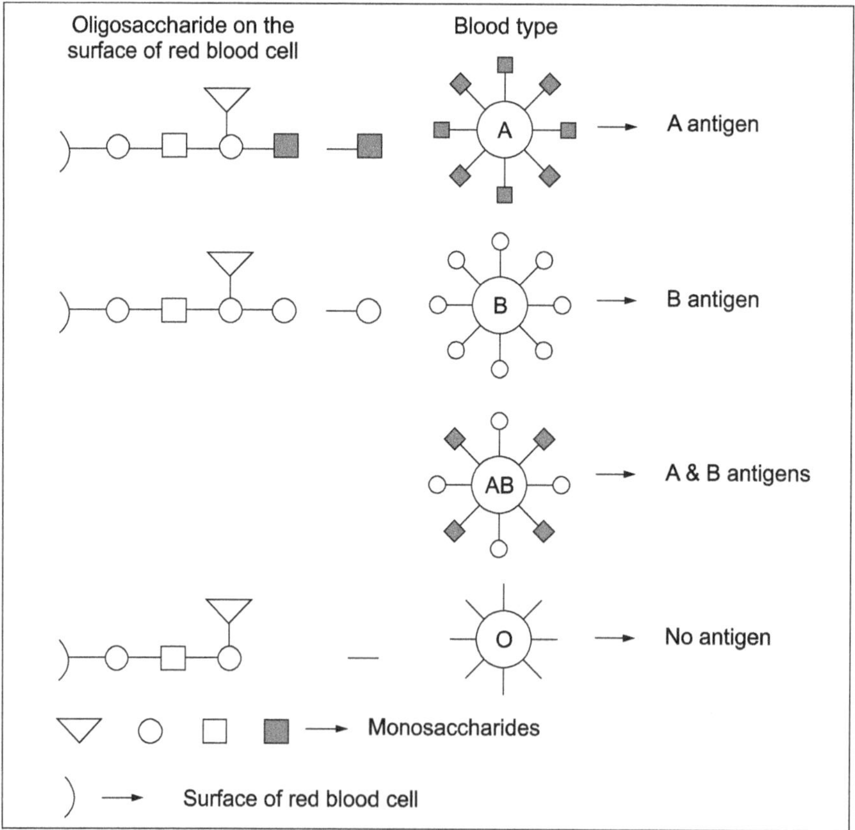

Figure 4.1: Oligosaccharides that are attached to the surface of the red blood cells differing in some of the monosaccharide constituents. *Blood groups are classified as A, B, AB or O depending on the type of oligosaccharide on the red blood cells. The oligosaccharide on the red blood cells of O type blood group does not act as an antigen and hence no antibodies are induced against them. The O blood group is therefore a universal donor and can be transfused to any blood group individual without adverse effects.*

Polysaccharides can form straight, or branched chains. The individual components in the polysaccharide, may all be similar when it is termed a homopolysaccharide; e.g. glycogen, starch, cellulose, inulin and chitin, or, different i.e., a heteropolysaccharide; e.g. peptidoglycan, hyaluronic acid (present in cartilage and tendon, the lubricant of joints in the body, and in the vitreous of the eye, where it provides a clear, elastic gel that holds the retina in its proper position).

4.2.1 Starch and Glycogen

The polymer of α glucose serves as a storage carbohydrate in living organisms – as starch in plants and glycogen in animals. These two complex carbohydrates, though chemically similar, have different structures. Starch gives a blue colour with iodine. Starch is made up of two components: amylose which is a chain of α glucose units without branches (15 to 20%) and amylopectin, which is branched (80 to 85%). Amylose is water soluble and made of 200 to 1,000 glucose molecules linked in a straight chain. Amylose acquires a spiral structure that contains six glucose units per turn. Amylopectin is water insoluble and is a branched molecule of 2000 to 20,00,000 glucose units (**Figure 4.2**).

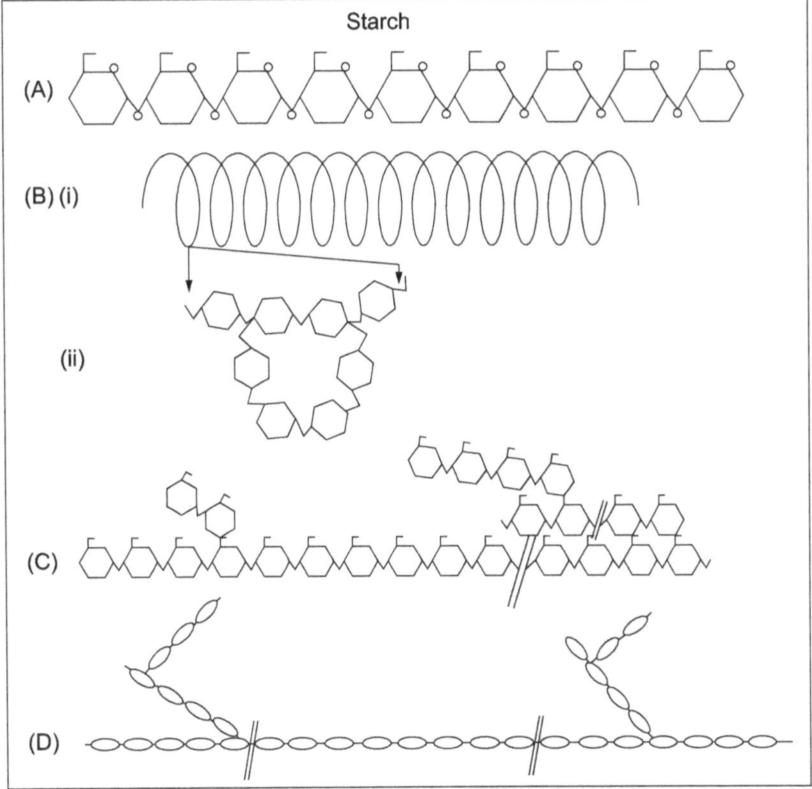

Figure 4.2: Structural components of starch – a polymer of α glucose molecules. *Starch contains two types of molecules: amylose and amylopectin.* **A& B**. *Amylose, a straight chain of α glucose molecules that takes a helical form with six glucose molecules per helix.* **C&D**. *Amylopectin, a branched chain of α glucose molecules || represents a break (to accommodate the large molecule in the figure).*

Glycogen is a larger polymer and has more frequently branched structure than amylopectin (**Figure 4.3**). It is present in liver, brain and muscles. Glycogen stains red with iodine.

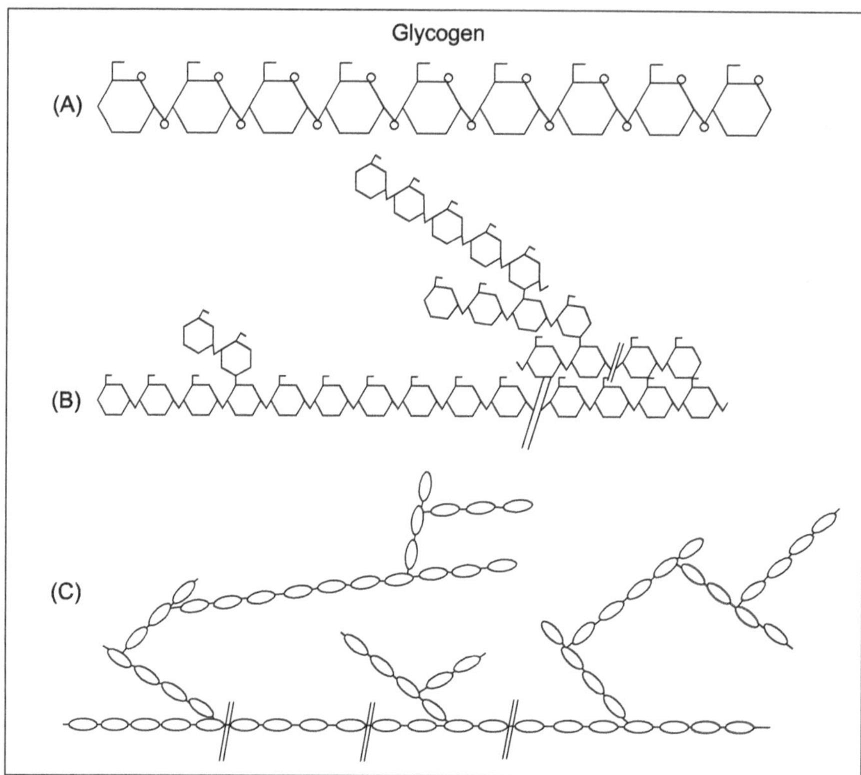

Figure 4.3: Structure of glycogen. *It is a highly branched chain of α glucose molecules. It is similar to amylopectin except that it is more frequently branched.* **A.** *Chain of α glucose molecules.* **B and C.** *Highly branched chain of α glucose molecules.* || *represents a break (to accommodate the large molecule in the figure).*

4.2.2 Cellulose

Cellulose is yet another polymer of glucose (the β form), being the main constituent of plant cell walls. It is only present in plants. It is the most abundant polysaccharide in nature. Paper made from wood pulp and cotton fibre are made of cellulose. A molecule of cellulose contains ~ 6000 β glucose units in a straight chain and has a very stable structure (**Figure 4.4**). Every alternate glucose molecule in cellulose is flipped over, thereby tightly packing the long chain. This gives cellulose, its rigidity and high tensile strength required by plant cells. The bonds between glucose molecules in cellulose cannot be broken down by human digestive enzymes. Therefore, it is termed dietary fibre and passed out as undigested matter. The herbivores such as cows, buffaloes, and horses can however break the glucose–glucose bonds in cellulose and thus use it as a source of food.

The second most abundant polysaccharide after cellulose is chitin. It is a long chain polymer of N-acetylglucosamine. Chitin is the main component of cell walls of fungi and forms the exoskeleton of crustaceans (such as crabs, shrimps, lobsters) and insects. Being strong and flexible and biodegradable, it is used as surgical thread, which wears out with time, as the wound heals.

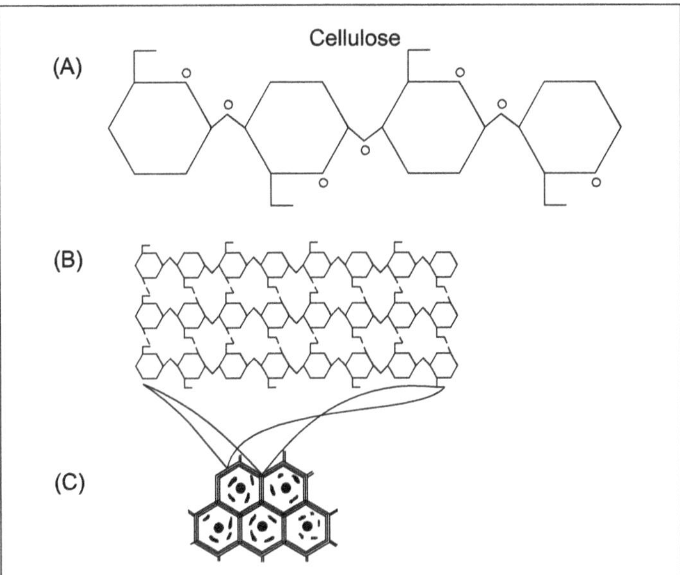

Figure 4.4: Structure of cellulose. *A. Cellulose is a straight chain of β glucose molecules.* **B.** *The individual chains of β glucose molecules; the molecules flip over alternatively and adopt a stiff rod like conformation; several chains run in parallel and are linked by hydrogen bonds.* **C.** *Plant tissue constituted by cells with thick cell wall. Cellulose is the chief constituent of plant cell walls.*

4.3 NUCLEIC ACIDS

Nucleic acids store, transmit and translate genetic information. They are two types – deoxyribo nucleic acid (DNA) and ribo nucleic acid (RNA). DNA is the hereditary material in most living beings. It is a self-replicating molecule. It is passed on from the parent to the offspring. RNA is the hereditary material in some viruses. There are several kinds of RNA, the main three being: messenger RNA (mRNA), transfer RNA (tRNA) and ribosomal RNA (rRNA). The mRNA, tRNA and rRNA facilitate the expression of the genetic material (described in module 3). The other RNAs have varied functions.

The nucleic acids are large molecules being polymers of nucleotides. The DNA is a longer polymer than RNA.

4.3.1 Nucleotides

The nucleotides have three components: a pentose (five carbon) sugar – deoxyribose in DNA and ribose in RNA, a phosphate, and a nitrogen base. The nitrogen base is attached to the first and the phosphate group to the fifth carbon atoms of the sugar (**Figure 4.5**). There are two kinds of nitrogen bases: purines (with two carbon rings) and pyrimidines (with one carbon ring). There are two kinds of purines: adenine and guanine and three kinds of pyrimidines: thymidine, uracil and cytosine. Cytosine is common to both DNA and RNA. Among the other two pyrimidines, thymidine is present in DNA and uracil in RNA. Depending on the nitrogen base attached to the sugar, nucleotides in DNA are of four kinds: deoxyribo thymidine tri phosphate dTTP, deoxyribo cytosine tri phosphate (dCTP), deoxyribo guanine tri phosphate (dGTP) and deoxy adenine tri phosphate (dATP). Adjacent nucleotides bind together by forming a bond between the phosphate group attached to the fifth carbon atom of ribose or deoxyribose in one nucleotide with the OH group attached to the third carbon atom of the sugar (ribose/deoxyribose) of the other nucleotide (**Figure 4.5**).

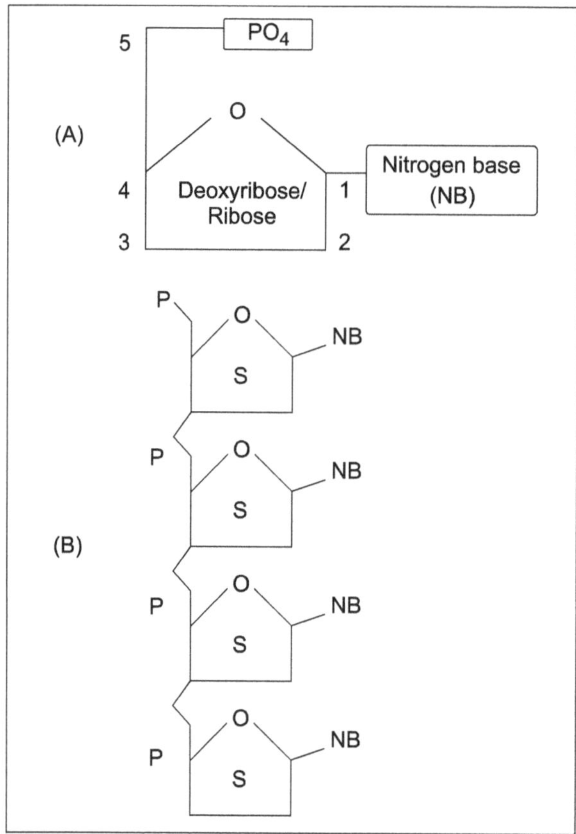

Figure 4.5: Molecular structure of a nucleotide and a polymer of nucleotides. A. *A nucleotide with pentose sugar with phosphate group attached to the fifth and a nitrogen base (NB) to the first carbon atom.* **B.** *A polymer of nucleotides with adjacent nucleotides attached through the phosphate group on the fifth carbon atom of the sugar of one with the hydroxy group on the third carbon atom on the sugar of the other. NB stands for nitrogen base (adenine/thymidine, guanine/cytosine/uracil)*

In the DNA molecule, a chain of nucleotides intertwines with another chain with the opposite nucleotides on the two chains bonding together. A pairs with T and G with C. Thus, a double helical structure is formed. The RNA molecule is single stranded and folds on itself in certain regions forming double strands at certain regions along its length.

Both the nucleic acids, DNA and RNA, have similar constituents in all living organisms – the difference lies only in the number and sequence of nucleotides. The sequence of nucleotides in the nucleic acid, carries the genetic information that translates into the overall form and characteristics of a living organism. The expression and replication of the nucleic acid, is facilitated by its structure. The detailed structure of DNA, its mode of replication and expression are described in chapter 6. The role of different types of RNA is also described in chapter 6.

4.4 PROTEINS

Proteins (from the Greek word 'proteios' meaning primary importance), are the most abundant of the organic molecules in the living system. All proteins contain C, H, O and N. It is the presence of N (nitrogen), that distinguishes them chemically from carbohydrates and fats. Some proteins have S in addition, and in a few proteins, P and other elements may be present. They are the work horses in the cell, responsible for various functions carried out by a cell. They also constitute the structural elements of a cell, being components of cell membranes. Proteins give energy to the body and also aid in building and repair.

Proteins are polymers of α amino acids. There are 20 different amino acids in proteins. The type, number and sequence of amino acids in different proteins differ. Proteins are the products of gene expression; the sequence of amino acids in the protein is determined by the nucleotide sequence of the gene which codes for it. They have a molecular weight of tens to several hundred kilo Daltons. They fold into specific shapes and their function is based on their structure. Protein structure is determined by the sequence of the amino acids in the polypeptide chain.

Proteins are classified as simple and conjugated proteins. Simple proteins consist of only amino acids e.g. albumins, globulins, glutelins, prolamines, histones, protamines and scleroproteins. Conjugated proteins contain a non-protein component e.g. nucleoproteins, glycoproteins, lipoproteins, phosphoproteins, chromoproteins and metalloproteins.

4.4.1 Amino Acids

Amino acids are small molecules with molecular weights of around 100–200 Da. They are the building blocks of all proteins. They are used in living systems to produce energy, to synthesize proteins and convert into other biologically functional substances.

Amino acids are bi-functional compounds that have a basic – an amino (NH_2), an acidic – a carboxyl (COOH), and a R group – carbon compound, attached to a carbon atom. They have the general formula of:

$$
\begin{array}{c}
NH_2 \\
| \\
R - C - COOH \\
| \\
H
\end{array}
$$

Because of the carboxyl group, an amino acid has a negative charge and due to its amino group, it is positively charged. Because amino acids contain groups with both charges, they have no net charge, and therefore, called zwitterions or dipolar ions. The net charge of the molecule is affected by the pH. The pH at which an amino acid exists as a zwitterion (equal positive and negative charge) is called the isoelectric point. Most of the amino acids, with the exception of a few, exist as zwitterions at a pH of 7. At this point, an amino acid does not migrate in an electric field. The proteins can be separated from each other based on their differences in isoelectric points.

The α amino acids have, R group attached to the carbon atom adjacent to COOH group. The amino acids differ in the R group they carry; e.g. when R is H, it is glycine, when it is methyl, the amino acid is alanine, when it is hydroxy methyl, it is serine. Amino acids are represented with the first three letters of their name or a single alphabet. Amino acids are classified based on their chemical constitution and structure (**Table 4.2**).

Table 4.2: Types of amino acids

Type	Nature	Example
Neutral	Have one carboxyl and one amino group.	Valine, Leucine, Glycine, Alanine, Isoleucine
Acidic	Contain an extra carboxyl group.	Glutamic acid, Aspartic acid
Basic	Contain an extra amino group.	Lysine, Arginine
Sulphur containing	Have a sulphur group.	Cysteine, Cystine, Methionine
Alcoholic	Have an OH group.	Serine, Threonine
Aromatic	Have a cyclic structure with a straight side chain with carboxyl and amino groups.	Tyrosine, Tryptophan, Phenylalanine
Heterocyclic	Have a cyclic structure with the N in the amino group within the cycle.	Proline, Hydroxyproline, Histidine

Of the twenty amino acids, humans cannot synthesize eight of them and synthesize two of them very slowly. The former are called essential amino acids. They are leucine, isoleucine, lysine, methionine, phenylalanine, threonine, tryptophan and valine. The latter two are arginine and histidine and termed semi-essential amino acids.

The amino acids besides being constituents of proteins are converted into other important compounds. For example, tyrosine is converted into the pigment melanin and also the hormones thyroxine and adrenaline. Glycine is involved in the synthesis of haem; tryptophan is involved in the formation of the vitamin nicotinamide and the plant hormone – indole acetic acid (IAA).

4.4.2 Polypeptides

Proteins are the most abundant and versatile biomolecules, having the most diverse range of functions among biomolecules. These are discussed in detail in chapters 5 and 7. A cell may contain several thousands of different kinds of proteins.

Proteins are made up of one or more peptide chains. Each peptide is a polymer – a chain of several amino acids with the amino acids linked together by a peptide bond. The peptide bond is formed between the COOH group of one amino acid and the NH_2 group of the neighbouring amino acid.

Simple proteins contain only polypeptide chains, complex proteins contain non-peptide prosthetic groups attached to them. They are called conjugated proteins. They are the lipoproteins (prosthetic group: lipid; e.g., chylomicron), phosphoprotein (prosthetic group: phosphate; e.g. casein of milk), metalloprotein (prosthetic group: metal; e.g. ferritin), glycoproteins (prosthetic group: carbohydrate; e.g. mucin), nucleoproteins (prosthetic group: nucleic acid; e.g. histones) and chromoproteins (prosthetic group: pigment; e.g. cytochrome).

Depending on the structure they assume, proteins are classified as fibrous or globular. The fibrous proteins are mostly present in animals. They have peptide chains running parallel held together by hydrogen or disulphide bonds; e.g. keratin (hair), myosin (in muscles), collagen (skin). The globular proteins have peptide chains coiling to form a spherical structure; e.g. albumin, insulin, haemoglobin.

Based on solubility in water, proteins can be differentiated as soluble, e.g. globular proteins, antibodies, hormones, serum albumin; or insoluble, e.g. myosin, fibrinogen.

When the protein is exposed to high temperature or unsuitable pH, its structure is disturbed. This is referred to as denaturation. A denatured protein loses its function. Denaturation is reversible in some cases.

The amino acid sequence in a peptide can be predicted from the nucleotide sequence of the gene coding it. The structure of a protein which is dependent on its amino acid sequence, can be predicted using bioinformatic tools. The function of a protein can be predicted from its structure. Determining the actual structure of a protein is technically difficult and expensive. It is done using X ray crystallography. Therefore, the structures of only important proteins have been determined.

The proteins by twisting and folding into complex three-dimensional shapes thus transform the informational content (sequence of nucleotides) in the gene into mechanically functional shapes; every turn and fold of a protein molecule is a geometrical key that carries out a specific function. The structural hierarchy of the proteins is described in chapter 7.

Since the informational content of a gene is translated into the protein, a central intriguing question in biology was as to which came first, the genetic code or the proteins? This is chicken or egg first? like puzzle. How could a genetic code work without proteins? How could there be proteins that were not coded for by genes? With the discovery of "self-catalytic" RNA (a nucleic acid which can do a function (catalysis) of a protein – technically a two in one i.e., both a nucleic acid and a protein at the same time), it is now settled, that early life began with a nucleic acid, i.e., the gene came first.

4.5 LIPIDS

Lipids (from Greek word 'lipos' meaning fat) constitute 2 to 3% of cell weight. They are chemically esters of alcohol and fatty acids. In true fats, the alcohol component is glycerol. Glycerol has three OH groups which can form ester bonds with fatty acids. A fatty acid has a carboxyl group linked to a carbon atom which is attached to an R group. The R group could be a methyl ($-CH_3$), or ethyl ($-C_2H_5$) or higher number of $-CH_2$ groups (1 to 23 carbons). Most biological fatty acids contain an even number of carbon atoms, because, the biosynthetic pathway common to all organisms involves chemically linking two-carbon units together. For example, palmitic acid has 16 carbons and arachidonic acid has 20 carbons including carboxyl carbon. Most naturally occurring fatty acids have an unbranched chain of carbon atoms, with a carboxyl group ($-COOH$) at one end, and a methyl group ($-CH_3$) at the other end. The R group may be saturated (all single bonds), e.g. stearic acid, palmitic acid; or unsaturated (with double bond/s). The R group in a fatty acid can be monounsaturated (single double bond) e.g. oleic, arachidonic, linoleic, linolenic acids or polyunsaturated (more than one double bond) e.g. eicosapentaenoic and docosapentaenoic acids. Animal fats have unsaturated fatty acids and are solids at room temperature, while those from plants and fish, have saturated fatty acids and are usually liquids. Liquid fats are called oils.

Fats in which one alcohol group of the glycerol is involved in ester bond formation with one fatty acid are called monoglycerides, when two alcohol groups bond with two fatty acids, it is a diglyceride and when all three alcohol groups are bonded by fatty acids, they are called triglyceride. Triglycerides with low melting point which remain as liquids at room temperature are called oils and those with high melting point and are solids at room temperature are called fats.

```
        H                   H                   H                   H
        |                   |                   |                   |
    H — C — OH          H — C — FA          H — C — FA          H — C — FA
        |                   |                   |                   |
    H — C — OH          H — C — OH          H — C — FA          H — C — FA
        |                   |                   |                   |
    H — C — OH          H — C — OH          H — C — OH          H — C — FA
        |                   |                   |                   |
        H                   H                   H                   H
     Glycerol          Monoglyceride        Diglyceride         Triglyceride
```

FA: Fatty acid

Lipids have only fatty acid and glycerol, e.g. oils, fats and waxes. Conjugated lipids have additional groups attached to the lipid, e.g. glycolipids, chromolipids (carotenes), lipoproteins, and phospholipids (important components of cell membranes). Phospholipids have a hydrophilic head and hydrophobic tail and these molecules line up to form a bilayer (**Figure 4.6**).

Lipids are small water-insoluble molecules with molecular weights of 750–1500 Da. Lipids show considerable solubility in organic solvents such as alcohol, ether and chloroform. Lipids do not polymerize to form macromolecules, but they can aggregate to form very large structures. Because they are defined by their water-insolubility, they are chemically more diverse than the other classes of biomolecules. There are about six major types of lipids. Lipids have several biological functions. Lipids serve both in storage and as sources for energy production. Triglycerides are storage compounds for the reserve energy of the cell. Lipids are in a highly reduced form and therefore release a lot of energy when burnt. Lipids act as carriers of natural fat-soluble vitamins such as vitamin A, D and E. They form structural elements of cell membranes. Lipids provide insulation for plants and animals. For example, they help keep aquatic birds and mammals dry because of their water-repelling nature. A large amount of fat is deposited in the subcutaneous layers of aquatic mammals such as whale and in animals living in cold climates. The myelin sheath of nerve cell which acts as an insulator, consists of lipids.

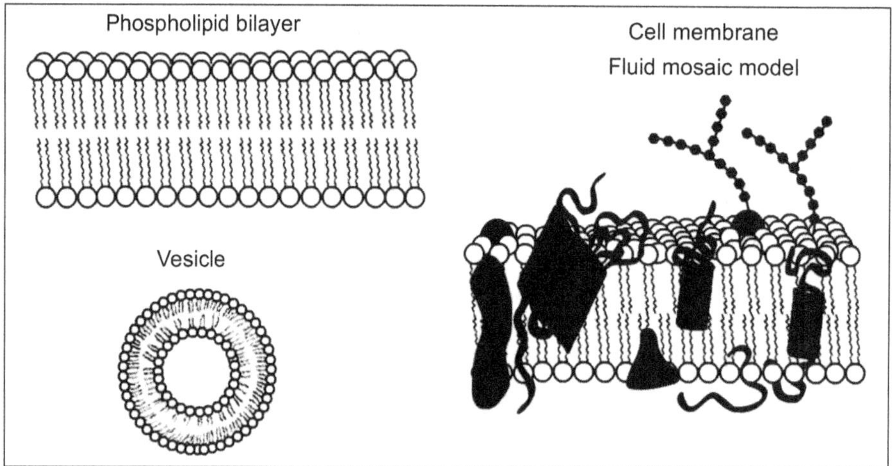

Figure 4.6: Phospholipids constituting the cell membrane and vesicles. *The hydrophilic head region of the lipid molecule faces water and the hydrophobic tail region is away from water forming a bilayer. In the fluid mosaic model of the organization of cell membranes, it is conceptualized that the proteins are either inserted in the lipid bilayer with their ends projecting out at both surfaces of the lipid bilayer (transmembrane proteins) or attached on the surface of the lipid bilayer (surface proteins).*

Lipids are classified as simple lipids, compound lipids and derived lipids.

4.5.1 Simple Lipids

They are esters of fatty acids with alcohols. They are of two kinds:

(*a*) **Fats:** esters of fatty acids with glycerol.

(*b*) **Waxes:** esters of high molecular weight fatty acids with alcohols other than glycerol (true waxes) or sterols.

4.5.2 Compound Lipids

Compound lipids on hydrolysis yield other groups along with alcohol and fatty acids. They are the following:

(*a*) Glycero phospholipids: phosphoric acid and two fatty acids are esterified to glycerol (e.g. cephalins, lecithins).

(*b*) Sphingo phospholipids: The amino alcohol sphingosine replaces glycerol. They have phosphoric acid and choline in addition to fatty acid.

(*c*) Glycolipids: They have a carbohydrate linked to the lipid. e.g. glycosphingo lipid – on hydrolysis, it yields sphingenine, a fatty acid and sugar (galactose).

4.5.3 Derived Lipids

Derived lipids are formed from oils and fats during metabolic reactions. Steroids are considered derived lipids, though they do not have a fatty acid, because of their fat like properties. An important steroid is cholesterol which is a component of cell membranes. Sex hormones such as adrenocorticoids, cholic acid and vitamin D are all synthesized from cholesterol.

4.6 SUMMARY

1. Carbon rich organic compounds present in living organisms are called biomolecules. Biomolecules are large in size and usually polymers of smaller biomolecular units.

2. The important biomolecules are carbohydrates, proteins, nucleic acids and lipids. These are the major types of biomolecules present in all living organisms. The large biomolecules have specific three-dimensional structures suitable for their function. Nucleic acids have the capability to self-replicate.

3. Carbohydrates are synthesized by plants, algae and cyanobacteria through photosynthesis. They are the chief source of energy for all living organisms. They are represented by the molecular formula $C_x(H_2O)_y$. Carbohydrates can be classified as monosaccharides, disaccharides, oligosaccharides and polysaccharides.

4. Monosaccharides contain three to six carbon atoms. The most common monosaccharides are glucose, fructose and galactose.

5. Disaccharides are formed from two monosaccharides; e.g. sucrose (fructose + glucose), maltose (two glucose molecules), lactose (glucose + galactose). The bond that holds the monosaccharides together is called a glycoside bond.

6. Micro carbohydrates, that are formed by condensation of three or more monosaccharides are termed oligosaccharides; e.g. raffinose (trisaccharide), trehalose (tetra saccharide).

7. Polysaccharides have definite shapes and are either constituents of structural elements (cellulose) or serve as source of stored metabolic energy (starch in plants and glycogen in animals).

8. Both starch and glycogen are polymers of α glucose molecules. Starch is made up of an unbranched glucose chain twisted into a helix called amylose and a branched glucose chain, called amylopectin. Glycogen is a larger molecule and has a more highly branched glucose chain than amylopectin.

9. Cellulose is a polymer of the β form of glucose. Several chains of glucose run in parallel and linked by hydrogen bonds. Cellulose is the chief substance in plant cell walls.

10. Nucleic acids are polymers of nucleotides. A nucleotide contains a pentose sugar (deoxyribose in DNA and ribose in RNA), a phosphate group and a nitrogen base. There are two classes of nitrogen bases: purines and pyrimidines. There are two purines: adenine and guanine and three pyrimidines: thymidine, cytosine and uracil. RNA has uracil in place of thymidine. Depending on the nitrogen base attached to it there are four kinds of nucleotides in DNA: dATP, dGTP, dCTP, dTTP. The sugar-phosphate forms the backbone strand with the nitrogen base projecting out. Two strands of DNA pair with the nitrogen bases between them bonding through hydrogen bonds: A with T and G with C. RNA is single stranded and folded in some regions to form double stranded regions.

11. Proteins are polymers of amino acids linked together by a peptide bond. A protein may be constituted by one or more polypeptides. The sequence of the amino acids determines the structure of a protein. The function of a protein is determined by its structure. The structure of a protein is determined by Xray crystallography which is an expensive technique. Bioinformatic tools are available to predict the structure of a protein from the sequence of its amino acids.

12. Lipids are small biomolecules. They are not polymers. Several lipid molecules aggregate. A lipid is formed by reaction between glycerol and fatty acids. A lipid with one fatty acid is termed monoglyceride, two fatty acids, a diglyceride and three fatty acids are triglycerides. Phospholipids constitute the membrane systems of cells. The phospholipid molecule has a hydrophobic head and a hydrophilic tail. These molecules aggregate with the heads facing the outside forming a bilayer. The tails of the lipid molecules in the two layers are in the centre. Proteins are both integrated in the lipid bilayer and present on its surface. This is the fluid mosaic model of arrangement of protein lipid bilayer of cell membranes and vesicles. Cholesterol is a derived lipid and plays an important role in synthesis of sex hormones and vitamin D.

4.7 SAMPLE QUESTIONS

4.7(a) Subjective Questions

Q.1. What are biomolecules? What are the main types of biomolecules? Describe briefly each of them.

Q.2. What are carbohydrates? How are they classified? Describe the structure of some important structural and storage carbohydrates.

Q.3. Describe the chemical constitution of nucleic acids and structural arrangement of their constituent units.

Q.4. Are lipids biomolecules? Why? Describe the chemical nature of lipids and their types.

4.7(b) Objective Questions

Q.1. Biomolecules are:

(*a*) Present in living organisms.

(*b*) Are organic carbon compounds.

(*c*) Have high molecular weight.

(*d*) Are polymers.

(*e*) All statements are true.

(*f*) Some statements are true.

Q.2. Carbohydrates are

(*a*) All water soluble.

(*b*) Some are water soluble.

(*c*) All have glycosidic bonds. (*d*) Can form complex polymers.

(*e*) All statements are true. (*f*) Some statements are true.

Q.3. Cholesterol is

(*a*) True lipid. (*b*) A steroid.

(*c*) Involved in production of thyroid hormone.

(*d*) A triglyceride.

Q.4. Blood groups in humans are determined based on the type of

(*a*) Lipid attached to the red blood cell. (*b*) Diglyceride attached to the red blood cell.

(*c*) Protein attached to the red blood cell. (*d*) Oligosaccharide attached to the red blood cell.

Q.5. The difference between cellulose and starch

(*a*) Cellulose is a polymer of α glucose units while starch is a polymer of β glucose.

(*b*) Cellulose is a polymer of β glucose units while starch is a polymer of γ glucose.

(*c*) Cellulose has a branched chain and an unbranched chain; starch has a highly branched chain.

(*d*) Starch has a branched chain and an unbranched chain; cellulose has a highly branched chain.

Q.6. Waxes

(*a*) Are derived lipids. (*b*) Have glycerol.

(*c*) Do not have glycerol. (*d*) Do not melt.

ANSWERS

1. (*f*) **2.** (*f*) **3.** (*b*) **4.** (*d*) **5.** (*d*) **6.** (*c*)

5 Enzymes

"Proteins hold the key to the whole subject of the molecular basis of biological reactions"

~ Linus Pauling

So many of the chemical reactions occurring in living systems have been shown to be catalytic processes occurring isothermally on the surface of specific proteins, referred to as enzymes, that it seems fairly safe to assume that all are of this nature and that the proteins are the necessary basis for carrying out the processes that we call life. **~ John Desmond Bernal**

Orgels first rule: *Whenever a spontaneous process is too slow or too inefficient, a protein will evolve to speed it up or make it more efficient.* **~ Leslie Eleazer Orgel**

5.1 INTRODUCTION

The perpetual physical and chemical changes that happen in living beings are the result of metabolism – a term used to describe the cocktail of biochemical reactions involving building (anabolic) and breaking (catabolic) substances. These biochemical reactions, occur under normal body temperatures, and narrow pH ranges, at an astounding pace. What is phenomenal about these reactions is that, they are accurate, mistake free processes (with very negligible errors), as against the natural law, of high speed, most often, deemed to be associated with disaster. The metabolic reactions happen at a speed, millions of times faster than, what would be expected, at the ambient conditions in the living systems. This is due to catalysis of these reactions by naturally occurring substances called enzymes.

5.2 PACE OF BIOCHEMICAL REACTIONS

Following are examples of the high speed with which some metabolic reactions are carried out.

DNA replication, that involves a series of phosphodiester bond formations between nucleotides, occurs in prokaryotes, at a rate of 1000 nucleotides per second; in eukaryotes, at 50 nucleotides per second (slower due to the dense packing of DNA).

Transcription – synthesis of RNA (also involving joining of ribonucleotides through phosphodiester bonds), happens at a rate of 10 to 100 nucleotides per second in prokaryotes, and 6 to 70 nucleotides per second in humans.

Translation – synthesis of polypeptides, that involves formation of peptide bond between amino acids, proceeds at a rate as fast as, joining of 20 amino acids per second in bacteria, and, 6 amino acids per second in rats.

Plant cells can perform the entire gamut of light and dark reactions of photosynthesis, for production of glucose molecule, in a matter of 30 seconds! One of the important reactions during photosynthesis – the splitting of water molecule by photosystem II, takes a few nanoseconds (one billionth of a second).

Breakdown of glucose to carbon di oxide (respiration), is estimated to happen within less than a millisecond. If respiration were to be measured as the number of ATP (energy currency) molecules generated, it is projected at, 70 ATP molecules per mitochondria per millisecond, or, about 14 microseconds per ATP produced, by a single mitochondrion.

The above described pace of various biochemical reactions, appears even more phenomenal, when the conditions in which they occur – body temperature and physiological pH, are considered. For example, the oxidation of a fatty acid to CO_2 and H_2O, occurs rapidly, within a narrow range of pH and temperature ambient in living systems, but, when carried out in a test tube, it requires extreme pH, high temperature, and corrosive chemicals. Breakdown of protein in living beings, takes place in less than four hours, at mild physiological temperature and pH. This would require 24 hours of boiling in 20% HCl if it were to be accomplished in the laboratory.

Enzymes, the biological catalysts derive their name from the word zymosis (= fermentation); the word meaning: en (in) zyme (yeast). Enzymes were recognized as the cause of fermentation by Louis Pasteur in late 1800. It was not until mid-1920's that an enzyme was extracted in pure form. The 1946 Noble prize in chemistry went to the scientists who isolated the enzymes in pure form and characterized them. Enzymes are globular proteins specialised to catalyse reactions. They have a site to comfortably fit in the reactants (substrate) of the reaction, called the binding site. More than 4000 enzyme catalysed reactions are known in living beings. Life activities would not happen, without the functioning of enzymes, which speedup biochemical reactions – a function they accomplish, through reducing activation energy required to initiate a reaction. Enzymes accelerate naturally possible reactions and are known to facilitate life processes in all life forms, from viruses to higher vertebrates. Enzyme catalysed reactions include, build up and replacement of tissues, conversion of food to energy and solar energy into food, disposal of toxic substances, reproduction etc.

Life activities are not possible without the play of the enzymes. Enzymes run the metabolism, and thus, life. Most enzymes are proteins.Of late (1980's onwards), RNA molecules with enzyme function (ribozymes) have been discovered. Enzymes being either proteins or RNA, are the expression products of genes. Enzymes are vital biomolecules synthesized by all living organisms (viruses get them synthesized through their hosts) to enable them to carry out life processes. This situation brings up the question of "chicken or egg first", whether it is the gene or the proteinaceous enzyme, that first arose, to make life possible. With the discovery of RNAs which function as enzymes, it is now settled that, the prediction of the maverick scientist, Francis Crick, that earliest life forms had RNA as the genetic material (the RNA world), is indeed logical. The RNA molecule has both the power of reproduction and catalysis – it has both information, and, the machinery to use the information.

Within the human body, enzymes are present in all fluids and tissues. Those present within cells, catalyse metabolic reactions. Enzymes in plasma membrane receive signals and transmit the signal, which triggers cell response through a cascade of reactions. Enzymes in the circulatory system regulate clotting of blood.

5.3 USE OF ENZYMES IN INDUSTRY AND MEDICINE

Understanding enzyme mediated reactions has helped in many practical applications in industry and medicine. Enzymes are extracted from, plant or animal sources, or, synthesized in microbes through fermentation, using recombinant gene constructs. Several types of microbes are used for enzyme synthesis like: yeast (*Saccharomyces cerevisiae*), fungi (*Aspergillus*), bacteria (*Bacillus subtilis*) and microalgae (*Dunaliella tertiolecta* and *Chlamydomonas reinhardtii*). Enzymes retain their activity even after extracting from cells and can be used in vitro.

5.3.1 Enzymes in Industry

A major chunk of enzymes used in industry are in food industry (45%); the others in: detergents (35%), textile (10%) and tanning (leather) (3%) (**Table 5.1**). Use of enzymes in these industries results in major reduction in costs. Moreover, enzymes are non-toxic and biodegradable.

Table 5.1: Industrial applications of enzymes

Enzyme	Use
Proteases, lipases, amylases	Detergents
Amylase, catalase	Textile industry
Protease, papain	Leather industry
Cellulase, xylanase, laccase and lipase	Paper industry
Esterase, papain, protease	Meat tenderizer
Bromelain, β amylase, glucose oxidase	Baking
β glucanase, αamylase, amiloglucosidase, protease	Brewing
Rennet, lactases, proteases, catalase	Dairy industry
Pectinase, β glucanase	Wine and fruit juice

5.3.2 Enzymes in Medicine

Enzymes are used in medicine in therapeutics (treatment) and diagnostics – as indicators of a disease.

5.3.2(a) Enzymes in Therapeutics

There are more than 300 marketed drugs, targeted to more than 70 enzymes, that work by inhibiting their activity. Blocking enzyme activity can, either kill a pathogen, or, correct a metabolic imbalance.

Enzymes make potent drugs, because of their specificity of the target, ability to act on numerous target molecules and their high efficiency in biological systems. They have very few, if any, side effects. The downside, however, is being proteinaceous, they are antigenic and induce antibody production which results in their quick clearing from the blood serum. There are many enzyme-based drugs for a wide range of medical conditions; some of the popular therapeutic enzymes are listed in **Table 5.2**. In some of the genetic disorders like SCIDS and Alkaptonuria, enzyme replacement therapy is given to provide the missing enzyme.

Table 5.2. Enzymes used as therapeutics

Enzyme	Used for
Streptokinase, Urokinase (thrombolytic enzyme)	Heart attack and blood clots in arteries of lung and other blood vessels.
β galactosidase (Ti lactase)	Lactose intolerance.

Enzyme	Used for
RNase (ranpirnase)	Many types of cancers.
Serine protease (thrombonin)	To stop bleeding.
Porcine, Bovine (pancreatic enzymes)	Pancreatic insufficiency.
Bromelain (from pine apples)	Improving digestion.
Proteases	Inflammation, arthritis and reduce symptoms of irritable bowel syndrome.
L asparaginase	Leukaemia
Streptodornase	Pain relief, remove debris in second- and third-degree burns, to clean dirty wounds and necrotic tissue.
Proteolytic enzymes (trypsin, chymotrypsin, papain)	Cancer
Lysozyme	Antibiotic – degrades bacterial cell wall.

Plant products like fruits and vegetables are taken raw. They contain the enzymes: proteases, amylases lipases and celluloses, which aid in digestion and absorption of nutrients in small intestines.

5.3.2(b) Enzymes in Diagnostics

The level of non-functional enzymes in the plasma (the straw-coloured clear fluid in the blood without red and white blood cells and platelets), can be monitored to diagnose certain diseases, and, in disease prognosis while treatment. An increased level of these enzymes is indicative of some diseases (**Table 5.3**).

Table 5.3: Elevated level of enzymes in disease diagnosis

Enzyme	Disease indicated
Creatine kinase	Muscle disorders, Heart attacks, Brain tumours.
Lactate dehydrogenase	Heart attacks
Aspartate transaminase	Heart attacks
Acid phosphatase	Prostate cancer
Amylase	Acute pancreatis, Lesion in salivary gland
Lipase	Pancreatis
Alkaline phosphatase	Cholestatic liver disease, Various bone diseases, Prostate cancer
Alanine transaminase	Acute liver disease, Hepatitis

5.4 CLASSIFICATION OF ENZYMES

Enzymes were formerly given the name of the scientist who discovered them. Now, for convenience and comprehension, they are given names based on the type of reaction they catalyse. International Union of Biochemists (IUB) classified enzymes into six classes. Translocases are yet another class of enzymes which help in movement of molecules across membranes. Enzymes catalysing similar type of reactions are grouped under one class (**Table 5.4**).

Table 5.4: Types of Enzymes: Classification of Enzymes

Class	Reaction catalysed	Examples
Oxidoreductase	Redox (oxidation-reduction) reactions where oxygen, hydrogen, or, electrons are transferred from the oxidant (electron donor) to the reductant (electron acceptor). $A^- + B \rightarrow A + B^-$ A is the reductant (electron donor) and B is the oxidant (electron acceptor) These enzymes use NADP or NAD+ as cofactors.	1. Reductases: NAD, NADP, FAD reductases 2. Oxidases: add electrons to oxygen. 3. Peroxidases: reduction of hydrogen peroxide. 4. Oxygenases: supply molecular oxygen to organic compounds. 5. Hydroxylases: add hydroxyl groups. 6. Transmembrane oxidoreductases: create electron transport chains in bacteria, chloroplasts and mitochondria.
Transferase	Transfer functional groups like phosphate, acetyl, amino etc. between donor and acceptor molecules. X group +Y= X+ Y group X is donor, Y is acceptor, group is the functional group.	More than 400 types. 1. Transacylases – acyl group 2. Transaminases – amino group 3. Transphosphorylases – phospho group 4. Transmethylases – methyl group
Isomerase	Intramolecular rearrangements through breaking and forming bonds without adding or removing atoms from the substrate resulting in change of shape of the molecule. $A \rightarrow B$ where A and B are isomers or A–B $\rightarrow$ B–A where A–B and B–A are stereoisomers Stereoisomers have the same ordering of individual bonds and the same connectivity, but, differ in the three-dimensional arrangement of bonded atoms.	1. Phosphoglucomutase: catalyses the conversion of glucose-1-phosphate to glucose-6-phosphate (transfer of a phosphate group from one position to another in the same compound). 2. Racemases 3. Cis-trans isomerases
Hydrolase	Addition or removal of water to break (cleave, hydrolyse) a bond. Usually dividing a large molecule into two smaller molecules.	More than 200 hydrolases known. 1. Proteinases 2. Carbohydrases 3. Lipases 4. Esterases 5. Nucleosidases
Lyase	Addition or removal of group of atoms by cleavage of bonds without water. Add water, carbon dioxide or ammonia across double bonds, or, eliminate these, to create double bonds or ring structure. The reverse reaction is also possible (called "Michael addition"). They require one substrate for the reaction in one direction but two substrates for the reverse reaction.	1. C-C Lyases 2. C-O Lyases 3. C-N Lyases 4. C-S Lyases

Class	Reaction catalysed	Examples
Ligase	Ligation process i.e., formation of single bonds with the help of ATP. Two large molecules are joined by forming a new chemical bond, usually, with accompanying hydrolysis of a small pendent chemical group on one of the larger molecules, or, the enzyme catalysing the linking together of two compounds. Ab + C → A–C + b or sometimes Ab + cD → A–D + b + c + d + e + f where the lowercase letters can signify the small, pendent groups.	**1.** Synthetases **2.** Synthases **3.** DNA ligase: Joins two complementary fragments of nucleic acid and repairs single stranded breaks, that arise in double stranded DNA during replication.
Translocase	Moving ions or molecules, usually, across a cell membrane	**1.** Translocase of the outer membrane of mitochondria (TOM) works in conjunction with translocase of the inner membrane (TIM) to transport proteins into the mitochondrion. **2.** ADP/ATP translocases enable the exchange of cytosolic adenosine diphosphate (ADP) and mitochondrial adenosine triphosphate (ATP) across the inner mitochondrial membrane.

5.5 COMPONENTS AND STRUCTURE OF A FUNCTIONAL ENZYME

5.5.1 Components

To date, more than 5,000 enzymes have been characterized. Except for a few (which are RNA), all of them are proteins. Their molecular weight ranges from 12 kDa to ~ 100 kDa. Only a small proportion of enzymes like lipases, proteinases and hydrolases, function with their protein part alone. A large majority of them, require non-protein molecules to function. An enzyme without its non-protein molecules is called an apoenzyme, and together with it, is described as a holoenzyme. Non-protein molecules are of two types: coenzymes and cofactors.

Holoenzyme (Total enzyme) $\longrightarrow$ Apoenzyme (Protein) + Cofactor (Non-protein)

Organic cofactors that are loosely bound to the enzyme, only during the reaction, are called coenzymes. Coenzymes are thermo-stable and related to the vitamins; e. g. Nicotinamide adenine dinucleotide (NAD), Flavin adenine dinucleotide (FAD). Many dehydrogenase enzymes like alcohol dehydrogenase, malate dehydrogenase and lactate dehydrogenase have NAD as their coenzyme. A coenzyme, unlike the enzyme, undergoes change (e.g. NAD is reduced to NADH) during the reaction, and therefore, can be considered, as a co-substrate.

Cofactors can be small organic molecules, or metal ions. Some enzymes have metal ions like Mg^{+2}, Mn^{+2}, Zn^{+2} as cofactors. Enzymes with a metal ion as cofactor are called metalloenzymes. While enzymes have great specificity, a cofactor can serve many apoenzymes

Some organic molecules (e.g. FAD) are tightly bound to the enzyme molecule. Such tightly bound organic molecules are called prosthetic groups.

5.5.2 Structure

Enzymes are globular proteins. The tertiary structure of the enzyme has a cleft, into which the substrate and cofactor fit in. This is called the active site of the enzyme. It constitutes less than 5% of the total surface area of the enzyme. The active site is constituted by the binding site and catalytic site (**Figure 5.1**).

The binding site of the enzyme has a complementary shape to the substrate. Therefore, the substrate fits in the enzyme binding site. The catalytic site facilitates the reaction to take place, by creating a suitable environment for the reaction, through arrangement of the atoms, in the molecules constituting the amino acid constituents in that region. Some enzymes, in addition to the active site, also have an allosteric site, to which, regulatory molecules (activators/inhibitors) bind, and affect their activity.

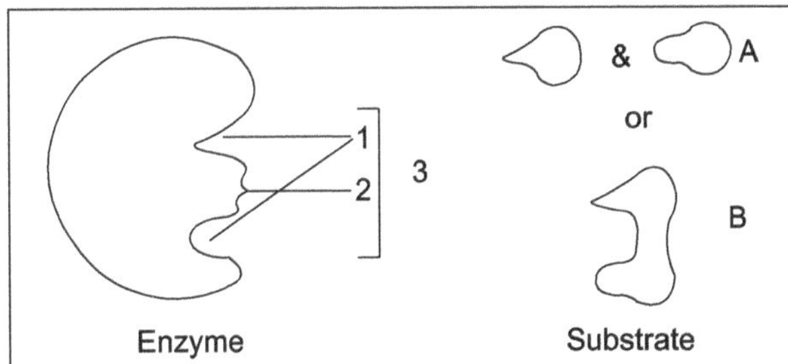

Figure 5.1: Enzyme and substrate molecules. 1. *Binding site.* **2.** *Catalytic site.* **3.** *Active site.* **A.** *Substrate molecules in anabolic reactions (substrate molecules combine together to form the product(s).* **B.** *Substrate molecule in catabolic reaction (substrate molecule is broken down to products).*

5.6 MECHANISM OF ENZYME FUNCTION

Enzymes bind to one or more substrates, forming an enzyme substrate complex, and facilitate the reaction, resulting in the formation of one, or, a few products.

5.6.1 Enzyme Specificity

Enzymes show specificity for substrates. Specificity of an enzyme may be:

1. Absolute specificity – the enzyme catalyses only one reaction.
2. Group specificity – the enzyme acts on molecules that have a specific functional group, like amino, phosphate, or, methyl groups.
3. Linkage specificity – the enzyme acts on a particular type of chemical bond like C-H, C-N, C-O etc.
4. Stereochemical specificity – the enzyme acts on a particular steric or optical isomer.

5.6.2 Binding of the Enzyme to Substrate

Enzymes are larger than the substrate. The active site of the enzyme has a shape that matches that of the substrate(s). The substrate fits into the active site and links with the enzyme, forming weak bonds with some of the residues in the binding site. The binding between the enzyme and substrate, can be either uniform, or differential binding. An enzyme substrate complex is thus formed. This intermediate state, where the substrate binds to the enzyme, is called the transition state. The reactant(s) converts into the product(s), and, later dislodges from the enzyme. The free enzyme, again binds to another substrate molecule(s), and the catalytic cycle continues, until, the reaction completes.

5.6.2(a) Enzyme Substrate Interaction

Since, the function of an enzyme is dependent on its structure, specially, that of the active site, it is logical, that the interaction between the enzyme and substrate(s), has to be through matching of the structures. Two models have been proposed to explain the interaction. One, assumes that the structure of the active site of the enzyme, is fixed (lock and key mechanism), while the other, considers it to be flexible, that undergoes conformational changes, to suit the substrate (induced fit model).

5.6.2(a)1 Lock and Key Model

In this model, the enzyme is considered the lock, with its active site being the key hole, and, the substrate(s) is the key. Enzyme substrate interaction happens only when the right key (substrate), inserts into the key hole (active site) of the enzyme (**Figure 5.2**).

5.6.2(a)2 Induced Fit Model

This is a more accepted model. In this, it is assumed that, initially the substrate interacts with the active site in the enzyme, forming weak bonds. This rapidly induces conformational changes in the enzyme, such that, the shape of the active site suits the substrate, and results in strengthening the binding between the two (**Figure 5.2**). In induced fit model, differential binding occurs between the enzyme and substrate.

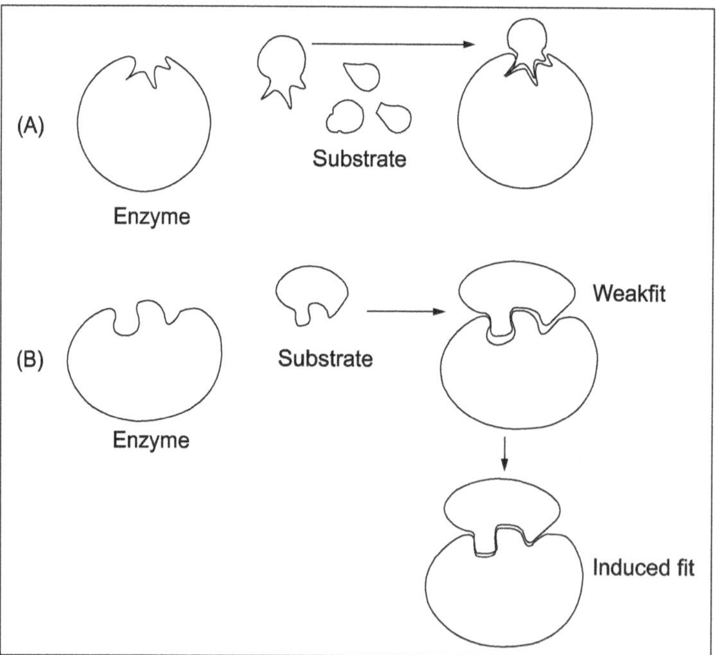

Figure 5.2: Enzyme-Substrate interaction. A. *Lock and key model – the binding between enzyme and substrate is like a lock and key. Enzyme is the lock with its active site constituting the key hole. The substrate is the key. Only the key (substrate) which fits into the keyhole (active site), can form the enzyme-substrate complex for the reaction to proceed.* **B.** *Induced fit model – the substrate molecule somewhat fits (weak fit) into the active site of the enzyme, which triggers a conformational change in the active site, enabling the substrate to fit properly in the active site (induced fit) to form the enzyme-substrate complex.*

5.6.3 Regulation of Enzyme Activity

Activity of the enzyme is sometimes affected by regulatory substances. The molecules of the regulatory substance, bind to the allosteric site, either enhancing, or, inhibiting its activity. Regulatory substances that enhance enzyme activity are called activators and those which prevent their activity are termed inhibitors. Both of them affect enzyme activity, by bringing a conformational change in the active site of the enzyme. The activator brings about a change that enables a good fit of the substrate molecule, while, the inhibitor causes disfiguring of the shape of the active site, so that, the substrate cannot bind to the enzyme (**Figure 5.3**) Inhibition in enzyme activity due to binding of the inhibitor to the allosteric site of the enzyme is termed allosteric inhibition. Enzyme activity can also be affected by inhibitors,

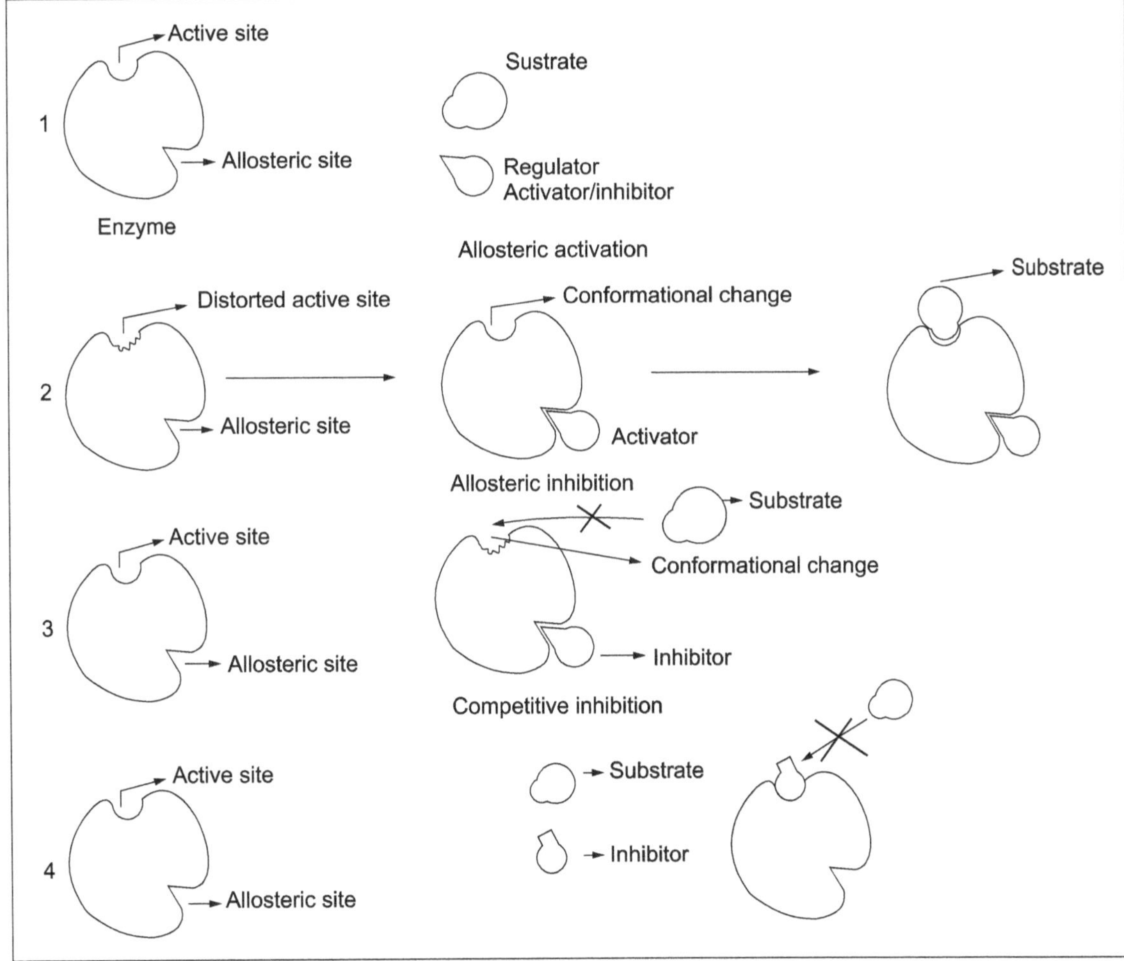

Figure 5.3: Regulation of enzyme activity. 1. *Enzyme has an allosteric site in addition to the active site. Molecules of the regulatory substance have a complementary shape to the allosteric site and can fit into it. If the regulatory molecule enhances enzyme activity, it is called an activator, if it inhibits enzyme activity, it is called an inhibitor.* **2&3.** *Allosteric activation and inhibition, both involve bringing about conformational change in the binding site. The activator, brings about a favorable change for binding of the substrate. The inhibitor causes loss of favorable shape for the binding of the substrate.* **4.** *Competitive inhibition results from inhibitor molecule competing with the substrate molecule for binding to the active site.*

which have a resemblance to the shape of the substrate molecule, and, can fit into its active site. Such inhibitors, compete with the substrate, to bind to the active site. When the active site is occupied by the inhibitor molecule, the substrate molecule cannot bind to the enzyme for the reaction to proceed. This mechanism of regulation of enzyme activity, is termed competitive inhibition (**Figure 5.3**).

5.6.4 Mechanism of Hastening Reaction Rate by an Enzyme

Chemical reactions need a certain minimum energy to activate atoms or molecules to a condition in which they can undergo chemical transformation or physical transport. This is called the 'Activation energy' of a reaction. The reactant molecules must possess sufficient energy to exceed the activation energy needed for the reaction. Activation energy is the potential energy barrier for a reaction. All molecules of the reactants possess energy, but, in varying amounts, depending on their recent collision history; only a few among them, have sufficient energy to participate in a reaction. The lower the potential energy barrier (activation energy) of a reaction, the more reactants have sufficient energy and, hence, the faster the reaction will occur. Even a very small reduction in this potential energy barrier, can result in an enormous increase in the rate of reaction. A catalyst hastens a reaction by lowering the activation energy, so that, a greater proportion of the reactant molecules have enough energy to react. Enzymes function, by forming a transition state, with the reactants of lower free energy, than would be found in the uncatalyzed reaction. This results in lowering of the activation energy of the reaction (**Figure 5.4**).

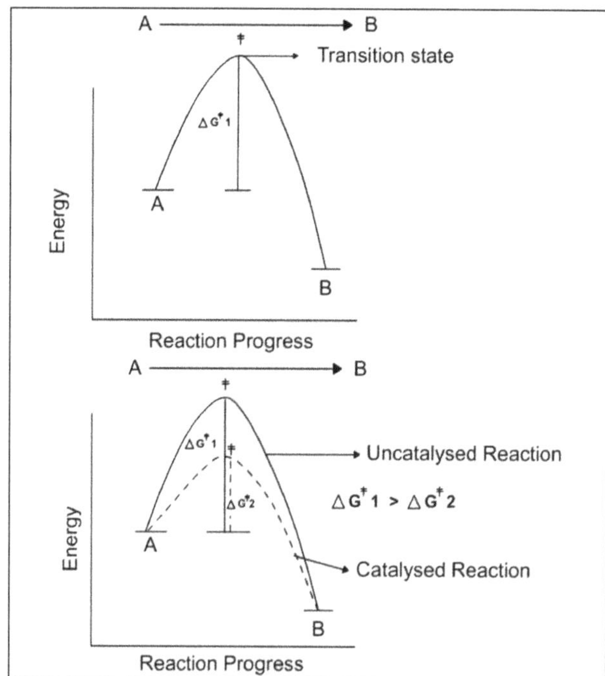

Figure 5.4: Hastening of an exergonic reaction (energy level of the energy substrate (A) is higher than that of the product (B)). *The energy level of the substrate should increase by $\Delta G^{\ddagger}$ to reach the high energy transition state (denoted by $\ddagger$), for the reaction to take place. $\Delta G^{\ddagger}$ is termed activation energy of a reaction. The enzyme lowers the activation energy, by which, more substrate molecules can reach the transition state to participate in the reaction.*

5.6.4(a) Mechanism of Lowering of Activation Energy of a Reaction by an Enzyme

Enzymes lower activation energy by bringing the reactants close to each other, overcome forces of repulsion, and, initiate breaking of bonds, to participate in a reaction. The various ways in which this is achieved are:

(*a*) Proximity and orientation: positioning the substrates together in the proper orientation.

(*b*) Applying torque (rotational force) on the substrates.

(*c*) Providing suitable pH or charge in microenvironment of the active site, e.g. creating a charge distribution, opposite to that of the enzyme substrate complex (electrostatic catalysis).

(*d*) Bound substrates undergo stretching, due to which, the bonds within their molecules are strained (bond strain), thereby facilitating the formation of the transition state of enzyme substrate complex. In addition to bond strain in the substrate, bond strain may also be induced within the enzyme itself to activate residues in the active site.

(*e*) Some enzymes lower activation energy by taking part in the chemical reaction themselves. The active site residues (NH_2 COOH or -SH), may form temporary covalent bonds with substrate molecules, as part of the reaction process (covalent catalysis).

(*f*) The interaction between the substrate and the enzyme, could be by hydrogen bonding or Van der Waals forces.

(*g*) The formation of the enzyme substrate complex is exergonic i.e., energy is released. This raises the energy level of the substrate molecule. Being at high energy state, the enzyme substrate transition state is unstable and spontaneously converts into the more stable product with lower energy. The thus formed product, has lower affinity with the enzyme, and is released from it.

(*h*) Just like inorganic catalysts, enzymes catalyse both reactions. They can neither, allow only forward reaction, or, only backward reaction to happen alone. They help the reaction to reach the equilibrium point of the forward and backward reactions, faster.

$$E + S \rightleftharpoons ES \rightleftharpoons EP \rightleftharpoons E + P$$

ES: Enzyme substrate complex; EP: Enzyme product complex

Examples of two types of reactions (catabolic and anabolic) that are catalyzed by an enzyme are described in chapter 7 under the enzyme section.

5.7 MONITORING ENZYME ACTIVITY: ENZYME ASSAYS

To monitor enzyme activity, enzyme assays are done. Assays essentially help, in estimating the rate of an enzyme catalyzed reaction. Enzyme assays, measure either the consumption (disappearance) of substrate, or, production (appearance) of product over time. The rate of a reaction is the concentration of substrate disappearing (or product, produced) per unit time (mol L^{-1} s^{-1}). Measuring the product is more accurate, because, to start with, product is nil (0) and even a small change in product concentration, is easier to observe and estimate.

Since in enzyme assays activity of the enzyme is quantified based on the number of active enzyme molecules present, it is important to mention, the conditions under which the assay is done. Assays are carried out under conditions optimal for enzyme activity. Therefore, an enzyme as assayed in the laboratory, may not have the same natural functional activity in the cell, in which, conditions may be different.

Enzyme activity is, the moles of substrate converted per unit time = rate × reaction volume. It is expressed in enzyme unit (U), which is equal to 1 μmol min^{-1}. It is the micro moles of product formed by an enzyme, in a given amount of time, (minutes) under given conditions, per milligram of total protein. Specific activity of the enzyme is equal to the rate of reaction multiplied by the volume of reaction divided by the mass of total protein. It is given as units/mg or nmol/min/mg.

Based on sampling method, assays are two types: continuous and discontinuous(stopped) assays. In continuous assay, as the reaction proceeds, several samples are taken at different time intervals; while, in the discontinuous assay, only one sample is taken, after the reaction is stopped, to estimate the concentration of substrate(s)/product(s).

The reaction is stopped in several ways: **1.** by denaturing the enzyme; or by using: **2.** strong acid; **3.** alkali; **4.** detergent; **5.** heavy metal ions which act as irreversible inhibitors; **6.** heat; **7.** by chilling on ice; **8.** by addition of a complexing agent such as EDTA (ethylene diamine tetra acetic acid).

Given the wide range of enzyme-controlled reactions, with different types of products formed, there is no single best method, for measuring reaction rates of all enzymes.

5.7.1 Continuous Assays

Continuous assays are more convenient. They monitor the enzyme activity as the reaction is occurring. Depending on the nature of the substrate/product, different methods of estimating their concentration are used

5.7.1(a) Colorimetric

If the substrate/product is coloured, or if it absorbs light at UV wavelength, absorbance is measured using a spectrophotometer. Estimation of the product concentration through this method, is based on the principle of Beer-Lambert law – the amount of light absorbed, is proportional to the concentration of the substance and the path length of light through the solution.

5.7.2 Coupled Reactions

In many reactions, changes in substrates or products, are not observable by spectrophotometric methods, because, they do not absorb light. These reactions can be measured by coupling the product of the reaction, to another enzyme, such that, it becomes a substrate to the second enzyme and the product released in the second reaction can be measured through absorbance using a spectrophotometer.

In some instances, for enzymes with substrates and products that do not absorb light, light absorbing non-physiological substrates, or a substrate that produces a light absorbing product, are synthesized to use in enzyme assay.

Based on the nature of the reactant/product, different methods are used for estimating enzyme activity.

5.7.2(a) Fluorometric

When the substrate/product is fluorescent, i.e. if it absorbs light of one wave length, and emits a light of a different wavelength, the amount of fluorescence emitted can be measured as an indicator of the amount of product. A fluorometer is used and estimations are made, based on the same principle of absorbance applying Beer-Lambert law.

5.7.2(b) Chemiluminescence

If light (photons) is emitted by the substrate/product, the emitted light is measured using a luminometer.

5.7.2(c) Manometric

If the product of the enzyme catalyzed reaction is a gas, the built up of pressure, due to the accumulation of gas, can be measured, to quantify the product formed, and, thus, the rate of the reaction. A manometer is used for the purpose.

5.7.2(d) Conductivity

If the reaction involves production or removal of ions, enzyme activity can be estimated by quantifying the difference in charge, using a conductivity meter.

5.7.2(e) Potentiometry

If the reaction involves change in pH, then it can be measured using a pH meter.

5.7.2(f) Titrimetric

The product is titrated with an acid or base, to estimate its quantity, and from it, enzyme activity can be estimated.

5.7.3 Discontinuous Assays

In these assays, enzyme activity is measured through radiometry (using radioisotopes) or chromatography – either HPLC (high performance liquid chromatography), or, TLC (thin layer chromatography).

5.8 ENZYME KINETICS

Enzyme kinetics is the quantitative measurement of rate of an enzyme catalysed reaction and a methodical study of factors that affect these rates. The catalytic properties of enzymes, and consequently their activity, are influenced by several factors: **a.** physical, like temperature and pressure; **b.** chemical properties of the reaction mixture, like pH and ionic strength and **c.** concentrations of the substrates, enzyme cofactors and its inhibitors. Enzyme activity is affected by binding of a small molecule, "allosteric effector", or, by covalent binding or release of molecules: e.g. phosphorylation and dephosphorylation.

A study of an enzyme's kinetics reveals its catalytic mechanism. Michaelis-Menten kinetics is one of the simplest and best-known models to study enzyme kinetics. The model serves to explain, how an enzyme can enhance the kinetic rate of a reaction, and, the reason, for the rate of a reaction depending on the concentration of enzyme present. Cells control the amount of an enzyme present, by regulating its rate of synthesis and degradation.

The principles of enzyme kinetics are similar to those of general chemical reaction kinetics. However, they show a distinctive feature of saturation. At lower substrate concentration, the initial reaction velocity is proportional to substrate concentration (1st order reaction). Further increase in substrate concentration, does not any more affect the reaction rate, with the reaction rate remaining constant. When substrate concentration is high, enzymes follow zero order kinetics, which means, with further increase in substrate concentration, no increase in the rate of the reaction happens.

The overall reaction catalysed by an enzyme is composed of two elementary reactions: **1.** bimolecular reaction – the substrate (S) forms a complex with the enzyme(E); **2.** unimolecular reaction – the enzyme substrate complex (ES), subsequently decomposes to product(s) (P), and, releases the freed enzyme(E). The first reaction is reversible while the second one is irreversible.

$$E + S \rightleftharpoons ES \rightarrow E + P$$

According to this equation, when the substrate concentration is sufficient to entirely convert the enzyme to the ES form, the second step of the reaction becomes rate limiting step. A further increase in substrate concentration, does not any more result, in increase in the rate of the reaction. Two rate equations are used in enzyme kinetics based on differences in assumption about the reaction: **1.** the rapid equilibrium assumption – Michaelis-Menten Equation; **2.** the steady-state assumption – Briggs and Haldane equation.

5.8.1 Michaelis-Menten Equation

This is best-known equation to study enzyme kinetics of one-substrate enzyme-catalysed reactions. It explains, the quantitative relationship in a reaction, between the initial velocity V_0, the maximum velocity V_{max} and the initial substrate concentration [S], all related through the Michaelis constant K_m.

When the rate of an enzymatic reaction with increasing substrate concentration is represented graphically the curve obtained is usually a hyperbola (**Figure 5.5**). For a short period after the reaction begins the curve is almost linear after which the rate of the reaction continuously slows and finally flattens off which means the maximum velocity (V_{max}) of the reaction is reached (**Figure 5.5**). The fact that at high substrate concentrations the velocity approaches a maximum provides support for the assumption that an intermediate enzyme-substrate complex is formed in the reaction and the rate does not further increase because the active site of all enzyme molecules are bound by the substrate. Because of this saturation of enzyme sites, the rate of enzyme catalysed reactions does not show a continuous linear slope.

The mathematical expression for this curve shown in figure 5.5 was developed by German biochemists, Michaelis and Menten. The Michaelis-Menten equation for the velocity (rate) of the reaction is given as

$$\text{Rate of the reaction} = \frac{V_{max}[S]}{K_m + [S]}$$

V_{max} is the maximal velocity of the reaction, [S] is the initial substrate concentration and K_m is called the Michaelis constant. K_m is the substrate concentration in moles per litre (M), at the point, where the reaction rate is half the maximum velocity ($1/2\ V_{max}$) (**Figure 5.5**). K_m gives a rough measure of the affinity of the substrate molecule for the surface of the enzyme, with its values, ranging between 10^{-8} to 10^{-2} M, for different enzymes. The V_{max} values, range from 10^5 to 10^9 molecules of product formed per molecule of enzyme per second.

The turnover number (V_M) of an enzyme is the number of reactions that take place per minute between the substrate molecule and the enzyme. It is expressed as, moles of product formed, per mole of enzyme per minute.

Where the linear slope of the reaction curve is very sharp i.e., the steady state of the reaction is very rapid, measurements are taken for the entire reaction duration and the data is fitted to a progressive curve.

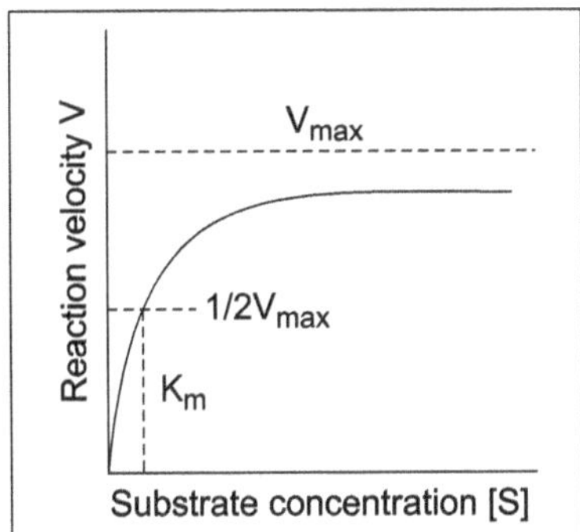

Figure 5.5: Enzyme kinetics with increasing concentration of the substrate. *The curve is a hyperbola. After an initial steep increase in rate with increase in substrate concentration, the rate flattens off at a particular substrate concentration with no further increase, even with increase in substrate concentration. V_{max} is the maximum rate of the reaction, $1/2V_{max}$ is the half of maximum rate of reaction and K_m represents the substrate concentration when $1/2V_{max}$ is reached. K_m is called Michaelis constant.*

5.8.2 Effect of pH and Temperature on Enzyme Kinetics

Enzymes have an optimum pH and temperature (or a range), at which they are maximally active. Effect of enzymes is strongly dependent on the pH. When enzyme activity is plotted against pH, a bell-shaped curve is usually obtained (**Figure 5.6A**). Bell shape of the 'activity–pH profile' results from the fact that, in the enzyme, presence of a amino acid residues with ionizable groups in the side chain are essential for catalysis. They have a basic group B, which has to be protonated (addition of H⁺) in order to become active, and, a second acidic amino acid AH, which is only active in a dissociated state (A⁻). At the optimum pH of 7, around 90% of both groups are present in the active form and at higher and lower values, one or the other of the groups, increasingly passes into the inactive state (**Figure 5.6A**).

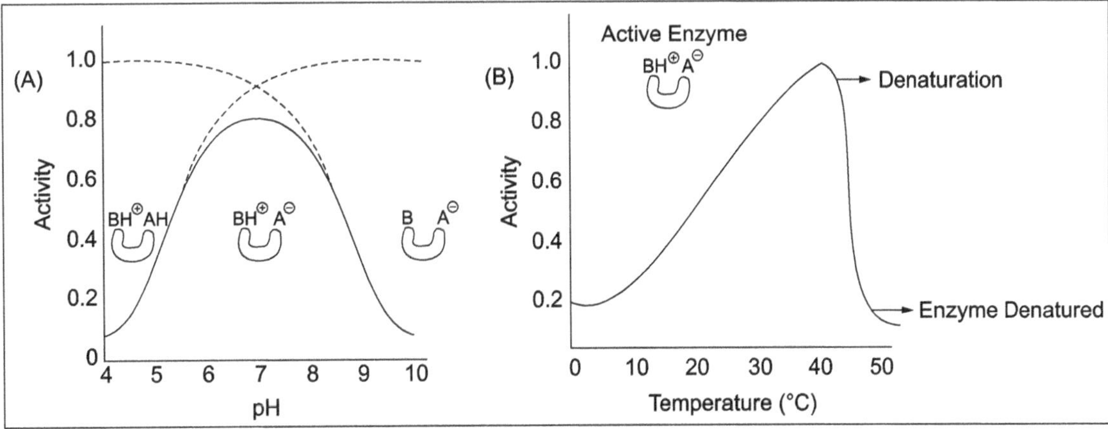

Figure 5.6: Effect of pH and temperature on enzyme kinetics. A. *Enzyme activity plotted with increasing pH results in a bell-shaped curve. Changes in the side chains of ionizable amino acids in the enzyme are required for enzyme activity. The basic group B has to be protonated to BH⁺ and the acidic group AH has to be dissociated to A⁻ for the enzyme to have maximum activity. At the optimum pH of 7, both the groups are in a condition that promote maximum enzyme activity, and, hence the highest reaction rate.* **B.** *Enzyme activity plotted with increasing temperature results in a steep slope which suddenly falls resulting in a skewed curve. With increasing temperature, the enzyme activity increases progressively, and at a critical temperature, at which enzyme denaturation begins, its activity steeply falls, and, finally becomes negligible when it is fully denatured.*

The temperature dependency of enzymatic activity is usually asymmetric. With increase in temperature, there is increased thermal movement of the molecules and initially, it leads to acceleration of the rate of the reaction. At a certain temperature, the enzyme then becomes unstable, and its activity is lost within a narrow temperature range, as a result of denaturation (**Figure 5.6B**).

Enzyme activity is also affected by hydrogen and other ion concentrations. Some enzymes require loosely bound Mg ions.

5.8.3 Allosteric Enzyme Kinetics

The kinetics of allosteric enzymes (enzymes which bind to regulatory molecules – activator/inhibitor), show different kinetics, with increasing concentration of the binding ligand (activator/inhibitor). Again, the kinetics between allosteric enzymes with competitive inhibition, can be differentiated, from that of allosteric enzymes with non-competitive inhibition. The competitive inhibitor behaviour systems are called K systems, because, here, with the increase in ligand concentration, they show difference in K_m (Michaelis constant), but no difference in V_{max} (maximum velocity) of the reaction (**Figure 5.7A**). The non-competitive inhibitor behaviour systems are called V systems, because, here, the maximum velocity that the reaction reaches, changes with the type of the ligand, being higher with an activator and lower with an inhibitor (**Figure 5.7B**).

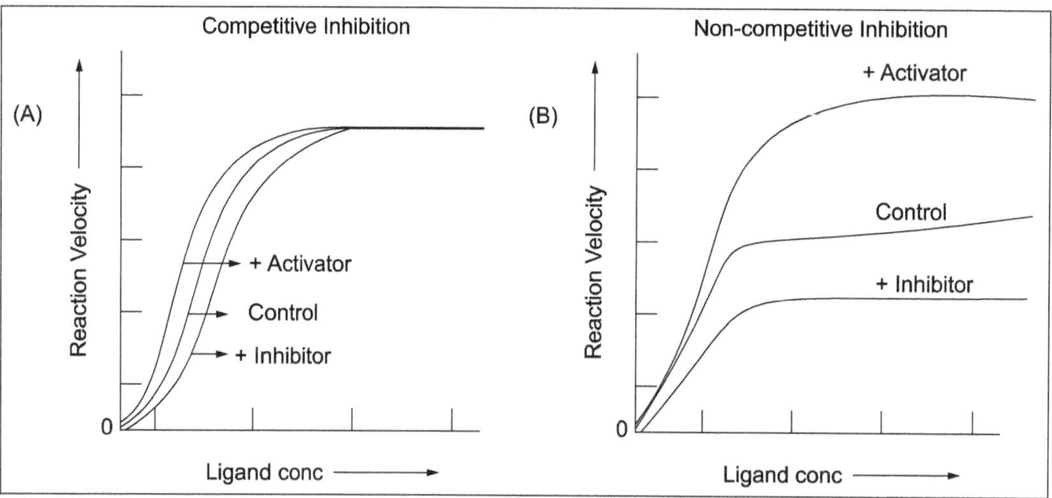

Figure 5.7: Enzyme kinetics of allosteric enzymes. A. *In competitive inhibition with increase in the ligand – activator/ inhibitor, the V_{max} does not change, only the K_m (Michaelis constant) differs between the control (with no ligand effector) and activator driven and inhibitor stopped reactions. Enzymes which show this behaviour are therefore termed K systems.* **B.** *In non-competitive inhibition, the V_{max} differs between the control (with no ligand effector), activator driven and inhibitor stopped reactions. Hence enzymes regulated by non-competing (with the substrate molecule) inhibitors are termed V systems.*

5.9 RNA CATALYSTS

In early 1980's, Thomas Cech and Sydney Altman discovered that RNA has catalytic properties for which they received the 1989 Nobel prize in Chemistry. Thus, arose the term ribozyme (ribo nucleic acid enzyme) for RNA with catalytic properties. This discovery of ribozyme demonstrated that a nucleic acid (with genetic information) can also serve as a biological catalyst, just like the protein enzymes. This lends credence to the idea that RNA could have been the elemental biological molecule. RNA with its replicating ability (not possible with protein), could have contributed to the RNA world, the forerunner of the modern biological world dominated by DNA directed organisms. The question of 'was it the nucleic acid (with life blue print), or, the protein (work horse), that came first in the origin of life?' (chicken or egg first dilemma) was put to rest, with the discovery of ribozyme which lends support to the hypothesis of the RNA world by Crick and his associates.

Ribozymes increase reaction rates by up to 10^{11}-fold which are still $\sim 10^3$-fold less than those provided by protein enzymes catalysing comparable reactions. All known ribozymes require Mg^{2+} for their activity and therefore considered metalloenzymes. Some of the important ribozymes are described in **Table 5.5.**

Table 5.5: Well studied ribozymes and their functions

Ribozyme	Function	
	Catalytic	Biological
Large ribozymes (100 to 3000 nucleotides)		
rRNA (ribosomal RNA in large subunit of ribosome providing peptidyl transferase function)	Peptide bond formation (between two adjacent amino acids).	Protein synthesis
Ribonuclease P	Hydrolysis of phosphodiester bond (between nucleotides).	Maturation of tRNA
Self-splicing Group 1 and Group 2 introns present in transcripts of bacterial and cell organelle genes.	Hydrolysis and ligation of phosphodiester bond (between nucleotides).	Remove introns in genes of bacteria, bacteriophages and organelle genes of eukaryotes.
Small ribozymes (30 to 150 nucleotides)		
Hairpin ribozyme, Hammer head ribozyme, Hepatitis Delta ribozyme, Varakud Satellite RNA	Hydrolysis of phosphodiester bond (between nucleotides).	RNA virus replication pathway.
Sn RNA's (Small nuclear RNA's)	Hydrolysis of phosphodiester bond between nucleotides and subsequent ligation.	Intron splicing of mRNA of nuclear genes of eukaryotes.

RNA catalyst chemistry is found in two complex ribonucleoprotein molecular machines – ribosome and spliceosome. Ribosome produces proteins within the cell. When the high-resolution structure of the ribosome was determined, it was found that a region within its large subunit provides the peptidyl transferase activity, required for the formation of peptide bond between the outgoing and incoming amino acids in the ribosome, during peptide synthesis. Hence, the ribosome can also be considered to be a ribozyme with its RNA and several proteins.

The group 1 and 2 introns (non-coding sequences) are part of the RNA transcript which cleave themselves out and the adjoining exons are ligated together. Since, no enzyme is involved in this cutting out of intron and joining of exons (cut and join = splice), these introns are described as self-splicing and considered as enzymes – the ribozymes. The mode of cleaving of the two introns differs slightly, though the outcome is the same (**Figure 5.8**). The introns in the mRNA of nuclear genes of eukaryotes are spliced with the help of spliceosome, which consists of five Sn RNAs (Small nuclear RNAs) – U1, U2, U4, U5 and

U6 (so named because they are rich in uridine) and several proteins. The components of the spliceosome get sequentially attached to the intron and help in cutting out the intron and joining the exons (**Figure 5.8**).

Ribonuclease P is a ubiquitous enzyme which contains an RNA and a basic protein. The protein is essential for its functioning in vivo but not in vitro. It processes the termini of precursor tRNA (**Figure 5.8**).

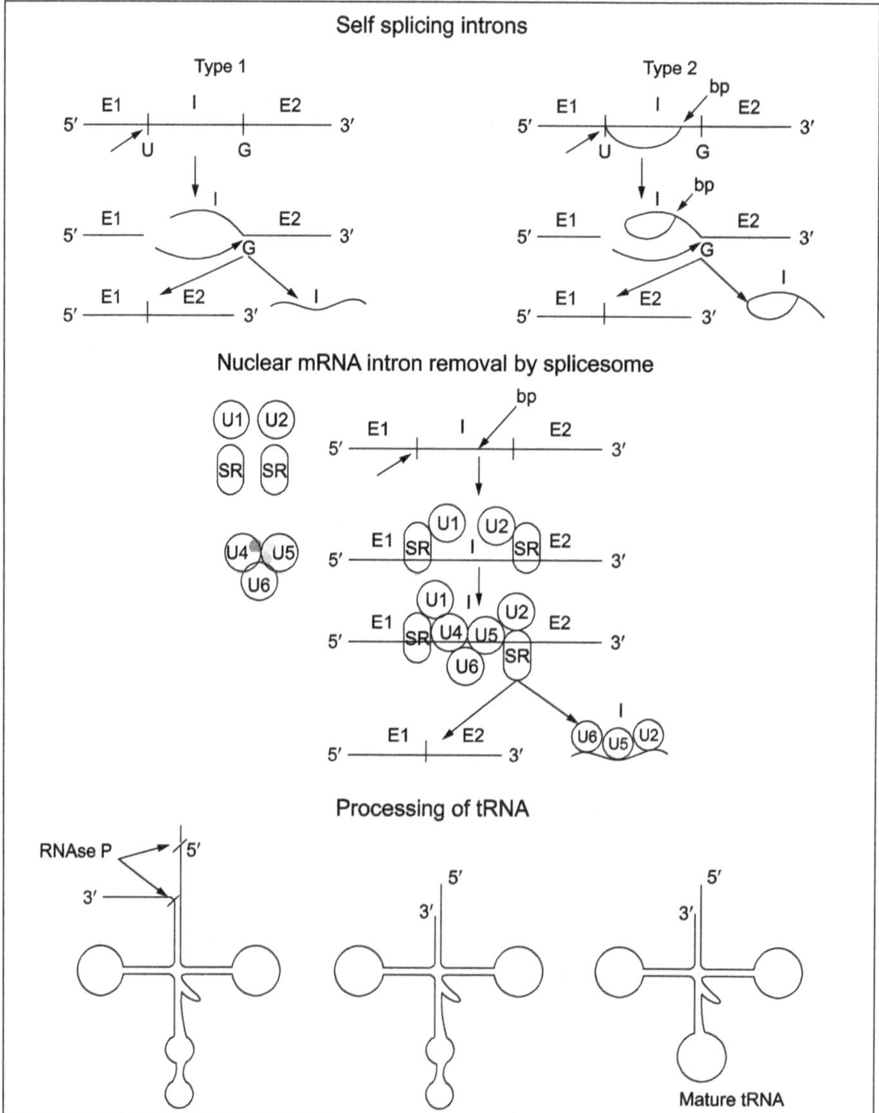

Figure 5.8: Ribozymes: Type 1, Type II and mRNA of nuclear gene introns and RNAse P. *Type I &II introns are present in bacterial and organelle genes and they are self-splicing. Type II intron has a short sequence within it, called the branch point (bp), which initiates the cleavage of the intron at the 5'end. The intron thus forms a lariat (a rope with a noose) like structure. Introns of mRNA of nuclear genes are spliced with the help of spliceosome. The spliceosome (an RNP- Ribo Nucleo Protein molecule) contains five SnRNA (Small nuclear RNAs) – U1 U2 U4 U5 U6 and other proteins (SR). RNAse P processes the 3' and 5' ends of tRNA removing some nucleotides to finally form the functional mature tRNA.E1 and E2 represent exons and I an intron.*

The hairpin, hammerhead, Hepatitis Delta virus, Varakud satellite ribozymes, all work in a similar manner. During replication of the RNA viruses or satellite RNA, the viral or satellite RNA is replicated as a long thread having multiple copies (concatemers). These ribozymes with their characteristic secondary structure (e.g. taking the form of hair pin or hammerhead), facilitate the cleavage of the multiple copy concatemer of nascent RNA, formed during viral replication, at specific sites (cos sites), into individual viral RNA copies. All these ribozymes are thus self-cleaving. It is now possible to synthesize ribozymes that cleave other RNA molecules at specific sequences. Such RNA catalysts have pharmaceutical applications – they can be used to fight against infectious RNA viruses like the HIV (AIDS) and Corona (COVID). Such RNA cleaving small ribozymes have a great potential for use in treatment of cancer. Ribozymes that can cleave oncogene (cancer gene) RNA, will prevent the development of cancer.

5.10 SUMMARY

1. The cocktail of biochemical reactions, involving building (anabolic) and breaking (catabolic) substances, happen in living organisms at normal body temperature and a narrow pH range, at an astounding pace. They owe their speed to the biological catalysts called enzymes.

2. Being natural catalysts not requiring high temperature or pressure, they are used in several industries. A major chunk of enzymes used in industry are in food industry (45%); the others are in: detergent (35%), textile (10%), and tanning (leather) (3%) industries. Enzymes are used in medicine in therapeutics (treatment) and diagnostics – as indicators of a disease.

3. International Union of Biochemists (IUB), classified enzymes into six classes: oxidoreductases, transferases, isomerases, hydrolases, lyases and ligases. Translocases are yet another recognized class of enzymes which help in movement of molecules across membranes.

4. With the exception of a few (RNA based), all enzymes are proteins. Only a small proportion of enzymes, like lipases, proteinases and hydrolases, function with their protein part alone. A large majority of them, require non-protein molecules, to function. An enzyme without its non-protein group is called an apoenzyme and together with it, is described as a holoenzyme. Non-protein groups are of two types: coenzymes and cofactors. Non-protein groups that attach tightly to the enzyme are called prosthetic groups.

5. Enzymes are globular proteins. The tertiary structure of the enzyme has a cleft into which the substrate and cofactor fit in. This is called the active site of the enzyme. The active site constitutes less than 5% of the total surface area of the enzyme and has both the binding site (for the substrate(s) and catalytic site.

6. Enzymes are very specific to the substrate on which they act. The 3D structure of the active site of the enzyme, is complementary to the shape of the substrate molecule. Association between the enzyme and substrate(s), has to be through matching of their structures. Two models have been proposed to explain the interaction. One, assumes that the structure of the active site of the enzyme, is fixed and matches exactly the shape of the substrate molecule (lock and key mechanism), while the other, considers it to be flexible, that undergoes conformational changes, to suit the shape of the substrate (induced fit model).

7. Activity of the enzyme is sometimes affected by regulatory substances. The molecules of the regulatory substance bind to the allosteric site, either enhancing (activator), or, inhibiting(inhibitor) its activity. The activator brings about a change in the structure of active site that enables a good fit of the substrate molecule, while, the inhibitor causes disfiguring of the shape of the active site, so that, the substrate cannot bind to the enzyme. This is described as non-competitive inhibition of enzyme activity. Enzyme activity can also be affected by

inhibitors, which have a resemblance to the shape of the substrate molecule, and, can fit into its active site. Such inhibitors, compete with the substrate, to bind to the active site. This is called competitive inhibition.

8. A catalyst hastens a reaction by lowering the activation energy so that a greater proportion of the reactant molecules have enough energy to react. Enzymes, function by forming a transition state (enzyme-substrate complex), with the reactants of lower free energy, than would be found in the uncatalyzed reaction. This results in lowering of the activation energy of the reaction.

9. To monitor enzyme activity, enzyme assays are done. Assays essentially help, in estimating the rate of an enzyme catalyzed reaction. Enzyme assays measure, either the consumption (disappearance) of substrate, or, production (appearance) of product, over time. The rate of a reaction, is the concentration of substrate disappearing (or product produced) per unit time (mol L^{-1} s^{-1}). Based on sampling method, assays are two types: continuous and discontinuous(stopped) assays. In a continuous assay, as the reaction proceeds, several samples are taken at different time intervals; while, in the discontinuous assay, only one sample is taken, after the reaction is stopped, to estimate the concentration of substrate(s)/product(s).

10. Based on the nature of the reactant/product, enzyme activity is monitored using different methods: spectrophotometry, calorimetry, fluorimetry, chemiluminescence, manometry, potentiometry, conductivity, titrimetry, radiometry or chromatography.

11. Enzyme kinetics is the quantitative measurement of the rate of an enzyme catalysed reaction and, a methodical study of factors, that affect these rates. Michaelis-Menten kinetics is one of the simplest and best-known models to study enzyme kinetics. The model serves to explain, how an enzyme can enhance the kinetic rate of a reaction, and, the reason, for the rate of a reaction depending on the concentration of enzyme present.

12. When enzyme activity is plotted against pH, a bell-shaped curve is obtained; with temperature, an asymmetric skewed curve results.

13. The kinetics of allosteric enzymes (enzymes which bind to regulatory molecules – activator/inhibitor), show different kinetics with increasing concentration of the binding ligand (activator/inhibitor). Again, the kinetics between allosteric enzymes with competitive inhibition, can be differentiated, from that of allosteric enzymes with non-competitive inhibition. The competitive inhibitor behaviour systems are called K systems, because Michaelis constant (K_m) is the same both in the presence or absence of the inhibitor/activator. The non-competitive inhibitor behaviour systems are called V systems, because, here the maximum velocity that the reaction reaches, changes with the presence of the ligand, being higher with an activator, and, lower with an inhibitor.

14. Ribozymes (RNA based enzymes) increase reaction rates by up to 10^{11}-fold which are still $\sim10^{3}$-fold less than those provided by protein enzymes catalysing comparable reactions. All known ribozymes require Mg^{2+} for their activity and therefore considered metalloenzymes. Some of the important ribozymes are intron 1 and intron 2, peptidyl transferase, tRNA, Sn RNAs.

5.11 SAMPLE QUESTIONS

5.11(a) Subjective Questions

Q.1. What is the chemical nature of enzymes? Describe the structure of an enzyme. Why are enzymes specific for a substrate? What are the models proposed to explain their specificity?

Q.2. How is enzyme activity regulated by regulatory molecules? Describe the various mechanisms by which regulatory molecules affect enzyme activity.

Q.3. What is enzyme kinetics? How is it influenced by temperature, pH, and concentration of regulatory molecules?

Q.4. How is enzyme activity monitored. Describe the various methods used for monitoring enzyme activity.

Q.5. Write a critical note on RNA enzymes.

5.11(b) Objective Questions

Q.1. All enzymes are biological catalysts and made up of:

(*a*) Only protein.

(*b*) Only protein, or protein and conjugate groups, or just RNA, or RNA and protein.

(*c*) Proteins and coenzymes and cofactors.

(*d*) Proteins and allosteric ligands.

Q.2. The binding site of an enzyme is

(*a*) A large region on the surface of the enzyme.

(*b*) Is adjacent to the catalytic site.

(*c*) Is a part of the active site of the enzyme.

(*d*) Some statements are true.

(*e*) All three statements (*a*), (*b*), (*c*) are true.

Q.3. Translocase enzymes

(*a*) Transfer functional groups from donor to acceptor.

(*b*) Add or remove water molecules.

(*c*) Help in moving ions or molecules across membranes.

(*d*) Add or remove a group of atoms without water.

Q.4. Coenzymes

(*a*) Like enzymes do not undergo change during the reaction.

(*b*) Undergo change during the reaction.

(*c*) Help the backward reaction to keep pace with forward reaction.

(*d*) Are more efficient than cofactors in increasing enzyme activity.

Q.5. Allosteric enzymes

(*a*) Do not have an active site.

(*b*) Have a ligand binding site instead of the active site.

(*c*) Have a ligand binding site as well as an active site.

(*d*) Function at a lower rate than normal enzymes.

Q.6. Competitive inhibition of enzyme activity is due to

(*a*) The ligand molecule binding to the ligand binding site.

(*b*) The ligand molecule having a structure similar to the reactant molecule.

  (*c*) The ligand molecule binds to the substrate and inactivates it.

  (*d*) The ligand competes with the activator.

Q.7. Compared to uncatalyzed reactions enzymes speed up the reactions by

  (*a*) Reacting with the substrate molecules at a lower energy level.

  (*b*) Reacting with substrate molecules at a higher energy level.

  (*c*) Undergoing change during chemical reaction.

  (*d*) Forming an enzyme substrate complex.

Q.8. When substrate concentration is very high

  (*a*) Enzyme activity is increased. (*b*) Enzyme activity is decreased.

  (*c*) The rate of reaction flattens off. (*d*) The rate of the reaction speeds up.

ANSWERS

1. (*b*) **2.** (*d*) **3.** (*a*) **4.** (*b*) **5.** (*c*) **6.** (*b*) **7.** (*a*) **8.** (*c*)

6 Information Transfer in Living Organisms

> *I'm fascinated by the idea that genetics is digital. A gene is a long sequence of coded letters, like computer information. Modern biology is becoming very much a branch of information technology.*
>
> **~ Richard Dawkins**
>
> *"Every time you understand something, religion becomes less likely. Only with the discovery of the double helix and the ensuing genetic revolution have we had grounds for thinking that the powers held traditionally to be the exclusive property of the gods might one day be ours."*
>
> **~ James D. Watson**
>
> *On 19th September 1957, Crick "permanently altered the logic of biology".* **~ Horace Judson**
>
> *The genes are the atoms of heredity.* **~ Seymour Benzer**

In living organisms, information transfer from one generation (parents) to the next (offspring), occurs through the duplication and passing of genetic material to the progeny. This is termed heredity. The physical and behavioural features of an organism (phenotype), are a consequence of the nature of its proteins, which are coded by the genes. This happens through information transfer from genes to proteins. Experiments through which the genetic material was discovered, its molecular structure and organization in cells, the process by which it is expressed i.e., information transfer of the genetic material to proteins which determine physical features, the process of duplication of genetic material and the basis for what constitutes a gene are described in this chapter.

6.1 THE GENETIC MATERIAL

Experiments on plants and animals in early 1900's, unravelled the mechanism of heredity, showing that, genes are responsible for the appearance of an organism. The inheritance pattern of genes was found to be correlated with the behaviour of chromosomes during cell division. Thus, chromosomes were concluded to be the carriers of genes. The chromosomes are constituted by two components: proteins and nucleic acid; which of them constituted the gene remained to be sorted out.

The chemical nature of the genes was discovered through a series of intriguing experiments on microorganisms (bacteria and viruses) between 1920–1950's.

6.1.1 Griffith's Experiments: Transforming Principle

In the early 1920's, Frederick Griffith, a British army medical officer, initiated studies on *Streptococcus* (*Diplococcus*) *pneumoniae*, a bacterium which caused pneumonia – a fatal lung infection, in several servicemen during the first world war.

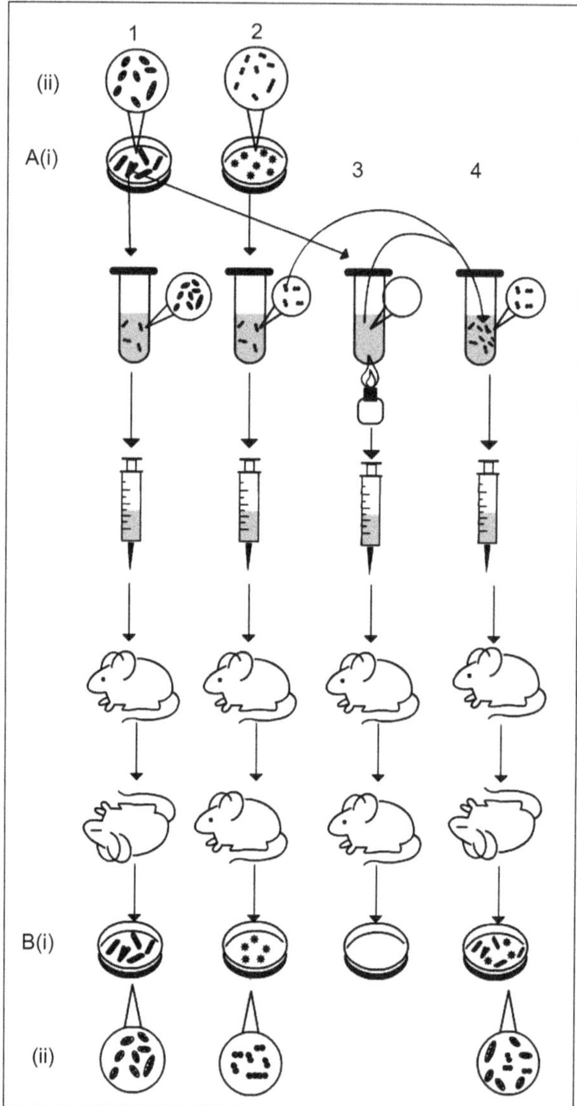

Figure 6.1: Griffith's Experiments: Injection of rats with strains of *Streptococcus* (*Diplococcus*) *pneumoniae*, **causative bacterium for a fatal lung disease – pneumonia. 1. A. i.** *Culture of virulent smooth strain (S) on solid culture medium in a Petri dish – S strain colonies with smooth, spreading outline.* **1. A. ii.** *Bacterial cells from smooth bacterial colony under the microscope – cells have an outer capsule.* **2. A. i.** *Culture of non-virulent rough strain (R) on solid culture medium in a Petri dish – small R strain colonies with rough outline.* **2. A. ii.** *Bacterial cells from rough bacterial colony under the microscope – cells with no capsule.* **3.** *S cells picked from solid culture, diluted with water and heated.* **4.** *Heat killed S cells from 3 and live R cells from 2.* **B. i.** *Culture from the tissue of the treated rat.* **B. ii.** *Bacterial cells from the colonies in B.* i. *under the microscope.* **Note:** *In treatment 4, in addition to avirulent R type, virulent S type colonies of Streptococcus are also found in cultures from treated dead rat. Griffith concluded that this is a result of transformation of R cells to S type, due to the transforming principle from the heat killed S cells in the mixture.*

Griffith isolated two strains of *S. pneumoniae* in his cultures. On solid culture medium, one of them formed pin head-like colonies with a rough texture and called Rough (R); the other developed smooth spreading colonies termed Smooth (S) (**Figure 6.1.**). The cells in the two colonies looked different under the microscope: the cells in the S strain colonies had an extra smooth coat on the cell wall, which was absent, in the cells of the R strain (**Figure 6.1**).

Griffith conducted experiments with S and R strains of *S. pneumoniae* on rats. He injected cells of the S and R strains separately into two different rats. The rat injected with S strain developed pneumonia and died, while the rat inoculated with R strain did not get pneumonia. He concluded that the S strain is virulent, causing disease, while, the R strain is non-virulent, and therefore, harmless (**Figure 6.1**).

In the follow up experiments, Griffith applied heat to the culture of S strain and injected it into a rat. The rat did not develop pneumonia – it survived the injection of heated cells of S strain. Griffith concluded that this was because, excessive heat killed the S strain bacteria (**Figure 6.1**).

Finally, Griffith did a curious experiment: he injected a rat, with a mixture of heat killed S cells, and live R cells, neither of which can cause pneumonia. This rat, developed pneumonial infection, and died (**Figure 6.1**). Griffith cultured samples from the rat, dead from the injection of the mixture of heat killed S and R strains – he observed both S and R type colonies in culture, which when examined under the microscope, showed living cells, of both the S and R strains (**Figure 6.1**). This meant that either the heat killed S bacteria came back to life (very unlikely), or some of the cells of the live R strain were somehow 'transformed' into the S strain. After repeating this experiment many times, Griffith concluded that it was actually due to the transformation of some bacteria of R strain to an S type in the presence of dead cells of S strain. He named this phenomenon "transformation".

A significant discovery was thus made and reported by Griffith in 1928 – organisms can somehow be genetically "re-programmed" into a different version of themselves. The R strain of *S. pneumoniae* was changed into S strain, presumably because of the transfer of genetic material from the heat-killed S strain; Griffith termed this material as "transforming principle". When S strain bacteria are heat treated, the cells break open, releasing many substances – carbohydrates, proteins, lipids and nucleic acids. One of these substances, is the transforming agent – it is absorbed by live bacteria of R strain leading to their transformation. The question thus arose – which molecule among the substances from the heat-treated S cells is the "transforming principle"?

The question generated from Griffith's experiments was examined by two groups of scientists: Oswald Avery, Colin MacLeod, and Maclyn McCarty (1944) and Alfred Hershey and Martha Chase (1952).

6.1.2 Experiments of Avery, MacLeod, and McCarty: Revelation of the Transforming Principle as DNA

Avery et al. repeated Griffith's experiment of treating a rat with a mixture of heat killed S and live R strains of *Streptococcus pneumoniae* but with some modifications. The mixture of heat-killed S strain and live R strain bacteria was separated into six test tubes. A different enzyme was added to each tube except one. The enzymes used were: RNase (RNA); DNase (DNA); protease (protein); lipase (lipids) and a combination of enzymes that break down carbohydrates. These enzymes differ in the substrate they act and break down. The tube which received no enzyme served as the 'control' (like the 4th one in Griffith's experiment – along with live R cells, it contained all the substances released from the heat killed S strain intact). The mixture of dead S and live R cells from each of the six tubes treated with different enzymes and the control was injected into six different rats (**Figure 6.2**).

The theory behind this experiment was that, the mixture (dead S+ live R cells + enzyme) among the five tubes, that does not cause pneumonia, and kill the injected rat, has the 'transforming principle' intact i.e. the substrate of the corresponding enzyme in the mixture. For example, if the rat injected with a cell mixture treated with protease survived, then, the transforming agent among the substances released from the dead S cells is a protein. The protease added to this cell mixture digests the protein and due to its absence, the R cells are not transformed to virulent S, to trigger pneumonia and kill the rat.

In the experiments of Avery et al., the rat treated with cell mixture with DNase was alive. This is because, DNA among the substances released from S cells was digested by DNase and the R cells remained untransformed to S type. This unequivocally showed that the transforming agent from the S strain bacteria is DNA (**Figure 6.2**).

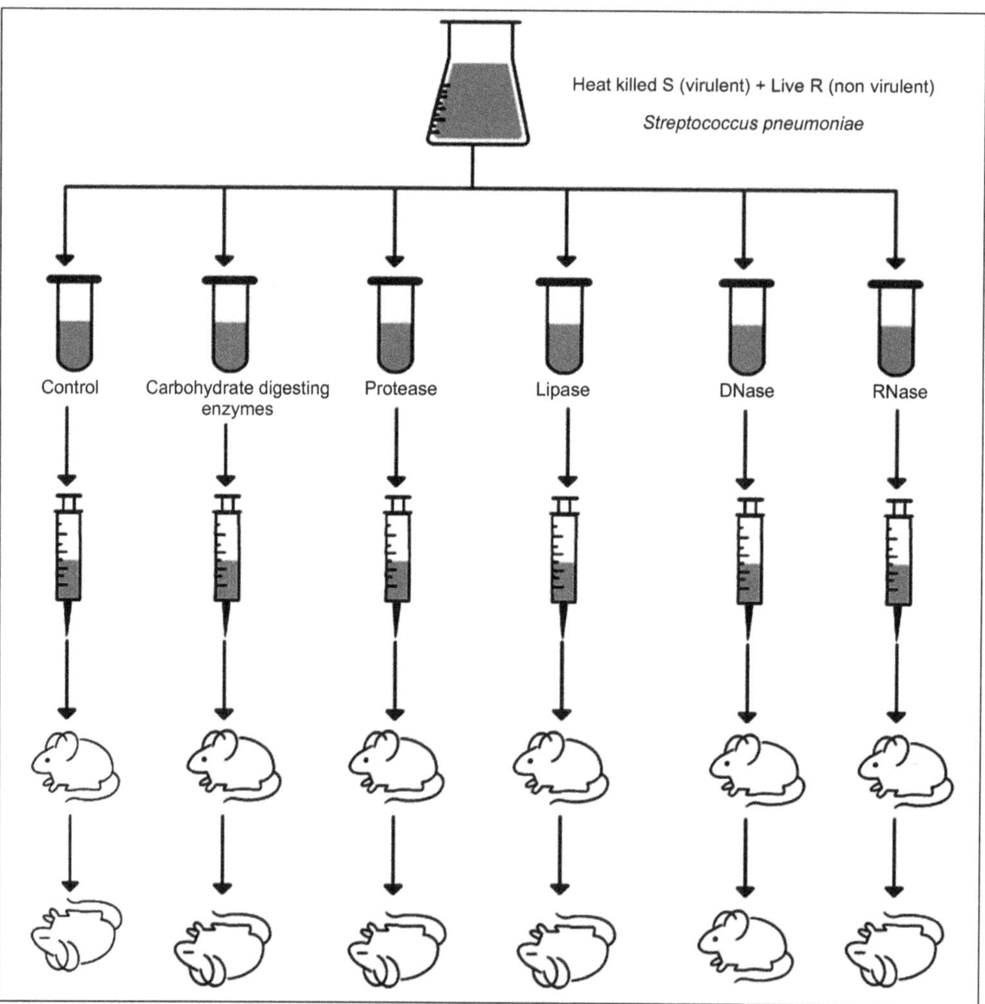

Figure 6.2: Avery, Mc Cleod and Mc Carty's experiment: Injection of rats with a mixture of heat killed virulent S strain and live R strain (non-virulent) of *Streptococcus pneumoniae* treated with different enzymes. *Only rat injected with DNase treated cell mixture was alive showing that DNA is the transforming principle. (The DNA of the smooth virulent strain was digested by DNAse and therefore the avirulent strain of rough strain was not transformed to virulent one. In all other treatments, the R strain was transformed by the DNA from the virulent S strain)*

To further test their conclusion, Avery et al. extracted pure DNA from heat-killed S strain and added it to a culture of live R cells. This mixture of pure DNA from S cells and live R cells also induced pneumonia when injected into a rat causing its death. Thus, the transforming agent was concluded as DNA.

6.1.3 Hershey and Chase Experiment: Proof of DNA as Genetic Material

Eight years after the famous Avery, MacLeod, and McCarty experiment was published, Alfred Hershey and Martha Chase, did experiments with a virus, to confirm the biochemical nature of the genetic material. An extremely small virus which infects only bacterial cells called a bacteriophage (or phage) was used in their experiments. A virus is made of only two components: a protein coat enclosing a nucleic acid (DNA/RNA).

Hershey and Chase used 'radioactive labelling' technique to determine which of the two components of the virus (phage): protein or DNA, constitutes the genetic material. The phage looks (under an electron microscope) like a space shuttle with a proteinaceous hexagonal head which encloses DNA, and a tail. For radioactive labelling, a radioactive isotope of a certain atom is used and its location is traced due to its radioactive emission using a simple laboratory instrument. Radioisotopes of phosphorous (P^{32}) and sulphur (S^{35}) were used in these experiments. Phosphorous is present in DNA, but not in the proteins of bacterium and phage; sulphur is in proteins, but not in DNA. One batch of phage culture was grown in the presence of P^{32}. This batch of phage will have radio-labelled DNA. Another batch of phage was cultured in the presence of S^{35}; these phages will have radio-labelled proteins which constitute its coat. The two batches of labelled (S^{35}/P^{32}) phages were used separately to infect bacteria and presence of radioactivity was tracked (**Figure 6.3**).

When the phage infects a bacterium, the protein coat is left outside the bacterial cell wall and only its DNA enters the bacterium. Within the bacterial cell, the phage DNA makes more copies of itself to reproduce phages. The liquid culture with bacteria and labelled bacteriophages was blended to separate the empty protein coat of the bacteriophage (referred to as ghost) from to the bacterium to which it is attached. The blended liquid was centrifuged. The bacterial cells being heavy settle at the bottom of the centrifuge tube as a pellet, the empty protein coats of the bacteriophage, being very small and light, compared to the bacterial cells, float in the top of the supernatant liquid. The location of radioactivity in the centrifuge tube was detected using a radioactivity detection system. In the tube with S^{35} labelled phages, radioactivity was detected on the top, which contained the empty protein coats of the phages, while, in the tube with P^{32} labelled phages, radioactivity was detected at the bottom in the bacterial pellet, which contained the phage DNA (**Figure 6.3**). Thus, Hershey and Chase concluded that, it is DNA that constitutes the genetic material of the phage – it entered the bacterial cell, and made more copies of itself, for phage reproduction.

The above three experiments in microbes, thus proved that, nucleic acid (DNA) is the genetic material. While the chemical nature of the genetic material was known from these experiments, the physical (molecular) structure of the DNA molecule remained to be elucidated. It is the molecular structure of the genetic material which helps to understand heredity (transfer of genetic information from parent to offspring) and expression of genetic material (information transfer from the DNA to proteins that determine the physical features of the organism).

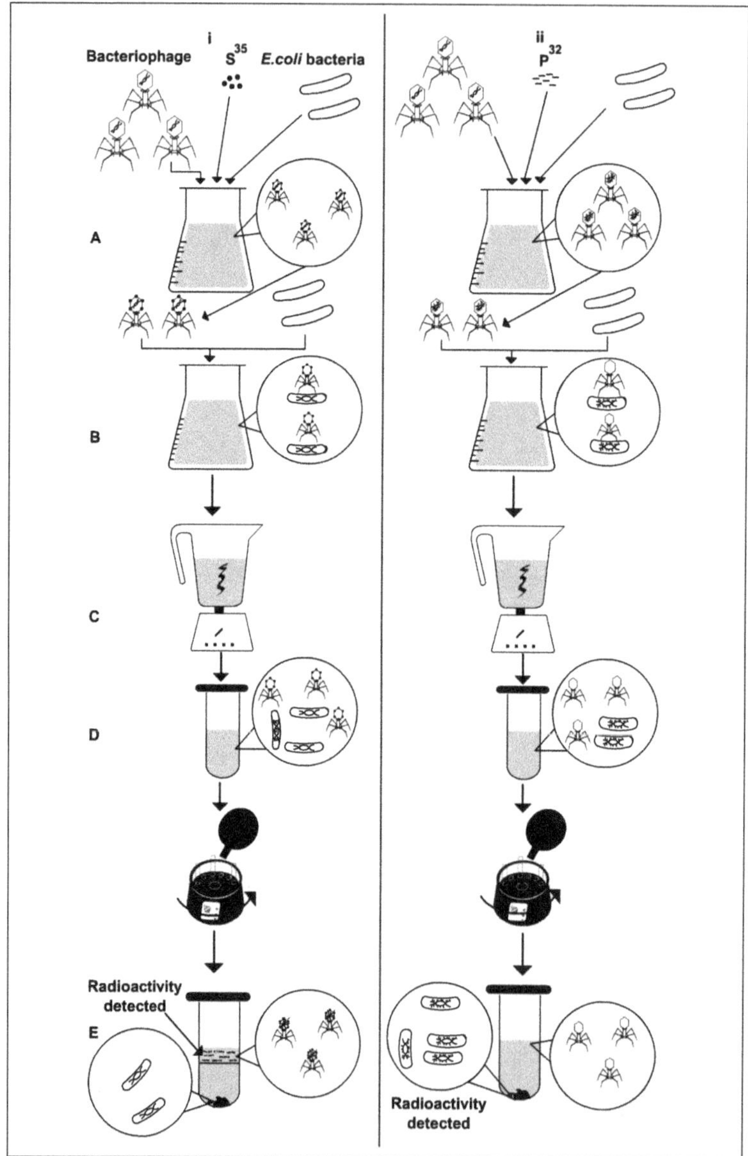

Figure 6.3: Hershey and Chase Experiment: Radioactive labelling of bacteriophage to find out whether protein or DNA is the genetic material. i. *Labelling outer protein coat of phage with radioactive sulphur(S^{35}).* **ii.** *Labelling phage DNA with radioactive phosphorous (P^{32}).* **A.** *Radio isotope labelling of phages:* **iA.** *Radioactive S^{35} incorporates into phage protein coat.* **iiA.** *Radioactive P^{32} incorporates into phage DNA.* **B.** *Inoculation of radio labelled bacteriophages harvested from cultures in 'A', and their host, E. coli bacteria, into culture medium (phage coat is left outside the bacterium; only its DNA enters the bacterium).* **iB.** *radioactive label is outside the bacterium.* **iiB.** *radioactive label is inside the bacterium).* **C.** *Blending the harvested culture from 'B' – bacterium and empty coat of phage attached to it are separated.* **D.** *Dispensing blended liquid into centrifuge tube and centrifugation – phage coats being small and light are in the top supernatant in the centrifuge tube, the large bacterial cells settle as a pellet in the bottom of the tube.* **E.** *Detection of radioactivity in the centrifuge tube.* **iiE.** *P^{32} label of phage DNA is detected in the bacterial cell pellet at the bottom of the tube. Therefore, it is concluded that genetic material of the phage is DNA (phage DNA enters the bacterium to make more copies and reproduce).*

6.2 HIERARCHY OF DNA STRUCTURE – FROM DOUBLE HELIX TO CHROMATIN

6.2.1 Structure of DNA: The Double Helix

The structure of DNA was deduced from chemical analysis and X ray crystallography images of the molecule.

Theodore Levene, chemically characterized the nucleic acids, both RNA and DNA. He found that DNA contained deoxyribose, a five carbon atom containing pentose sugar and four nitrogen bases – adenine, guanine, thymidine, cytosine, and a phosphate group. Thymidine and cytosine have a single carbon ring while adenine and guanine have two carbon rings in their structure; the former two nitrogen bases are termed pyrimidines and the latter, purines. Levene also found that sugar, phosphate and nitrogen base were linked together to form a unit which he termed a 'nucleotide'. Depending on the nitrogenous base component, nucleotides are four types. Levene stated that the DNA molecule consisted of a string of nucleotide units linked together through phosphate groups. He formulated the 'tetranucleotide hypothesis 'stating that the DNA molecule was a tetranucleotide with four types of nucleotides. He ruled out its role as the genetic material.

Inspired by the discovery of Avery et al. that DNA molecule is the genetic material, Erwin Chargaff, an American biochemist, analysed it. Chargaff discovered that in a DNA molecule, amount of guanine was equal to that of cytosine and amount of adenine was equal to thymidine. From studies on DNA samples from several organisms, Chargaff in 1949, stated unequivocally that in all DNAs tested, the molar ratios of purines to pyrimidines, of adenine to thymidine and guanine to cytosine approached 1.0. He also made a land mark discovery – that the proportion of the different nitrogen bases varied between different species. The molecular diversity of DNA discovered by Chargaff, led credibility to its role as the genetic material. Chargaff disproved tetranucleotide hypothesis of DNA structure.

During 1951–53, Rosalind Franklin working with Maurice Wilkins in England, through X ray crystallography, produced high-resolution photographs of DNA fibres and very accurately described the arrangement of various components of the DNA molecule.

The structure of DNA was unravelled in 1953, though ingenious model building, by James Watson and Francis Crick, through an insight of chemical data of Chargaff, and, the X ray crystallography images of Rosalind and Wilkins (**Figure 6.4**). The ingenuity of the discovery of DNA structure was rewarded with a Nobel prize to Watson, Crick and Wilkins in 1962.

Watson-Crick's model of DNA structure, became a celebrity, as it explains the mechanism of both inheritance (information flow from generation to generation), as well as its expression (information transfer from gene to protein that determines the appearance of an organism). As per this model, DNA is a macromolecule – a polymer of nucleotides. The phosphate group in one nucleotide that is attached to the fifth (5′) carbon atom of deoxyribose sugar covalently bonds with the OH group at the third (3′) carbon atom of deoxyribose of the next nucleotide forming a polymer of a long chain of monomer nucleotides. The sugar–phosphate chain forms the backbone of the DNA strand with the nitrogen bases sticking out. The DNA molecule is made of two strands which twist in the from a helix. The double helix makes a right-handed twist at regular intervals of ten nucleotides. The nitrogen bases sticking out from the opposite strands of DNA are linked by hydrogen bonds. The linking is specific, adenine always pairing with thymidine with two hydrogen bonds, and guanine always with cytosine, with three hydrogen bonds.

Since a purine (double carbon ring nitrogen base) pairs with a pyrimidine (single carbon ring nitrogen base) the width between the two strands is uniformly constant. The two strands of DNA which form the double helix are antiparallel to each other; that is, the phosphate bearing end at the 5th carbon position of deoxyribose (5′ prime end) of one strand aligns with the hydroxyl group bearing end on the 3rd carbon of deoxyribose (3′ prime end) of the opposite strand and vice versa (**Figure 6.4**).

The DNA molecule is like a helical staircase with the paired and linked nitrogen bases forming the stairs and the sugar phosphate links making the banister (**Figure 6.4**).

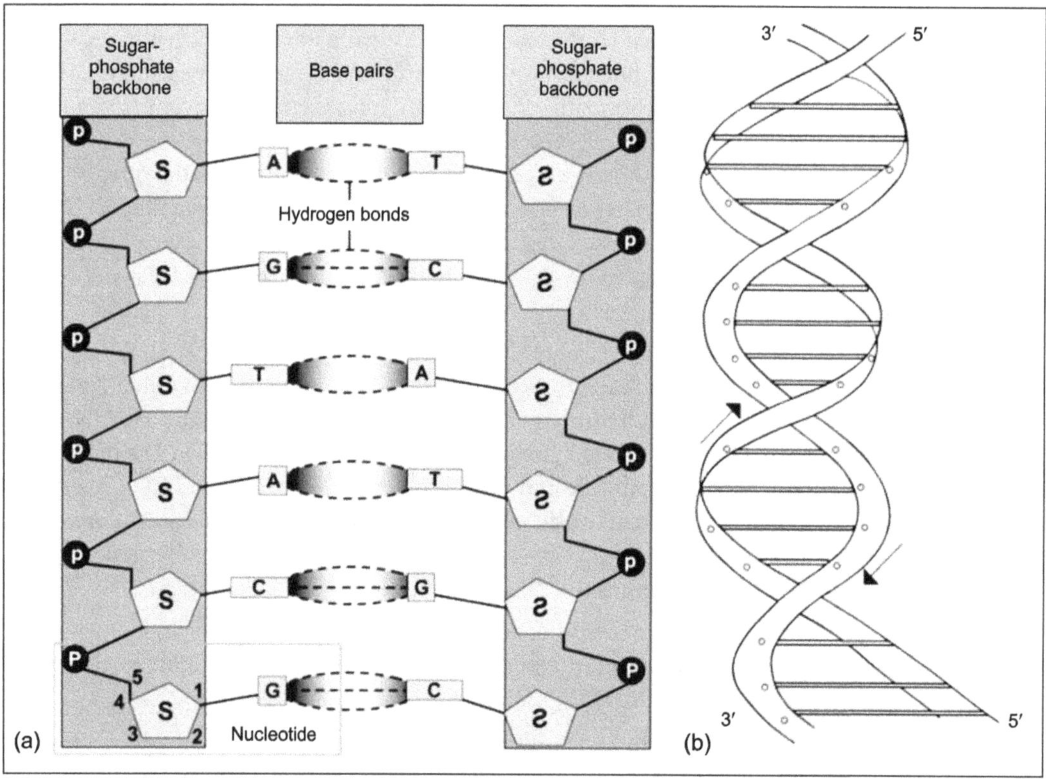

Figure 6.4: DNA structure. a. *Molecular structure.* b. *Double helix- the two strands are antiparallel*

6.2.2 Packaging of DNA in Cells

The amount of DNA present per cell differs among different organisms. It ranges from ~ 4.6 million base pairs in a simple bacterium like *Escherichia coli* to 3 billion in humans to even much more in organisms like some salamanders. Stretched end to end, the DNA molecule from a single *E. coli* cell extends to 1.6 mm; in human cells, to 2 meters. The cells are too small (few nm to µ) to be visible to naked eye – they are microscopic. Thus, the DNA must be packaged, much beyond the coiling of double helix, to fit within the cell.

The double helical DNA is differently packaged in prokaryotes (bacteria and viruses) that do not have a nucleus, and the eukaryotes (plants and animals) which have a nucleus. In bacteria, the two extreme ends of the DNA molecule are attached to form a circle. The circular DNA is supercoiled and folds into loops averaging a length of 20,000 bp. The supercoiling of DNA in the bacterial cell is facilitated and maintained by proteins. The specific region of the bacterial cell in which the densely looped supercoiled DNA is present is called the nucleoid (**Figure 6.5**). In eukaryotes, the quantity of DNA being much more than prokaryotes, it requires to be more tightly packed. Moreover, it needs to fit into the nucleus, the membrane bound organelle within the cell (**Figure 6.5**). The DNA is even more condensed in chromosomes formed during cell division (**Figure 6.5**).

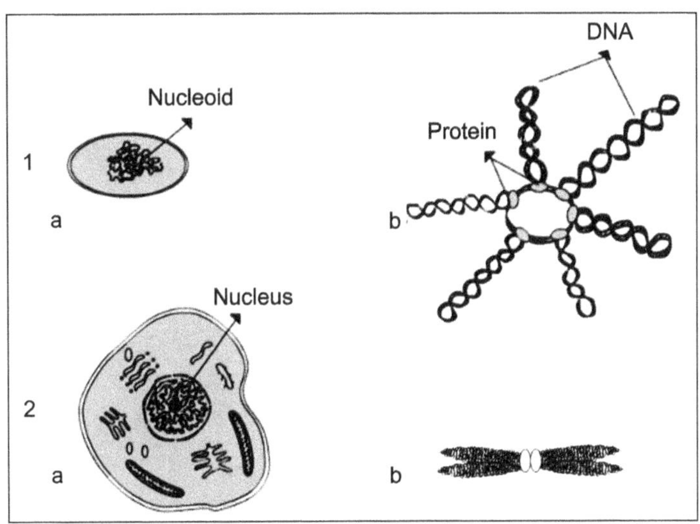

Figure 6.5*:* **DNA in prokaryote (bacteria) and eukaryote. 1.** *Prokaryote –* **a.** *Bacterial cell showing nucleoid – the region containing DNA.* **b.** *Magnification of a part of supercoiled DNA in the nucleoid; the supercoils are held together with protein molecules.* **2.***Eukaryote –* **a.** *Animal cell with nucleus in interphase stage (non- dividing stage) – DNA complexed with proteins (chromatin) is present within the nucleus bound by a double membrane with pores.* **b**. *A chromosome as seen during cell division containing extremely condensed chromatin.*

The required high packaging ratio of DNA is achieved through association with several proteins. The DNA package proteins are two kinds – basic proteins called histones and acidic proteins termed the non-histone proteins. Five kinds of histones: H1, H2A, H2B, H3 and H4 and several types of non-histone proteins are associated with DNA. The packed DNA protein complex in eukaryotes is called chromatin. The chromatin is less densely packed during the growing phase of a cell termed interphase. The chromatin progressively gets more and more condensed to form chromosomes during division phase of the cell. Different organisms have different number of chromosomes. Each chromosome has a single molecule of DNA (**Figure 6.5**).

The chromatin in eukaryotes is formed by packaging of DNA in four tiers – a 10 nm and 30 nm fibre and 300 nm and 700 nm thick bundle of folds and loops.

6.2.2(a) The 10nm Fibre – Nucleosome Chain

The DNA double helix, winds twice around a disc like protein core, called the nucleosome. The nucleosome is an octamer of eight histone molecules, two each of H2A, H2B, H3 and H4. About 150 base pairs(bp) of DNA is wound round a nucleosome; the linker DNA between two adjacent nucleosomes is about 50bp. This gives DNA a packing ratio of about 6 i.e., the double helix is shortened to $1/6^{th}$ of its length and it is 10nm (11nm to be exact) in diameter (**Figure 6.6**).

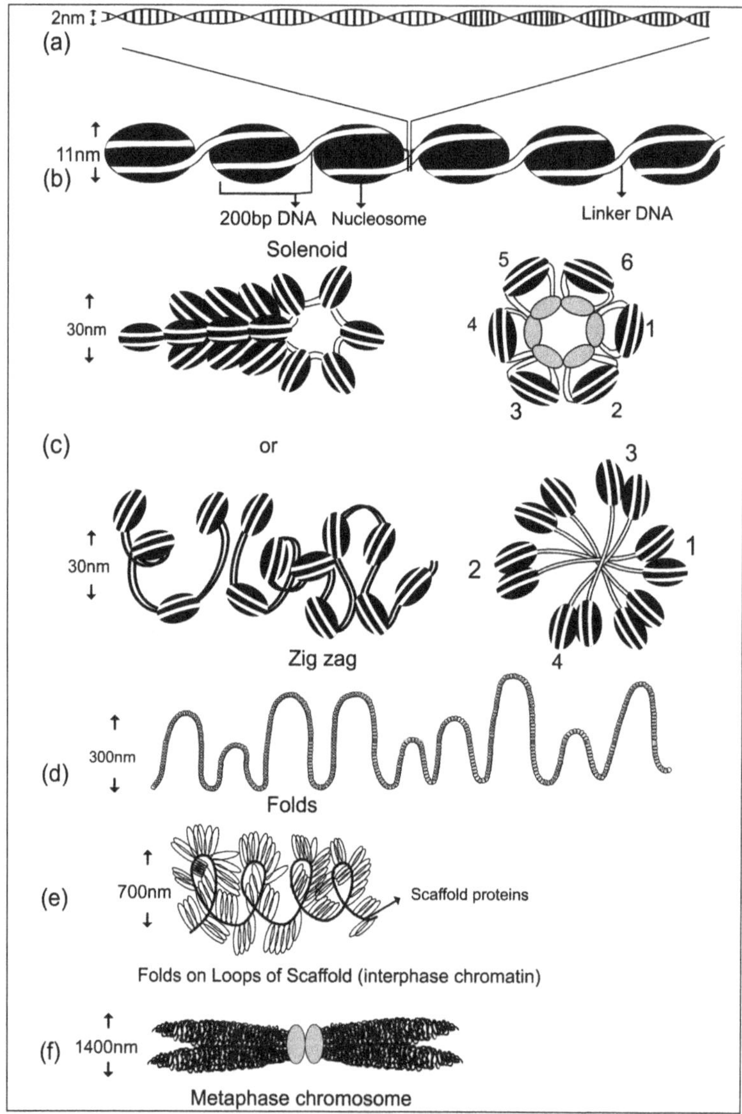

Figure 6.6: Packing of DNA in Eukaryotes. a. *The double helical DNA is 2 nm wide.* **b.** *The first level of packing involves winding of DNA twice around the nucleosome – a disc shaped structure formed from the aggregation of 8 molecules of four types of histone proteins. The structure formed from the winding of DNA around the nucleosomes is 11 nm wide.* **c.** *Second level of packing involves the twirling of the 11 nm fibre to form a 30 nm fibre –the twirling can be either like a solenoid or in a zig zag fashion.* **d.** *Folds: the 30 nm fibre folds with the help of non-histone proteins to form a 300 nm fibre.* **e.** *Loops: The folds get attached to a matrix of scaffold proteins in the form of loops to form a 700 nm fibre. This constitutes the chromatin in a cell in interphase stage.* **f.** *During cell division, the chromatin gets even more densely aggregated due to condensing of the scaffold protein matrix resulting in the 1400 nm chromatid of metaphase chromosome.*

6.2.2(b) The 30nm Fibre

The second level of packing of DNA is achieved, by coiling of the 10 nm fibre of nucleosome beads, to form a 30 nm fibre. The pattern of packing in the 30 nm is not fully resolved. Two models have been proposed to explain the pattern of coiling: solenoid, zigzag. The solenoid model assumes that, the 10 nm fibre of DNA wound round nucleosomes, gets coiled into a helical structure, with six nucleosomes per turn. In the second model, nucleosomes are assumed to be arranged in a zigzag pattern, such that, two rows of nucleosomes are formed, and the linker DNA criss crosses, between each stack of nucleosomes. This produces a double-helical structure (a two-start helix), while a solenoid, is a one-start helix. H1 protein molecule, binds to the nucleosome and the linker DNA, to stabilize the 30 nm fibre. The coiling of the 10 nm to 30 nm fibre accomplishes a packing ratio of 40, i.e., every 1 µm along the axis of the coil has 40 µm of DNA. The coils are held in place with the association of H1 histone protein (**Figure 6.6**).

6.2.2(c) Folds – 300 nm Structure

The next level of packing occurs through a series of radial folds of 30 nm fibres held in place by scaffold proteins to form a 300 nm thick structure.

6.2.2(d) Folds as Loops over Scaffold Protein Matrix – 700 nm Structure

The 300 nm folds are anchored on loops of a protein matrix of non-histone proteins which serves as the scaffold. Each loop has 5 to 200 kb of DNA and a length of 10-30 μm. The overall thickness of the thus formed structure is 700nm. The loops emanating from a central scaffold have been actually observed in metaphase chromosomes. The extreme condensation of interphase chromatin to metaphase chromosomes with 1400 nm thick chromatids is accomplished through tighter compaction of scaffold protein matrix. Thus, a final packing ratio of about 1000, in interphase chromatin and 10,000 in mitotic chromosomes is attained (**Figure 6.6**).

6.3 MOLECULAR BASIS OF INFORMATION TRANSFER FROM GENETIC MATERIAL

DNA has the genetic information. The external, anatomical and biochemical features of an organism are governed by proteins. Proteins are the work horses in a cell. Information transfer from DNA in the nucleoid or nucleus, to proteins is an intriguing process of an orderly series of events. The mechanism is universal, being similar in prokaryotes and eukaryotes with a few differences.

Watson and Crick, while proposing the structure of DNA in 1953, predicted that the sequence of the nucleotides is the determinant of information in the gene. Crick also foresaw the existence of an 'adapter' that helps in the translation of information from the nucleic acid to a protein. It was later discovered that the adapter was another nucleic acid – an RNA – a tRNA (transfer RNA).

While a gene constituted by DNA has its information as a set sequence of four nucleotides: dATP, dGTP, dTTP and dCTP, the proteins are polypeptide chains with a set length and sequence of all or some of the twenty known amino acids. Information transfer from genes involves, paraphrasing the nucleotide sequence to amino acid sequence. There is a collinear relationship between nucleotide sequence of DNA and the amino acid sequence of the protein. Different genes, differ from each other in their DNA sequence, resulting in proteins with different amino acid sequence. The amino acid sequence determines the structure of the protein molecule. The structure of the protein governs its functions. Thus, proteins with different amino acids have different structure and therefore, different function.

6.3.1 Information Transfer from Genetic Material (DNA – Gene) to Protein

6.3.1(a) The Central Dogma

Without any experimental evidence, but with intuition from the proposed DNA structure, Francis Crick, in 1958 made a bold statement that, the information transfer between DNA and protein is only 'one way' i.e., it can occur from DNA to the protein but not vice versa; also, information transfer is not possible between proteins. This statement of Crick is famously known as 'Central dogma of molecular biology' and diagrammatically represented in **Figure 6.7**. In a field like science, where hypothesis is the basis for any theory, and hypothesis only stands until proven wrong (possible as more information becomes available), a dogmatic (incontrovertibly true) statement, from a scientist giant, is indeed the result of great insight, on the nature of working of the DNA molecule, whose structure he predicted. The main theme of central dogma stands true to date, with some additions (**Figure 6.7**).

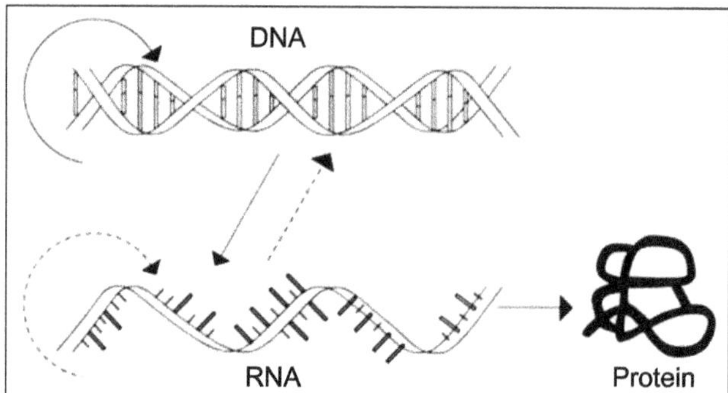

Figure 6.7: Central dogma of molecular biology stated by Francis Crick. *Information flow from nucleic acid to protein is unidirectional; information in protein cannot be transferred to nucleic acid. The dotted lines (synthesis of DNA with information from RNA and replication of RNA) were discovered in viruses later (after Crick's statement).*

6.3.1(b) RNA in Information Transfer

Experiments by several scientists in late 1950's led to the discovery of three kinds of RNA (ribo nucleic acid) that mediate information transfer from DNA to protein: the messenger RNA (mRNA), transfer RNA (tRNA) and ribosomal RNA (rRNA). Messenger RNA carries the information from DNA in the form of its nucleotide sequence which is translated to amino acid sequence of a peptide, hence the name messenger RNA. Ribosomal RNA forms the skeleton of ribosomes which are the protein synthesizing machines of the cell. There are four kinds of rRNA in the ribosomes: 16S, 23S, 5.8S and 5S in prokaryotes; 18S, 28S, 5.8S and 5S in eukaryotes. Transfer RNA serves as an adaptor to mediate conversion of nucleotide sequence to the amino acid sequence. It carries amino acid to the site of protein synthesis, hence called transfer RNA. The rRNA and tRNA though single stranded, due to folding, form double strands in some regions. RNA has ribose sugar and has the nitrogen base uracil instead of thymidine of DNA.

6.3.1(c) Genetic Code

What remained unclear was – in what form the nucleotide sequence is decoded to the amino acid sequence. The seminal work of Marshall Nirenberg and Hargobind Khorana, explained the basis of decoding, for which, they were awarded Nobel prize in 1968. They showed that a sequence of three nucleotides (a triplet), coded for an amino acid. All the 64 possible triplet sequences with the four types of nucleotides were spelt out – which triplet nucleotide sequence coded for which amino acid (**Figure 6.8**). Since there are only 20 types of amino acids, all the amino acids except two, are coded by more than one triplet codon. Three of the 64 codons, code for no amino acid; these three codons were initially called non-sense codons; they are now termed 'stop codons'. The triple nucleotide code was christened the 'genetic code'.

6.3.1(c)1 Features of the Genetic Code

The genetic code is read from RNA which has been synthesized using a DNA strand as its template. In RNA, thymidine of DNA is replaced by uracil. Therefore, the genetic code has U in place of T. The genetic code is a triplet – a sequence of three nucleotides. It is universal – a codon codes for the same amino acid in all living beings with very few exceptions. It is non-overlapping and comma less – the triplet codons on a stretch of DNA are read as a set of three starting from a particular nucleotide triplet set (start codon) in a continuous manner with no intervals of unread nucleotides. Genetic code is degenerate – more than one codon codes for the same amino acid. The genetic code is ordered – multiple codons for a given amino acid and codons for amino acids with similar chemical properties are closely related, usually differing by a single nucleotide. The genetic code is unambiguous – a triplet codon always codes for the same amino acid (**Figure 6.8**).

6.3.1(d) The Process of Information Transfer from DNA to Protein (Transcription and Translation)

The information which is in the form of nucleotide sequence in DNA, is first transferred to RNA through copying – a process termed 'transcription'. Transcription involves synthesis of RNA, using one of the DNA strands of a gene, as a template. The RNA thus synthesized, has a sequence, complementary to the DNA strand that served as the template strand, and similar sequence, to the other DNA strand of the gene. The sequence of DNA (non-template) strand which is similar to mRNA is called coding sequence or sense strand. The template strand is called non-coding or antisense strand.

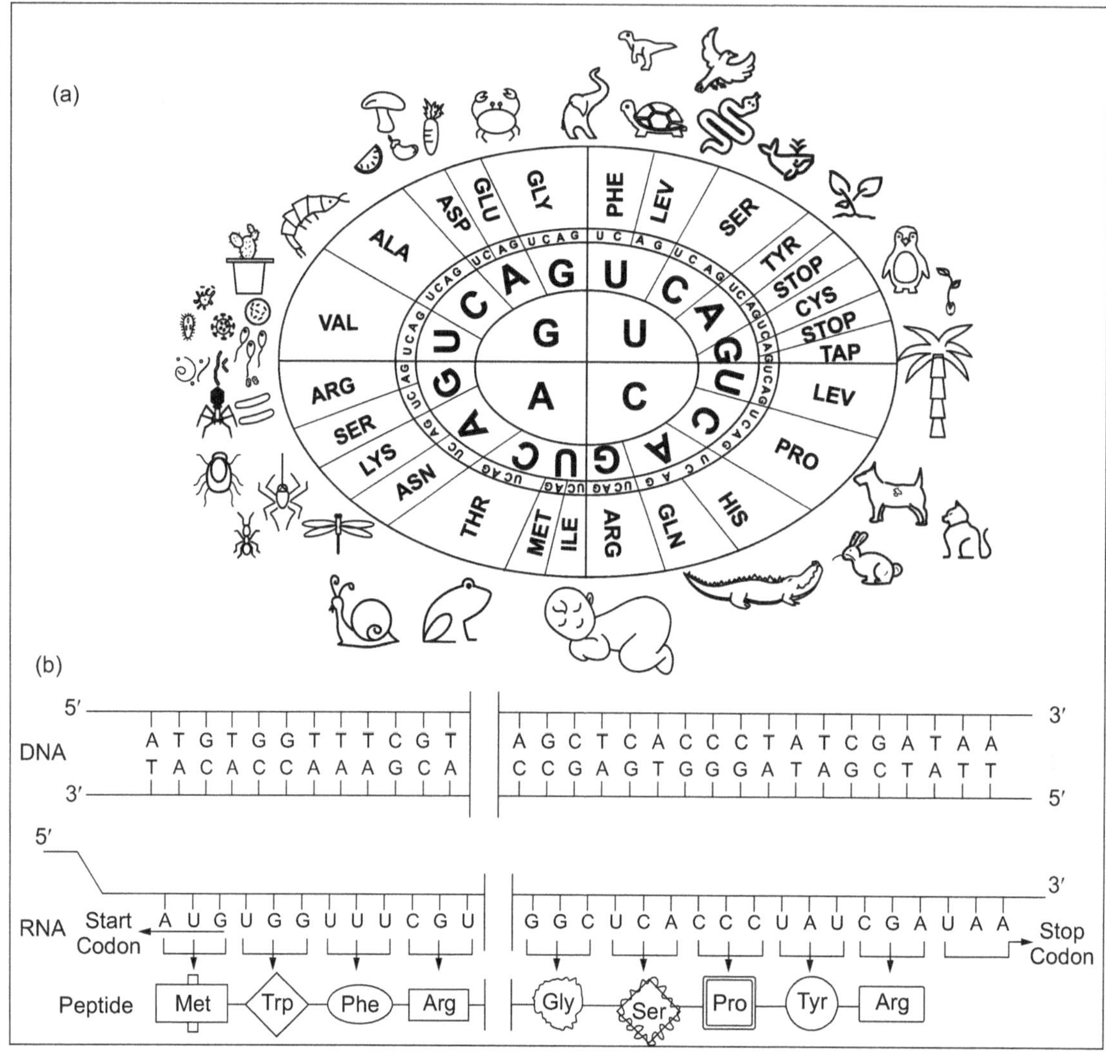

Figure 6.8: The genetic code: a. *Universal triplet genetic code for the 20 amino acids that constitute proteins. Abbreviations of amino acids are given. Three of the codons do not code for any amino acid, when such a codon is encountered on the RNA, protein synthesis stops, therefore are called stop codons. The degeneracy (one amino acid being coded by more than one triplet codon), ordered nature (related codons coding for the same amino acid) and unambiguity (one triplet code coding for only one amino acid) are evident.* **b.** *The process of decoding the genetic code showing its comma less, non-overlapping nature. Note: The DNA, RNA and protein are given a break: || to indicate the long length of the DNA of the protein coding DNA and the RNA and protein that emerge from it*

The rRNAs and tRNAs which facilitate information transfer from DNA to protein are transcribed from rRNA and tRNA genes respectively. The tRNA has a clover leaf like secondary structure with arms and loops which folds into an L shaped form in its tertiary state (**Figure 6.9**). One of the arms of tRNA carries the amino acid during protein synthesis. The loop opposite to this arm has a triplet sequence which is termed 'anticodon' – so called since it is complementary to one of the triplet genetic code on the mRNA. To match the 61 codons, there are about 40 different tRNAs. A mismatch between the genetic code and anticodon in the third nucleotide position is tolerated in some instances through wobbling, therefore there are fewer tRNAs than are required for all the 61 codons.

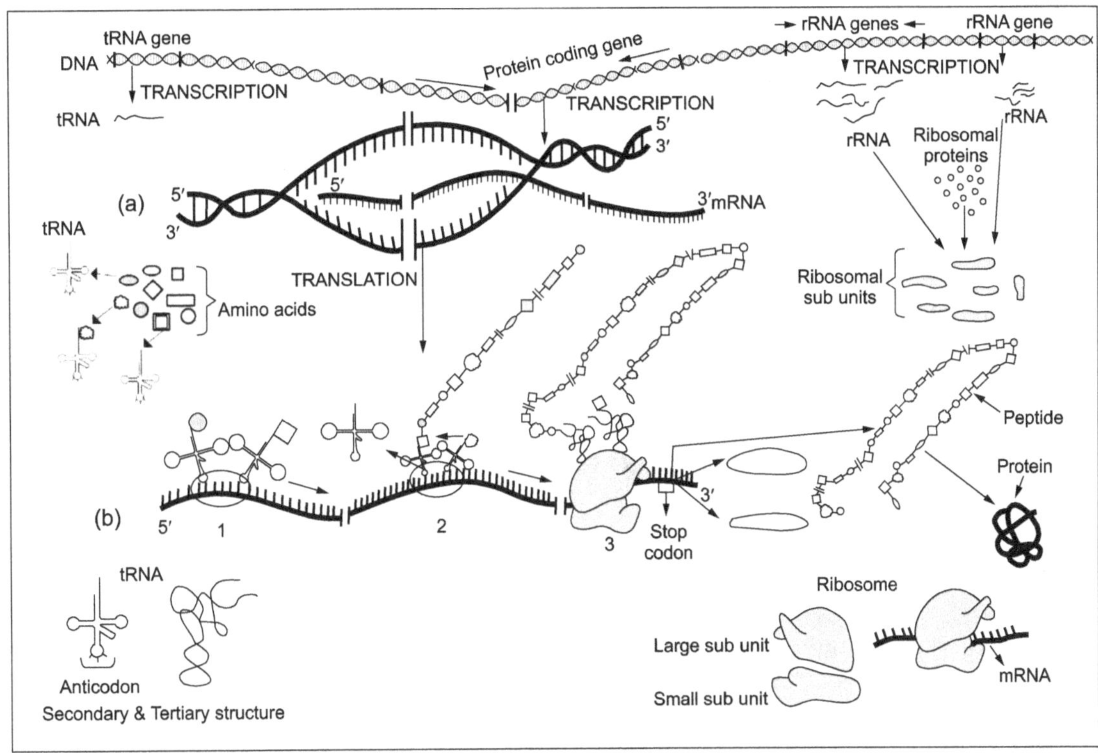

Figure 6.9: Process of Information transfer from DNA to protein (protein synthesis). *Three types of RNAs are involved – mRNA, tRNA, rRNA; all three are transcribed from DNA using one of the DNA strands as a template. mRNA is transcribed from protein coding genes, tRNA from tRNA genes and rRNA from rRNA genes. The different t RNAs differ in the anticodon and carry different amino acids – they are transcribed from different t RNA genes.* **a.** *Transcription – information transfer from DNA to RNA in a protein coding gene. The messenger RNA (mRNA) is synthesized on the DNA using one of its strands as a template.* **b.** *Translation – information transfer from mRNA to peptide.* **b1.** *Ribosome is assembled on the mRNA with two tRNAs carrying amino acids with their anticodon sequence aligned with two triplet codons on the mRNA.* **b 2 & 3.** *Ribosomes engaged in the process of peptide synthesis, as they move from the 5' end of mRNA, a long peptide chain is synthesized.* **b3.** *shows the actual shape of ribosome and tRNA. The ribosome and the peptide are disengaged from the mRNA when encountered with a stop codon. The peptide folds to form a protein. Note: The DNA, RNA and protein are given a break:* ‖ *to indicate the long length of the DNA of the protein coding DNA and the RNA and peptide that emerge from it.*

The RNA synthesized on the DNA representing a protein coding gene is termed messenger RNA (mRNA), since having complementary sequence to the DNA strand that served as its template, it contains the genetic code. The mRNA moves to a ribosome where its sequence is used in the form of triplet codes to direct the linking of particular sequence of amino acids to form the polypeptide chain. This process is termed 'translation'. The information transfer from DNA to protein is facilitated by genetic code.

Ribosome which is the protein synthesizing organelle of a cell is a ribonucleoprotein complex not bound by a membrane consisting of three kinds of rRNA and several types of proteins. It has two subunits one large (50 S and 60 S in pro and eukaryotes respectively) and the other small (30 S and 40 S in pro and eukaryotes respectively), both of which come together on the mRNA, while synthesizing protein. The mRNA strand passes through the two ribosomal sub units. Several ribosomes are engaged along the length of mRNA each producing a peptide chain; the unit of mRNA with several ribosomes is called a polysome.

The mRNA along its length, has a series of triplet codons starting with a start codon (usually AUG, sometimes GUG) and ending with a stop codon (UAA/UGA/UAG). The ribosome moves starting from the start codon, three nucleotides at a time, along the mRNA, as a codon is translated into an amino acid. tRNA – the adapter predicted by Crick, carries the amino acid appropriate for the triplet code in the mRNA, that is in contact with the ribosome. The anticodon of tRNA plugs to the corresponding codon on the mRNA. Two neighbouring amino acids bound to tRNAs that are mounted on an mRNA through codon-anticodon pairing, are fastened with a bond – the peptide bond. The tRNAs now unstrip from the mRNA and exit from the ribosome. The ribosome contains the enzyme required for peptide bond formation. It is the peptidyl transferase which is an rRNA molecule with an enzymatic function. When a stop codon is encountered on the mRNA, no tRNA arrives with an amino acid as a stop codon is a nonsense codon, coding for no amino acid. The peptide chain is thus terminated, followed by the dismantling of the ribosome, by separation of its sub units from the mRNA. The ribosome is thus a mobile protein synthesizing factory which moves along the mRNA molecule to synthesize protein (**Figure 6.9**).

6.3.2 Information Transfer to Next Generation (Cell Reproduction Through DNA Replication)

During cell reproduction, DNA which constitutes the genetic material, is duplicated, to pass on to the daughter cell. The duplication of DNA is a very precise mistake free process aided by several enzymes. The two strands of DNA separate during duplication. Each of the separated strands, acts as a template for synthesis of the new strand. The separation of the DNA strands occurs progressively beginning from

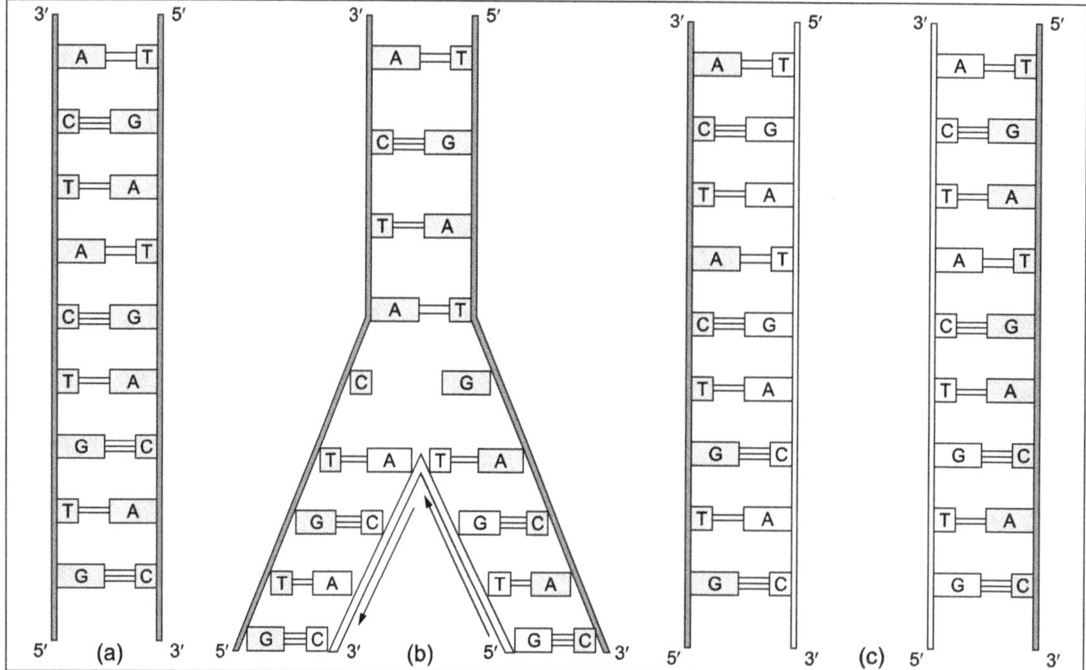

Figure 6.10: Information transfer to next generation (heredity): Semi-conservative replication of DNA.
a. *DNA molecule.* **b.** *DNA during replication – the two DNA strands separate. The new strands are synthesized in opposite directions using the old separated strands as template.* **c.** *Two daughter DNA molecules with one strand each of new and old.*

the start of replication. Synthesis of the new DNA strands occurs in the 3′-5′ direction, with sequential addition of each nucleotide, and the enzyme, DNA polymerase, aiding in the linking of adjacent nucleotides. Since the two DNA strands of a DNA molecule are antiparallel, the two new DNA strands are synthesized in opposite direction. The new strand that is synthesized on the template parental strand has a base sequence complementary to it, i.e., when the nucleotide on the template strand is A, the new DNA strand has a matching T in that position and if it is G on the template, then, a nucleotide with C is assembled in the new strand. In this manner, the two separated parental DNA strands reproduce identical copies. The resulting two DNA molecules have one old (parental) and one newly synthesized strand in the double helix. This type of replication is described as semi-conservative as there is one old and one new strand in the daughter DNA molecule (**Figure 6.10**). Watson and Crick while proposing the double helical structure of DNA, mentioned that, such a structure gives scope, for faithful copying (replication for reproduction) of information to the next generation.

6.4 DEFINITION OF A GENE IN TERMS OF COMPLEMENTATION AND RECOMBINATION

The entire set of genes of an organism is present as a circular DNA molecule in prokaryotes; in eukaryotes the total DNA in the nucleus is fragmented with one fragment per chromosome. A chromosome thus contains, a proportion of the total genes, of the organism. A gene is constituted of a certain length of DNA. What length of DNA can be considered a gene and what constitutes its structural and functional parts was deciphered in mid-1950's by Seymour Benzer through his inventive experiments with a bacterial virus (bacteriophage). Referred to as the 'fine structure analysis of the gene', Benzer's work, stands out as an example of an elegant science experiment, with profound findings, likened to the ground-breaking discoveries during early 1900's, of the sub-parts (electron, proton and neutron) of an atom which was hitherto believed to be the smallest particle of an element and unsplittable.

Benzer worked with a bacteriophage strain – T_4 which infects *Escherichia coli* bacterium. Two strains of *E. coli* – the B and K strains were used in his studies.

T_4, being a virus that is equivalent to an inert particle outside a live host, can reproduce only within the bacterial cell. T_4 is therefore, multiplied in lab cultures, by inoculating it on an *E. coli* culture. *E. coli* culture appears as a white opaque mat on a solid medium in a Petri dish. When it is inoculated with T_4, the region on the culture where the bacterial cells are infected and destroyed (lysed) by the phage appear as clear zones called 'plaques' (**Figure 6.11.i**). In his *E. coli* cultures inoculated with T_4 phage, Benzer sometimes observed a large (2-2.5mm diameter) plaque among the several normal sized (1-1.5 mm diameter) plaques. These large plaques had irregular (mottled) outline compared to the round outline of the usual plaques. The large plaques were formed due to destruction of more bacteria in that zone. The larger plaques were inferred to be formed by a mutant of T_4; it was named r mutant (r for rapid lysis – killing of more bacteria) (**Figure 6.11.ii**).

Benzer identified two types of r mutants: r I and r II. He chose r II for his studies. The common wild type T_4 phage which formed small plaques is represented as R II (capital R for wild type). Benzer collected more than 3000 r II mutants in his cultures. Though all the r II mutant plaques looked alike, Benzer numbered them serially.

The r II mutants, besides producing large mottled plaques, differed from the normal (wild) T_4 phages in their host range. The wild type T_4 phage, can infect both the B and K strains of *E. coli,* while the r II mutants, can infect only the B strain but not the K strain. The B strain is thus a permissive host and K strain, a non-permissive i.e., restrictive host for r II mutants (**Figure 6.11.iii**).

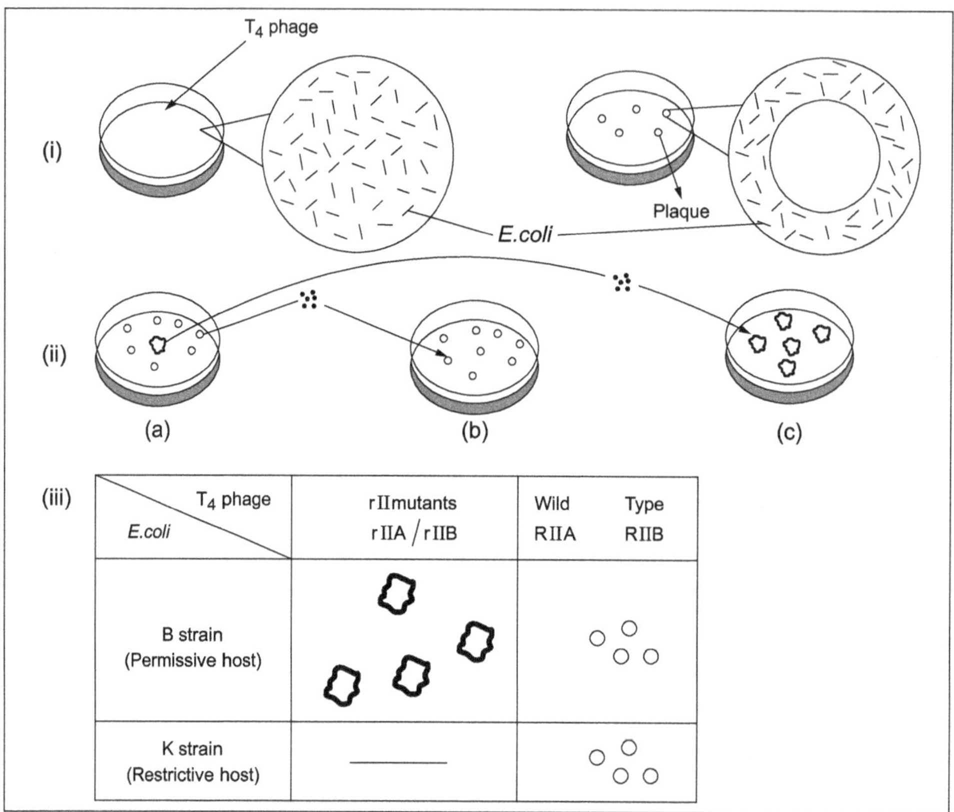

Figure 6.11: rII mutants of T₄ phage. i. *Inoculation of E. coli culture (as a mat) on solid medium with T₄ phages results in death of E. coli cells in the phage infected zone, resulting in clear zones (plaques) on the bacterial cell mat.* **iia.** *Mutant (rapid lysis) r II T₄ phages: an occasional R II mutant plaque spotted on an E. coli culture inoculated with T₄ phage with several small round wild type plaques.* **iib.** *normal wild type T₄ phage produces small round plaques.* **iic.** *r II mutant phage produces large plaques with an irregular outline.* **iii.** *Characteristics of wild type (R II A & B) and mutant (r II A & B) phages.*

Benzer conducted two sets of experiments. Both involved inoculation of *E. coli* cultures simultaneously with two r II mutant phages. In the first set of experiments, K strain *E. coli* was infected with, two different r II mutants at a time, in all pairwise combinations of the ~3000 r II mutants. K strain, being a restrictive host to r II mutants, no plaques are expected to be formed. But, some pairwise combinations of r II mutants, did result in the formation of large mottled plaques on the K strain cultures. Through these experiments Benzer classified the ~ 3000 r II mutants into two groups – A and B groups. Any r II mutant of the A group, when paired with any r II mutant member of the B group, produced plaques on the K strain *E. coli*, which is restrictive host to these mutant phages. Double infections of K strain *E. coli* cultures, with r II mutants of the same group i.e., two r II mutants of A group, or two r II mutants of B group, did not produce plaques. Benzer, deduced from these results that the phage infection starting from entry of the phage into the bacterium, its multiplication within the bacterial cell and final lysis (breakage) of the bacterial cell, is controlled by two phage genes – R IIA and R IIB. Thus, when a K strain bacterial culture, is simultaneously infected by an r II A and r II B mutants, since they represent mutations in two different genes influencing different stages in the infection cycle, they compensate for each other's deficiency, in the infection cycle i.e., they are mutually complementary. Two mutants with mutations in the same gene cannot complement each other because both are deficient for the same function (**Figure 6.12**).

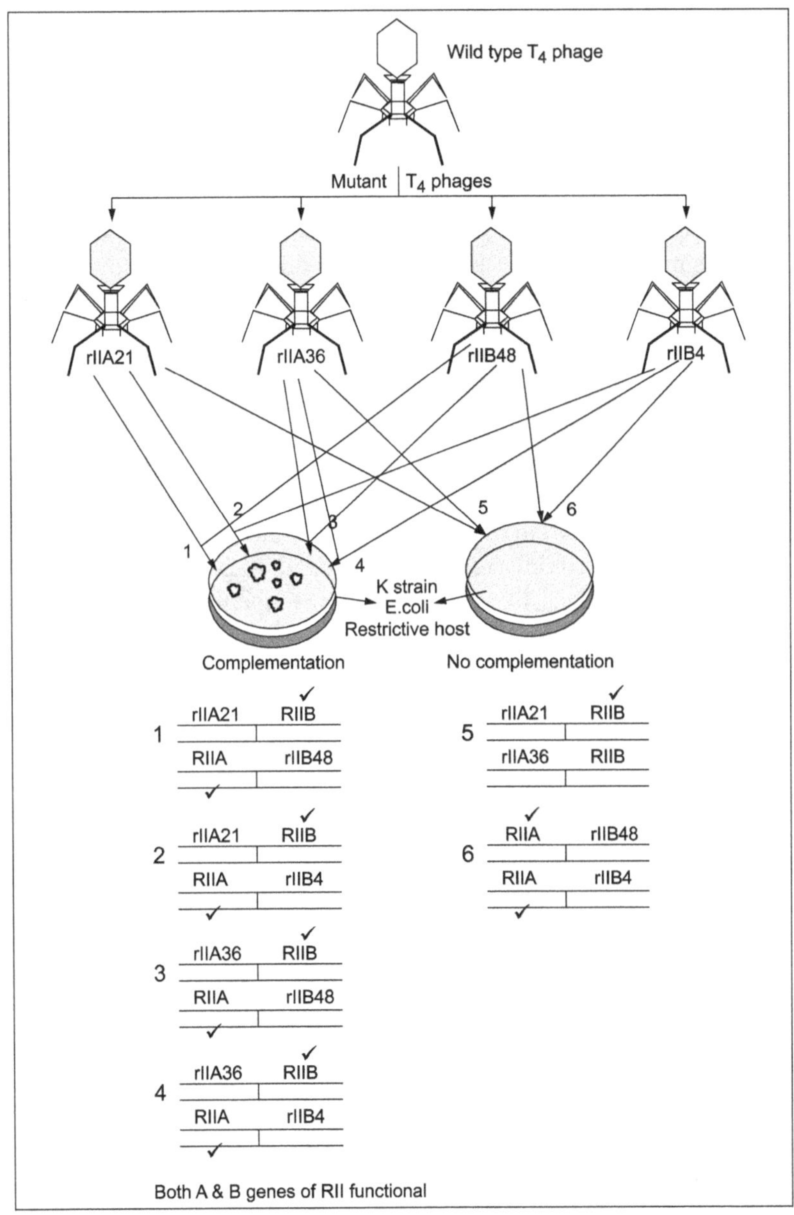

Figure 6.12. Benzer's experiments: Test of complementation between different r II mutants. *Double inoculation with two r II mutants on E. coli K strain, a restrictive host for r II mutants, led to the identification of two groups of r II mutants – A and B. Members from the same group do not complement each other; when simultaneously inoculated on the restrictive host, no plaques were formed. Double inoculation with r II mutants, one from each group (A and B), resulted in complementation and formation of r II type plaques. Complementation between r II mutant of A group and r II mutant of B group occurred because plaque formation (infection of E. coli by phage) was due to the function of two genes R II A and R II B. The r II A mutant was wild type for R II B and r II B mutant was wild type R II A. The deficiency of function in one mutant was compensated by the normal function of its wild type gene in the other and vice versa. Note: Arbitrary numbers are given to the r II A and r II B mutants to aid in comprehension.*

Complementation can be likened to a walking journey, taken up by a blind and limbless man together. While neither of them can make the journey alone, together, they can conduct the journey, as the deficiencies they possess are mutually complementary – the limbless, though cannot walk, can show the way to the blind; the blind, though cannot see, can walk, and carry the limbless, to take up and complete the walking journey. This set of experiments of Benzer that served to identify genes based on function is termed complementation test

Benzer's second set of experiments involved the following steps. The B strain (permissive host for r II mutants) culture was simultaneously inoculated with two different r II mutants of the same group. All possible combinations of r II mutants within each group were used in several such experiments. Since B strain *E. coli* is a permissive host, both the r II mutants multiply. The resulting r II mutant phage progeny produced in the B strain *E. coli* inoculated with two different r II mutants within the same group (A/B), were collected, and next, inoculated on the culture of K strain *E. coli* on solid medium. K strain being a restrictive host and the two r II mutants used are non-complementary, no plaques are expected on the K strain. But occasionally, plaques were formed on the K strain and the plaques were small, like the normal wild type phages. The frequency with which such plaques appeared differed in the different pair-wise r II mutant combinations (**Figure 6.13**). Both these unexpected results i.e., plaques being formed from two r II mutants of the same group on K strain, a restrictive host, and the plaque appearance being not similar to the mutant r II type, but being small, like the wild type was explained by Benzer.

Benzer, concluded that the R IIA and R IIB genes involved in the infection cycle of the phage each constitute a locus (region) on the total DNA of the phage. The two R II loci, A and B, are stretches of a certain length of DNA and were termed 'cistrons'. A cistron represents the functional unit of a gene i.e., if the cistron is disturbed (with respect to its sequence), then, its function is lost – it has functional integrity, only when it has a particular length and sequence of DNA. The r II mutants in the A and B group, represent mutations in two genes controlling different steps in the infection cycle of the phage. When inoculated together on K strain *E. coli* which is a restrictive host, the function lost in a member of r II mutant of A group, is compensated by that function by the r II mutant of B group, and vice versa – they complement each other. Therefore, together they cause infection and lysis of the K strain. From the more than 3,000 r II mutants isolated, Benzer concluded that each represents a mutation (change) at different site along the length of DNA that constitute R II genes. The smallest unit within each of the R II genes (A & B) that can undergo mutation was termed a "muton", the size of the muton being one nucleotide (**Figure 6.12**).

When the phages harvested from the double infection of B strain (permissive host), with two r II mutants of the same group (either A or B), are used to inoculate the culture of the K strain (restrictive host), the formation of a wild type (small) plaque is due to recombination between DNA of the two r II mutants resulting in a wild type DNA (**Figure 6.13**). Based on the frequency with which such wild type plaques were recovered in different combinations of mutants, Benzer estimated that the size of the smallest unit of a gene that can participate in recombination is a single nucleotide – he termed this the 'recon'.

Thus, the gene can be defined as a unit of function (cistron) with sub-components (nucleotides) which have the capacity to mutate (muton) and recombine (recon).

The complementation test of Benzer is a way to determine if two mutant genes which affect the same character represent mutations in the same gene or two different genes with related functions. The intragenic recombination observed by Benzer in T_4 phage, is very difficult to identify in higher organisms such as plants and animals, because, it is a very rare event, and a very large number of progeny are to be screened, to spot the occasional recombinant.

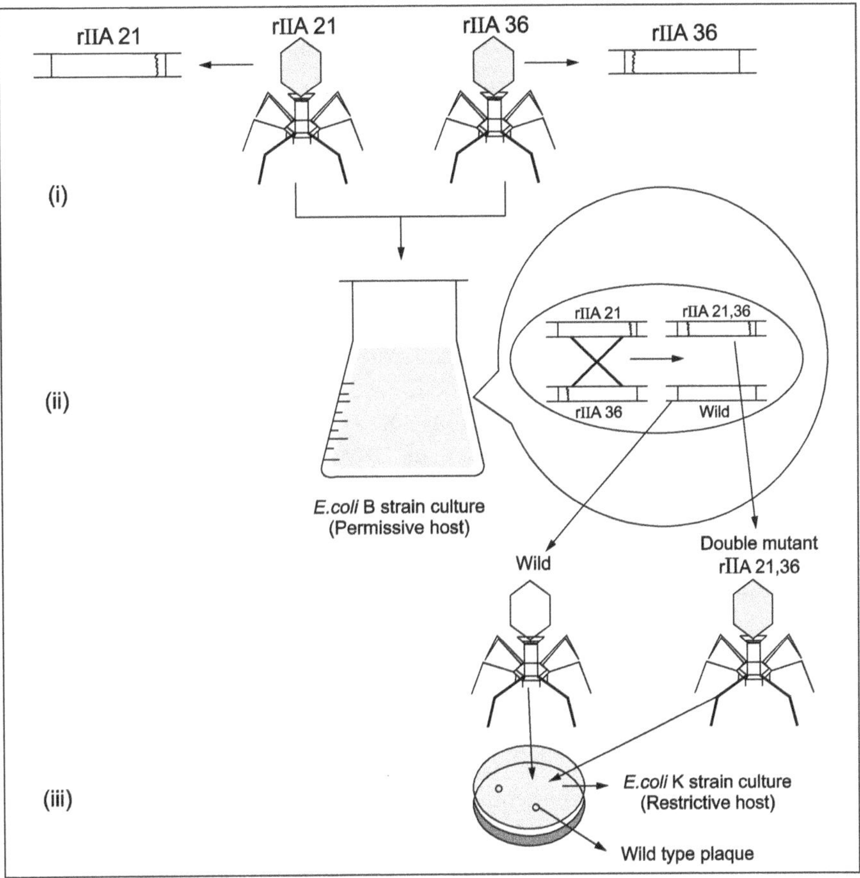

Figure 6.13. Benzer's experiments: Observation of intragenic recombination in RII gene of T 4 phage.
*i. Two r IIA mutants – r IIA21 and r IIA36. The site of mutation (marked as a wavy line) in the two mutants is different. **ii.** Inoculation of r IIA21 and r II A36 into the liquid culture of E. coli B strain, a permissive host for r II mutant phages. Some of the E. coli cells may be simultaneously infected by the rIIA21 and rIIA36 mutants and occasionally, recombination in the r IIA region can occur. Recombination results in formation of double mutant rIIA21-36 and wild type. **iii.** Inoculation of phages harvested from cultures in 'ii' on the culture mat of E. coli K strain (restrictive host) on solid medium results in spotting the occasional wild type phages formed due to intragenic (within r IIA gene) recombination as small round plaques. The double mutant rIIA21,36 cannot form plaques on culture of K strain since it is a restrictive host. Note: Arbitrary numbers are given to the r IIA and r IIB mutants to aid in comprehension.*

6.5 SUMMARY

1. The chemical nature of genes was discovered through a series of intriguing experiments on microorganisms (bacteria and viruses) between 1920 – 1950's. Fredrick Griffith, with his experiments on *Diplococcus pneumoniae*, discovered that, the virulent strain of this bacterium could transform the avirulent strain to a virulent one. Not knowing the chemical constituent of the virulent strain that was involved in transformation, Griffith called it the 'transforming principle'. Avery, MacLeod and McCarty, from their investigations with the same bacterial strains that Griffith worked, identified the transforming principle to be the DNA. Hershey and

Chase, through their work on a bacterial virus (phage), using radioisotopes of sulphur and phosphorous, demonstrated unequivocally, that DNA is the genetic material.

2. Theodore Levene, chemically characterized the nucleic acids, both RNA and DNA. He found that DNA contained deoxyribose, a five-carbon atom pentose sugar, a phosphate group, and four nitrogen bases – adenine, guanine, thymidine and cytosine. Thymidine and cytosine have a single carbon ring, while, adenine and guanine have two carbon rings in their structure; the former two nitrogen bases are termed pyrimidines and the latter, purines. Levene also found that sugar, phosphate and nitrogen base were linked together to form a unit which he termed a 'nucleotide'. Depending on the nitrogenous base component, nucleotides are four types. Chargaff, through his biochemical analysis, discovered that the ratio of purines to pyrimidines in DNA is 1, i.e., purines and pyrimidines are present in equal proportion in DNA. Translated to an equation, Chargaff's discovery can be represented as A+G/T+C = 1. Rosalind Franklin obtained the image of DNA molecule through X ray crystallography and described the details of image. Putting together all this information on DNA, Watson and Crick proposed the molecular structure of DNA, for which, they were awarded a Noble prize in 1953. RNA has the sugar ribose and uracil instead of thymidine.

3. The DNA is a double helical structure. The sugar and phosphate form the backbone strands of DNA. The nitrogen bases on opposite strands of DNA bond: A bonds with T (two hydrogen bonds) and G links to C (three hydrogen bonds). The DNA strand is formed from linking of several nucleotides – the link being formed between the phosphate group on the fifth carbon atom of the sugar (deoxy ribose) to the OH group on the third carbon of the sugar on the adjacent nucleotide. The end of the DNA molecule with the free phosphate group is called 5′ end and the other end with the free hydroxyl group is the 3′ end. The two strands of the double helix are antiparallel. The sequence of nucleotides in the DNA varies between different genes and between different individuals within the same species as well as different species.

4. The DNA (meters long) is very tightly packed to fit into the minute microscopic cell. In prokaryotic cell, there is a single DNA molecule which is super coiled. The supercoiled nature is retained, by protein molecules, plugged into the coils. The amount of DNA in eukaryotes is much higher than in prokaryotes and many linear DNA molecules are present. The DNA packs in a multi-tier fashion to fit the nucleus in the cell. There are five levels of packing. The double helical DNA which is 2nm in diameter winds round disc shaped protein aggregates called nucleosomes to form a 11 nm fibre. Nucleosomes are made of four kinds of basic proteins called histones. The 11 nm fibre packs (as a solenoid or in a zig zag configuration) to form a 30nm fibre. Next, the 30nm fibre folds, forming a 300 nm structure. The folds, then attach themselves on loops of a scaffold protein matrix made up of acidic non-histone proteins, to form a 700 nm fibre. This DNA protein complex fibre is called chromatin. During cell division, the protein scaffold condenses, bringing the loops closer together, forming structures which are 1400 nm thick. These structures are called chromosomes.

5. The gene (DNA) expresses itself as a protein. The information on DNA, is in the form of its nucleotide sequence. Three types of RNA: messenger RNA (mRNA), transfer RNA (tRNA) and ribosomal RNA (rRNA) aid in the expression of DNA. The nucleotide sequence in the DNA is copied on to RNA using one of its strands as a template, a process called transcription. This RNA with the information of the DNA is the mRNA. Ribosome, a ribonucleoprotein complex, is the cell organelle that conducts protein synthesis. The ribosomes bind to the mRNA and travel

along its length as the peptide is synthesized. The nucleotide sequence in the mRNA is read as code – the genetic code. A sequence of three nucleotides forms a code with the mRNA having several, non-overlapping comma less codes. With four types of nucleotides, 64 triplet codons are possible. Each codon is specific for a particular amino acid. There are twenty different types of amino acids in proteins. The amino acid specific for each codon on the mRNA is carried to it by the tRNA. The tRNA plugs to the codon through a complementary sequence to the codon called the anticodon. The rRNA in the ribosome provides enzymatic activity (peptidyl transferase function) for linking adjacent amino acids with a peptide bond formation. This process of decoding the genetic code to the sequence of amino acids by tRNA and ribosomes is called translation. Translation stops on reaching a stop codon on the mRNA. Three codons of the 64 codons do not code for any amino acid and act as stop codons.

6. DNA replicates itself very faithfully, through semi conservative replication. The two strands of DNA separate at different points along its length. At these points, each strand synthesizes a new partner strand, which has base pairs complementary to it. The two strands replicate in opposite direction with the help of DNA polymerase and several other enzymes. The two daughter DNA molecules that are formed have one old strand of the original and one new strand. This is called semi-conservative type of replication.

7. Seymour Benzer, through his elegant experiments with bacteriophages, demonstrated that the gene is not an unsplittable particle, but is a stretch of DNA, with individual functional components. He worked with phage mutants called R II mutants and showed that, the gene can be seen at three functional levels: cistron, recon and muton. The entire length of DNA, which has the information for a particular function is termed the cistron. The region within the cistron (gene) which can undergo mutation was named muton; that which can participate in recombination, the recon. Through his fine structure analysis of the gene, Benzer showed that the size of muton and recon is one nucleotide. Thus, a cistron contains as many mutons and recons as its number of nucleotides.

6.6 SAMPLE QUESTIONS

6.6(a) Subjective Questions

Q.1. Describe the series of experiments that led to the unequivocal conclusion that DNA is the genetic material.

Q.2. Describe the structure of DNA.

Q.3. How is DNA expressed? Explain the several steps in DNA expression showing the role of various RNAs in the process.

Q.4. How is DNA packed in eukaryotes?

Q.5. How were the observation of complementation between genes and recombination within a gene in R II mutant phages help in deciphering the fine structure of the DNA?

6.6(b) Objective Questions

Q.1. The width of the DNA double helix is maintained uniformly because:

(*a*) There are hydrogen bonds between the nitrogen bases.

(*b*) The nucleotides are uniformly placed.

(c) The nitrogen bases are equally distributed.

(d) Always a single carbon ring pyrimidine pairs with a double carbon ring purine.

Q.2. Griffith observed that rats died when infected by

(a) Smooth strain of *Streptococcus.*

(b) Heat treated smooth strain of *Streptococcus.*

(c) Rough strain of *Streptococcus.*

(d) A mixture of heat-treated rough strain and live rough strain of *Streptococcus.*

Q.3. In the experiment of Hershey and Chase, the isotopes used were of the following elements

(a) C & S   (b) S & H

(c) C & P   (d) S & P

Q.4. Genetic code is

(a) A triplet nucleotide sequence on the mRNA that specifies for a particular amino acid.

(b) A sequence of nucleotides that codes for a group of related amino acids.

(c) An iterative code on the gene.

(d) Present in the ribosome which is the protein synthesizing cell organelle.

Q.5. Translation is

(a) Synthesis of RNA on DNA.   (b) Replication of DNA to form a duplicate.

(c) Transfer of information from nucleus to cytoplasm.

(d) Deciphering the genetic code to facilitate assembling of relevant amino acids to enable the synthesis of a polypeptide chain.

Q.6. R II mutants of T_4 phage investigated by Seymour Benzer resulted from

(a) A mutation in one gene.   (b) A mutation in two genes.

(c) Several mutations in one gene.   (d) Several mutations in two genes.

Q.7. DNA replication of the two strands happens

(a) Simultaneously on the two strands in the same direction.

(b) Simultaneously on the two strands in opposite directions.

(c) Sequentially one strand after another.

(d) On only one strand.

ANSWERS

1. (d)   **2.** (a)   **3.** (d)   **4.** (a)   **5.** (d)   **6.** (d)   **7.** (b)

7 Protein Structure and Function

As can be seen even by this limited number of examples proteins carry out amazingly diverse functions. In many biological structures' proteins are simply components of larger molecular machines.

~ **Michael Behe**

7.1 INTRODUCTION

Proteins essentially drive almost every life process. These biomolecules constitute ~50% of the dry mass of a cell. They are a very diverse class of macro biomolecules. A cell can contain thousands of different kinds of proteins, each with a different function. Despite their extreme diversity, proteins are all made of chains of amino acids – the polypeptides. The difference lies in the composition and sequence of the 20 types of amino acids that constitute proteins. While the functional property of DNA, another important polymer biomolecule, is determined by the sequence of the nucleotides along a particular length, it is not the same with proteins. The overall structure of DNA remains uniform and is not much altered by the sequence of the nucleotides. Different proteins with different sequence combination of amino acids however, have very different structures.

The structure that a protein conforms to, is to a great extent, determined by the sequence of the amino acids in the polypeptide chain. The function of a protein is dependent on its structure – a particular structure facilitates a specific function. Proteins have an extensive diversity of structures resulting in a wide range of functions. Sometimes, a functional protein contains, more than one polypeptide chain, with the constituent polypeptide chains being similar, or dissimilar. To conclude about the function of a protein, it is thus crucial to know its structure.

To determine the exact structure of a protein, Xray crystallography technique is employed. For this, the protein should be available in pure form in sufficient quantities and needs to be crystallized. This technique is thus, a difficult and expensive process. Given the huge number of different proteins, determining the structure of every protein, through this method is not possible. Based on the information of the structure of the proteins determined from Xray crystallography and their amino acid sequence, computer algorithms have been developed to predict the structure of a polypeptide, from its amino acid sequence. Though not completely accurate, the bioinformatic tools developed with these algorithms, are useful in predicting the structure of a protein and therefore its function, when the amino acid sequence of the polypeptide is known.

The sequence of the amino acids of a polypeptide can be determined by mass spectrometry methods. Again, this procedure is cumbersome and therefore not possible to do for all known proteins. If the gene coding for the protein is known, the amino acid sequence of the polypeptide can be determined from the nucleotide sequence of the gene. This is because, the nucleotide sequence of DNA is read as a non-overlapping comma less triplet genetic code with each code being specific for a particular amino acid, resulting in a direct correlation between, nucleic acid sequence, to the sequence of amino acids. Determining the nucleotide sequence of DNA that constitutes a gene, is relatively much simpler. With the nucleotide sequence on hand, it is just the press of a button on a computer with a bioinformatic tool 'Translate' (available at NCBI), to get the amino acid sequence. The relation of the sequence of amino acids in a polypeptide, to its function can be represented as:

$$\text{Sequence} \longrightarrow \text{Structure} \longrightarrow \text{Function}$$

A functional protein consists of, one or more polypeptides, precisely twisted, folded, and coiled, into a unique shape, with a three-dimensional form. Polypeptides are unbranched polymers of amino acids, with a length ranging from, a few, to more than a thousand, monomers (amino acids). The folding, and the consequent 3D structure they achieve, is due to the formation of chemical bonds (hydrogen and disulphide bonds), between the atoms of different amino acid molecules, such that, the shape is held in a particular form.

The polypeptide chain starts folding in its nascent form, simultaneously as it is being synthesized in the ribosome, even before its synthesis is complete. Random folding of the long polypeptide chain, is prevented and a precise pattern of folding, resulting in the right conformation, which brings it to a functional form (native form), is made possible, with the help of a group of proteins termed chaperones. These chaperone proteins, associate with the nascent polypeptide, as it is being synthesized, and, prevent it, from improper folding. The chaperone proteins however, do not have the information about the functional native form of a protein. They only prevent random folding of the long polypeptide, and thus, allow it to fold into its

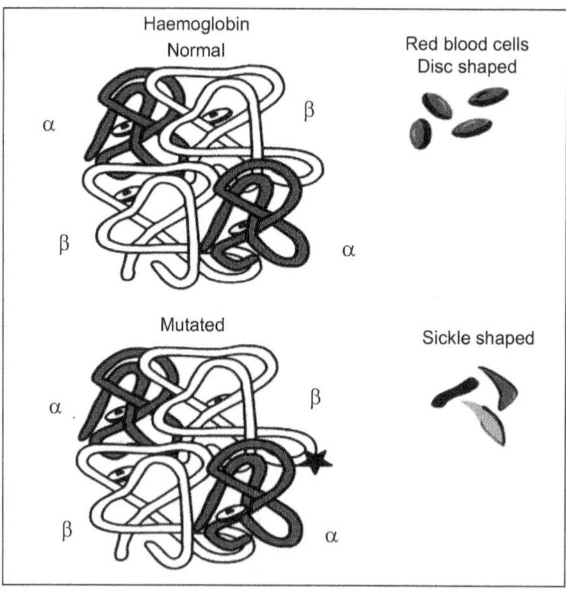

Figure 7.1: Normal and mutated haemoglobin forming normal biconcave disc and sickle shaped red blood cells respectively. *A change in a single amino acid in one of the polypeptide chains of haemoglobin (marked as star) causes the protein to become sticky. The sticky haemoglobin molecules aggregate as clumps, resulting in sickle shaped erythrocytes. The sickle celled red blood cells cannot flow freely in blood vessels, causing a disease called sickle cell anaemia. There are four iron (hae me) groups ◉ in a haemoglobin molecule.*

native structural form, which is determined by the sequence of the amino acids. The folding of the polypeptide is determined by, the form the polypeptide takes, based on the restricted bending angles or conformations, that are possible with the type and sequence of amino acids in it. The allowable angles of protein folding are described in a two-dimensional plot, popularly known as the 'Ramachandran plot'. The folding of the protein is a quick process, taking milli to microseconds. This is possible due to some enzymes – the folding enzymes.

The biological activity of a protein depends on its three-dimensional structure. This native structure, under normal physiological conditions, is quite stable. A particular protein, most often has the same sequence of amino acids and the same structure, in all living organisms. Even a very minor change of one amino acid in a protein, in some instances, can lead to a very different structure. For example, in haemoglobin, the replacement of one amino acid – valine with glutamate, results in a change of its property, resulting in aggregation of haemoglobin molecules (**Figure 7.1**).

The erythrocyte (red blood cell) that contains mutant haemoglobin molecules, become sickle or crescent shaped, instead of the normal disc shape (**Figure 7.1**). The sickle shaped erythrocytes, cannot freely flow in the blood vessels like the normal biconcave or disc shaped erythrocytes, thereby, clogging the blood vessels. This leads to a disease in humans, called the sickle cell anaemia.

7.2 ANALYSIS OF PROTEIN STRUCTURE IN A REDUCTIONIST APPROACH

The reductionist method of breaking down biological systems into their constituent parts has been helpful in explaining the chemical basis of several living processes. The famous molecular biologist, Francis Crick (1966) proclaimed that, "the ultimate aim of the modern movement in biology is to explain all biology in terms of physics and chemistry". The structure of the protein can be analysed with a reductional approach – to break the system down to its pieces, to reason about it, from properties of these pieces. It is assumed that, the isolated molecules, and, their structure, have sufficient explanatory power, to provide an understanding of the whole system. Now, many biologists have realized that, this approach has reached its limit.

The problem with reductionism is that, it is based on the concept that, the properties of a system, are the sum of the reduced properties of its components. Today, it is clear that, the specificity of a complex biological activity, does not arise from the properties of the individual molecules that are involved, as these components, frequently function in many different processes. Therefore, a new concept called 'emergence' has developed, that complements 'reductionism'. Emergent properties cannot be predicted or deduced, by precise calculation or other means. Emergent properties differ from resultant properties predicted from information of sub components – the way saltiness of sodium chloride is not reducible to the properties of sodium and chlorine gas. Similarly, proteins are extremely complex, and, the reductionist approach underestimates this complexity. Proteins have emergent properties that cannot be explained, or even predicted, by studying their individual parts. However, for sake of basic understanding, protein structure is explained based on a methodological reductionist approach.

7.2.1 Four Levels of Protein Structure

The complexity of protein structure, is best analysed, by considering the molecule at four organizational levels – primary, secondary, tertiary, and quaternary (**Figure 7.2**). An examination of these hierarchies of increasing complexity, has revealed that, certain structural elements are repeated in a wide variety of proteins, suggesting that, there are general "rules", regarding the ways in which, proteins achieve their native, functional form. These repeated structural elements, range from, simple combinations of α helices and β sheets forming small motifs, to the complex folding of polypeptide domains of multifunctional proteins.

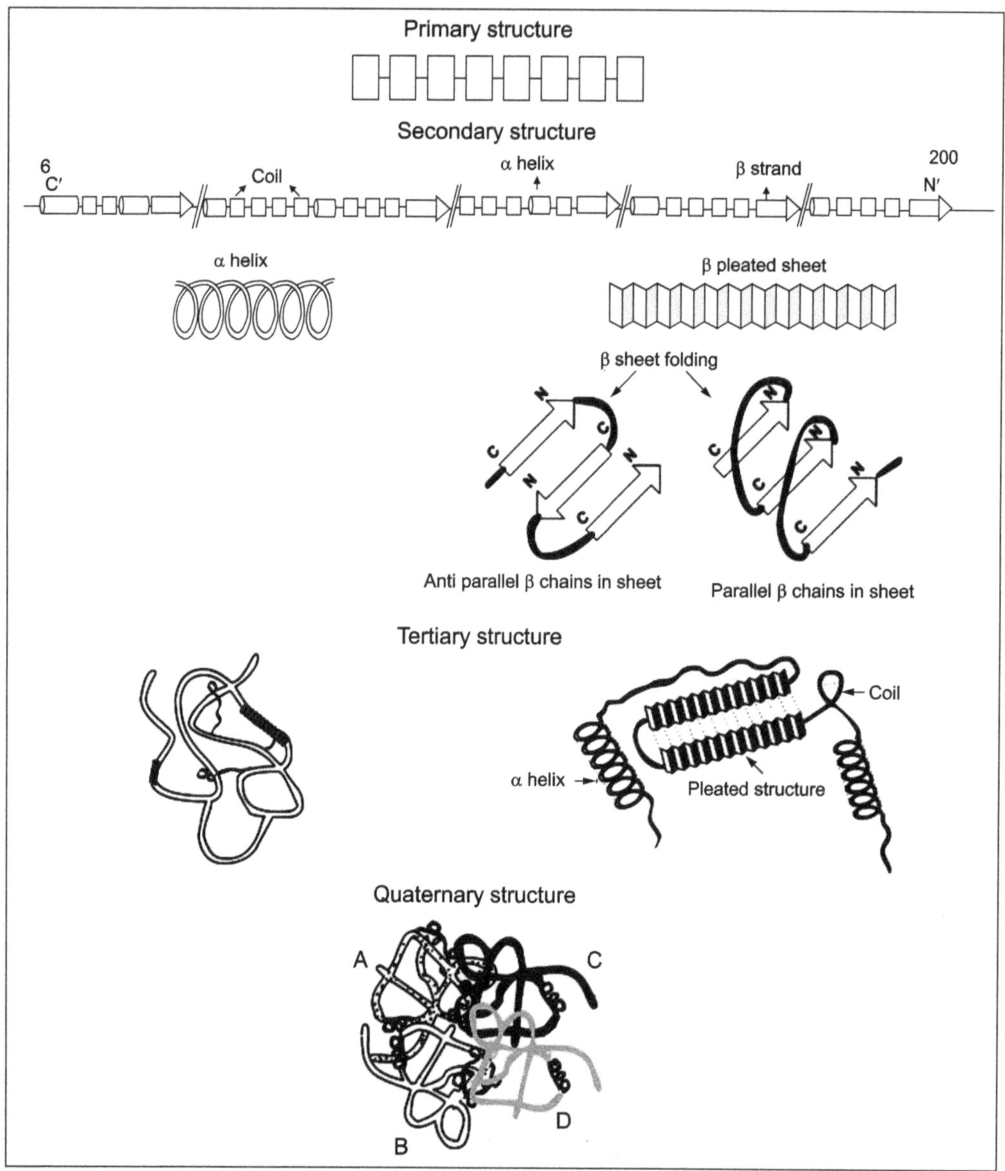

Figure 7.2: Various levels of protein structure. *Primary – constitutes the sequence of amino acids.* **Secondary** *– depending on the amino acid sequence, certain lengths of the polypeptide chain take a right-handed helical shape (α helix) or β pleats; β pleats in neighbouring regions can fold in parallel or antiparallel way into sheets; regions of the polypeptide which neither form α helix or β pleats, remain as coils.* **Tertiary** *– a specific 3D structure attained by folding of the different kinds of secondary structures in a specific manner.* **Quaternary** *– in complex proteins with multiple polypeptides, the 3D forms of different polypeptides assemble in a specific manner to attain a quaternary structure . The tertiary and quaternary structures are the functional forms of a simple and complex protein respectively.*

7.2.1(a) Primary Structure

The linear sequence of amino acids in the polypeptide chain, constitutes its primary structure. In the polypeptide chain, the amino acid in the extreme left is the first and that on the extreme right is the last. A peptide bond (called amide bond), is formed by linking of the carboxyl group of one amino acid, with the amino group of the next amino acid. Since peptide bond formation is accompanied by loss of a water molecule, this process is called condensation. The amino group in the first amino acid and the carboxyl group in the last amino acid in a polypeptide chain are free. Therefore, the left end is called the N′ end and the right, the C′ end.

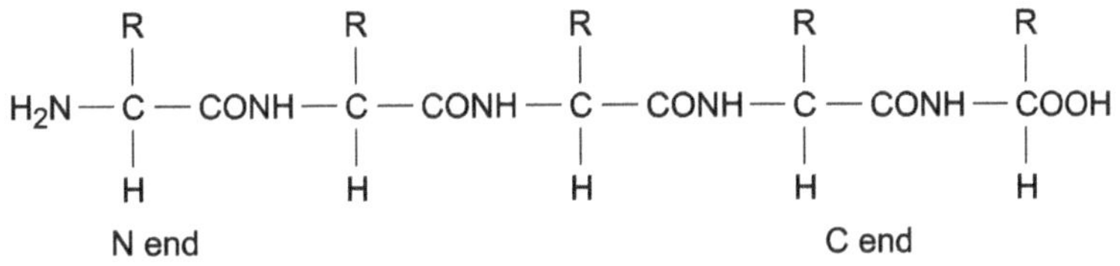

$$H_2N-\overset{\displaystyle R}{\underset{\displaystyle H}{C}}-CONH-\overset{\displaystyle R}{\underset{\displaystyle H}{C}}-CONH-\overset{\displaystyle R}{\underset{\displaystyle H}{C}}-CONH-\overset{\displaystyle R}{\underset{\displaystyle H}{C}}-CONH-\overset{\displaystyle R}{\underset{\displaystyle H}{COOH}}$$

N end   C end

Polypeptide chain

7.2.1(b) Secondary Structure

The polypeptide chain does not stay straight, but folds. The folds are of two types: the α helix and β sheet. The type of fold depends on the sequence of amino acids in the stretch of a polypeptide. The α helix is like a right-handed helical staircase (like a coiled spring), being formed by a series of hydrogen bonds, between an amino acid with the fourth amino acid in the chain. The spiral has about 3.6 amino acids per turn, and the amino acid side chains stick out from the helix. A β sheet is formed by intramolecular hydrogen bonding, holding two or more polypeptide chains. The β sheet structure can be formed between, polypeptide chains that run parallel (all N-terminal ends on one side), or, antiparallel (neighbouring N-terminal ends on opposite sides) (**Figure 7.3**). The β sheet structure is formed, when in the polypeptide chain, there are a stretch of bulky (with long side chain (R-group)) amino acids.

The peptide bond between two amino acids is rigid, with no flexibility. The alpha carbon (Cα) (carbon atom adjacent to the carboxyl carbon), of each amino acid, is held in the polypeptide main chain, by two rotatable bonds (**Figure 7.3**). The dihedral (torsion) angles of these bonds are called phi and psi (in Greek letters, ϕ and ψ). These two torsion angles in the polypeptide chain are also called Ramachandran angles (after the Indian physicist who worked on modelling the interactions in polypeptide chains). The torsion angles can be between $-180°$ to $+180°$: the negative value angles representing anticlockwise rotation, and, the positive value angles indicate, clockwise rotation. It is the flexibility of the alpha carbon atom to rotate that enables a polypeptide to fold.

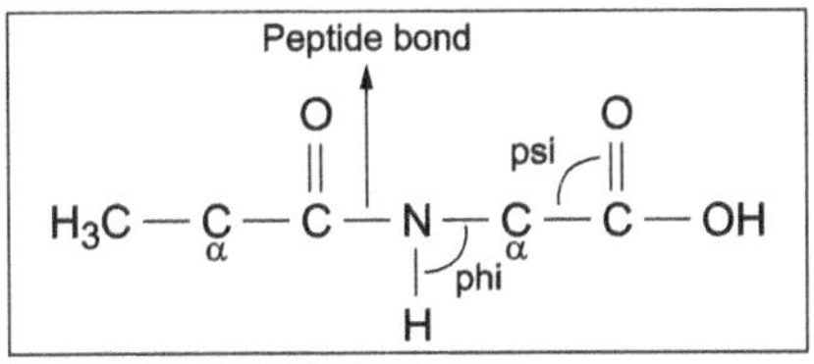

Figure 7.3: Linkage between two amino acids by a peptide bond showing the two (phi and psi) torsion angles of the alpha carbon atom (carbon adjacent to COOH) in the amino acid. *The flexibility (depending on the value of the torsion angles which range from $-180°$ to $+180°$) of alpha carbon atom enables the polypeptide to fold in different ways.*

Ramachandran et al. plotted the torsion angles, taking the phi angles on the X axis and the psi angles on Y axis, of several polypeptides and came up with the range of values of torsion angles, that result in formation of an alpha helix, or, beta sheet. This plot, is termed the Ramachandran plot. From the Ramachandran plot, it is derived that, the average phi and psi values for α helices are between – 57 and – 47 and β sheets are approximately –80 and +150.

Haemoglobin has only helices, natural silk is constituted by only beta pleated sheets, and many enzymes have a pattern of alternating helices and beta strands. The secondary structure elements are connected by "loop" or "coil".

7.2.1(c) Tertiary Structure

The tertiary structure of the protein is attained by the folding of the polypeptide on itself, bringing distant amino acids closer together at specific points, giving a third dimension (3D) to the protein structure. Large proteins use molecular chaperones for folding. The folding of the peptide is kept in place, through several types of binding, like – hydrogen bonding, disulfide bridges and Van der Waals force between the atoms of the different amino acids (**Figure 7.3**).

Globular proteins have a very compact 3-dimensional structure. Fibrous proteins have long, extended structures. The functional properties of a protein are attained with a proper tertiary structure, the function being lost, when the structure is dismantled.

7.2.1(d) Quaternary Structure

The highest level of protein organization is the quaternary structure, which applies to oligomeric proteins (proteins with more than one polypeptide chain). Quaternary structure, refers to the non-covalent interactions of protein subunits and their spatial arrangement. The arrangement of the different polypeptides, with their acquired tertiary structure, gives the quaternary structure of a protein (**Figure 7.3**). The subunits are packed and held together by hydrogen bonds, salt bridges, and hydrophobic interactions – the same forces that operate within tertiary structures.

Complex proteins (oligomeric proteins), can have, from two identical subunits, to ten different subunits, to very large protein complexes with 102 subunits of three different types. There are two classes of oligomeric proteins, namely, homo-oligomers and hetero-oligomers. The former are composed of identical subunits, and the latter, with different subunits. Haemoglobin has multiple subunits. It is a hetero tetramer consisting of two α and two β subunits. Each subunit has one iron (haem) group, which can bind to an oxygen molecule (**Figure 7.1**).

Most proteins have a limited lifespan, after which, they are degraded. The time from synthesis to degradation, is termed its turn-over time. It involves: synthesis, folding, modification, function and degradation. A newly synthesized polypeptide chain undergoes folding and often, chemical modification, to generate the final protein. All molecules of a protein species attain the same shape.

7.3 FUNCTIONS OF PROTEINS

Proteins are workhorses of the cell, playing a very important role both in its structure and function. They carry out a myriad functions (**Table 7.2**)

Table 7.1: Proteins with different functions

Kind	Function	Examples	About
Structural	Constituents of cytoskeleton.	Actin	Constituent of microfilaments.
		Tubulin	Constituent of microtubules.
		Keratin	Constituent of intermediate filaments.
	Components of different structures.	Collagen	Present in muscle, skin, bone and tendon.
		Keratin	Present in hair and nails.
		Elastin	Present in skin and bladder.
Integral	Constituents of membranes of the cell.		
A. Transport (Carrier)	Movement of ions and other substances.	Ion channels	Transport Na+, K^+, Ca^+ ions in neurotransmitters.
		Carrier proteins	Transport glucose etc. in intestinal cells.
B. Receptor	Receiving and transducing signals to a cell.	Ion gated channels	Present in cardiac muscle cells.
		G protein coupled	Present in nasal and ear cells.
		Enzyme linked	Present in cell membranes.
Non-membrane-bound transport	Carry hormones, oxygen etc. in the blood.	Serum albumin	Carries steroids, fatty acids and thyroid hormones in the blood.
		Haemoglobin	Carries oxygen.
Enzyme	Catalyse biochemical reactions.	Trypsin	Present in the pancreatic juice; digests protein.
		RUDP carboxylase	Present in chloroplast; facilitates fixing atmospheric CO_2.
Hormone	Regulate metabolism.	Insulin	Synthesized in the pancreas; maintains blood glucose levels.
		Somatotropin	Synthesized in pituitary gland; stimulates growth.
Immunoglobulin	Defence proteins	Antibodies	Synthesized by white blood cells called B lymphocytes.
Motor	Muscle cell proteins	Myosin α actin	Help in contraction and relaxation.
Storage	Growth of embryo or infant.	Ovalbumin	Present in egg.
		Casein	Present in milk.
		Gliadin	Present in wheat grain.

7.3.1 Structural Proteins

Structural proteins are responsible for cell shape; they are part of the cartilage and bone in vertebrate animals. They constitute the hair, feathers, quills, nails and hoofs, and are important components of the skin cells. There are many kinds of structural proteins the most important are described below.

7.3.1(a) Cytoskeletal Proteins

The cell has a definite shape because of a cytoskeleton. The cytoskeleton is composed of filaments of different thickness and protein constitution: microfilaments, intermediate filaments and the microtubules of up to 7, 15 and 25 nm width respectively. They form a network and scaffold which gives support to the cell and holds the cell organelles in position. It is a flexible structure and dynamic, undergoing frequent changes, in response to the environment and cell activity. The cytoskeleton is visible under light microscope, only when, fluorescent dye tagged antibodies of the proteins of the cytoskeleton are used. For its conceptualisation, a cartoon diagram depicting the organization of cytoskeleton is shown in **Figure 7.4**.

7.3.1(a)1 Actin

Actin is one of the most abundant protein in cells, comprising 1 to 5% and up to 20% in muscle cells, of the total cellular protein by weight. In vertebrate animals, there are three forms of actin: α, β and γ isoforms. The α-actins are found in muscle tissues, while β and γ, form the microfilaments of the cytoskeleton. There are several variants of each type of actin, with slight differences in flexibility. Actin makes up microfilaments. The actin molecules polymerise to form filaments with a double helical structure (**Figure 7.5**). A network of actin filaments (microfilaments), lie below the cell membrane, which are linked to it by connector protein, and give a specific shape to the cell (**Figure 7.4**).

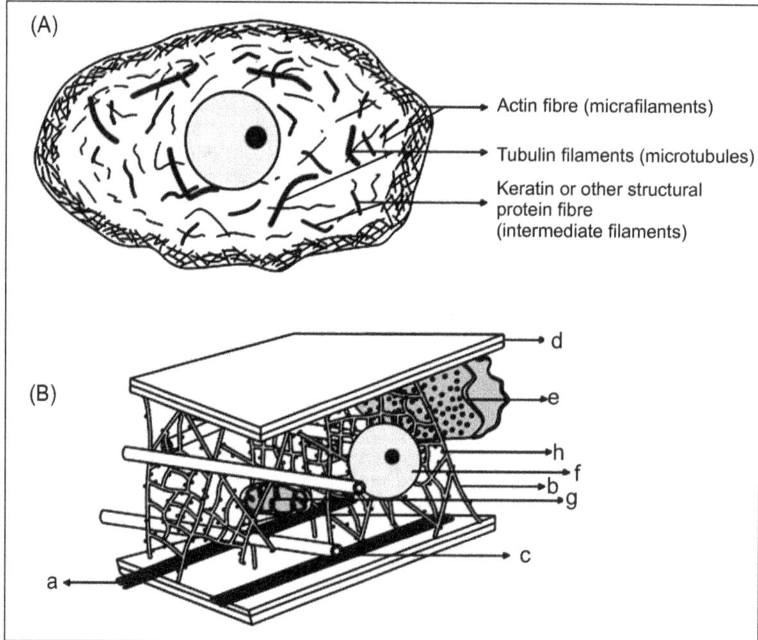

Figure 7.4: Cytoskeleton formed from structural proteins. Cartoon diagram showing organization of the cytoskeleton. *A. Distribution of the different components of the cytoskeleton – microfilaments, intermediate filaments and microtubules within the cell. **B.** 3D view of the cell showing the support and scaffold afforded by the cytoskeleton. a – microfilaments (actin fibres), b – microtubules (tubulin filaments), c – intermediate filaments (keratin or other structural protein fibres), d – cell membrane, e – endoplasmic reticulum, f – nucleus, g – mitochondrion, h – ribosome.*

7.3.1(a)2 Tubulin

The microtubules of the cytoskeleton, that form a scaffold for support of the cell organelles in the cytoplasm, are formed by tubulin protein. There are two types of tubulin proteins: α and β, in the microtubules. Each of this type, has again many variant forms, in different cells, just like the actin proteins, of the microfilaments. The α and β tubulin molecules dimerise; the dimers coil in a spiral pattern, with thirteen dimers per spiral, forming a hollow straw like tube (**Figure 7.5**). Microtubules play the role of

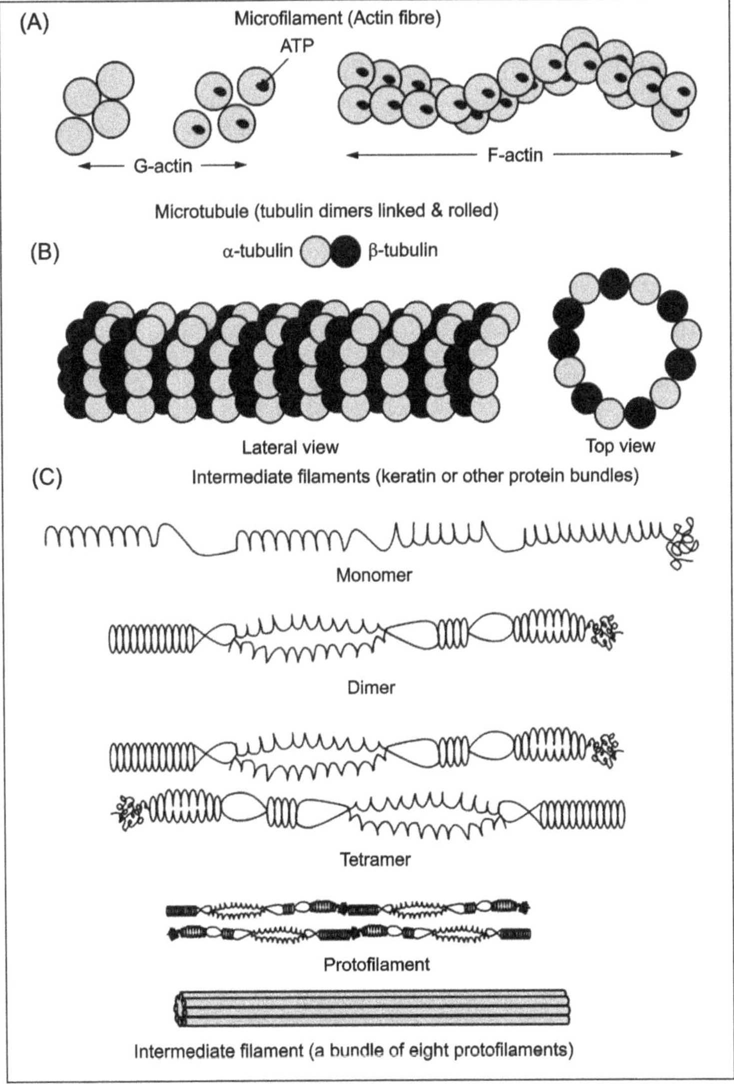

Figure 7.5: Structural proteins and their arrangement to constitute different components of the cytoskeleton. A. *Actin* *forms microfilaments – globular actin molecules (G actin) associate with ATP and polymerise to form double helical actin fibres (F actin) which are the microfilaments.* B. *Tubulin* *is a constituent of microtubules – a molecule of* α *tubulin and one of* β *tubulin dimerise, the dimers role around with 13 tubulin molecules per turn forming the microtubules.* C. *Keratin* *constitutes intermediate filaments – Keratin molecules are in the form of* α *helices with a coiled head. Two helices together form dimers; two dimers join in opposite orientation to form a tetramer; tetramers join together to form the protofilament; eight bundles of protofilaments form the intermediate filament.*

a scaffold, holding cell organelles in position (**Figure 7.4**). They help the cell to resist compressive forces and retain its shape. During cell division, the cytoskeleton is dismantled, the microtubules reassemble, to form the spindle apparatus, which helps the movement of chromosomes to opposite poles.

7.3.1(a)3 Keratin and Other Structural Proteins

The intermediate filaments are made of a variety of proteins different in different kinds of cells. One of the important proteins is keratin, a fibrillar protein which is also found in the epidermis and epidermal appendages like hair, wool, nails, quills, scales etc. Keratin helps to attach the cells to each other. A keratin molecule has an alpha helical structure, with the amino and carboxy terminals, sticking out as coils. These molecules dimerise; two dimers join in opposite orientation to form a tetramer, two of which together form a protofilament. A bundle of eight protofilaments form an intermediate filament (**Figure 7.5**). The intermediate filaments have high tensile strength and specialised to bear tension thereby maintaining the shape of the cell, and, anchoring the nucleus and other cell organelles, in the cytoplasm (**Figure 7.4**). Beta keratins are present in silk and spider webs. They are flexible, but do not stretch. The structure is an antiparallel β sheet.

7.3.1(b) Other Structural Proteins: Collagen and Elastin

Collagen and elastin are two other important structural proteins. Collagen is the most abundant mammalian protein. It is present in bones, cartilage, nails, hair and skin. A collagen molecule consists of three subunits – one α_1 helix and two α_2 helices. These molecules, cross link with each other to form collagen microfibrils that spiral around an elongated straight axis, to form collagen fibrils. Several such fibrils bundle up to form a collagen fibre. Several collagen fibres bundle up to form a tendon fibre (**Figure 7.6**). Collagen undergoes aging process with crosslinks being formed between its fibres.

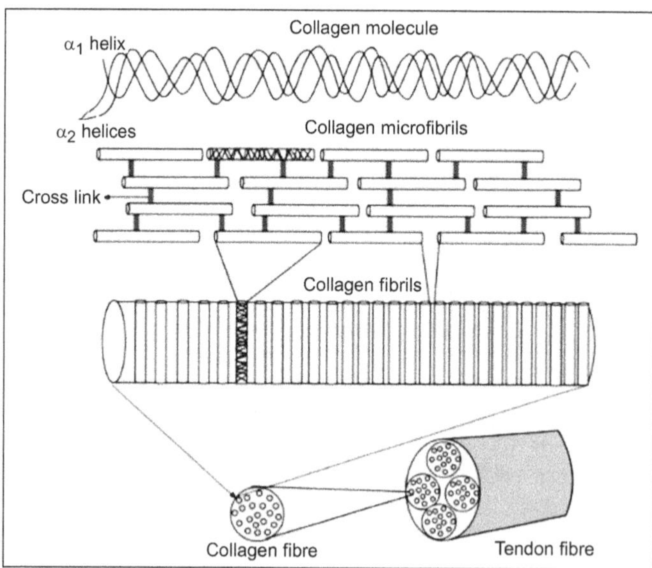

Figure 7.6: Collagen protein forming fibres. *Collagen molecules have three α helices – one of α1 and two of α2 type.* **Collagen microfibrils** *are formed due to crosslinks between collagen molecules.* **Collagen fibrils** *are formed by spiral packing of the microfibrils.* **Collagen fibre** *is formed when several collagen fibrils bundle together.* **Tendon fibre** *consists of several collagen fibres organized into a bundle.*

Elastin is present in ligaments, tendons and skin. It is present in walls of arteries, bladder, lungs, skin, cartilage and uterus, which need elasticity. Elastin can stretch and therefore give elasticity. Elastin molecule takes a lot of irregular conformations, and sometimes, is in the form of a β-spiral. The elastin molecules cross link and can undergo stretching (**Figure 7.7**).

Fibroin is a structural protein present in silk. Myosin is a contractile structural protein present in muscle cells in association with α actin fibres.

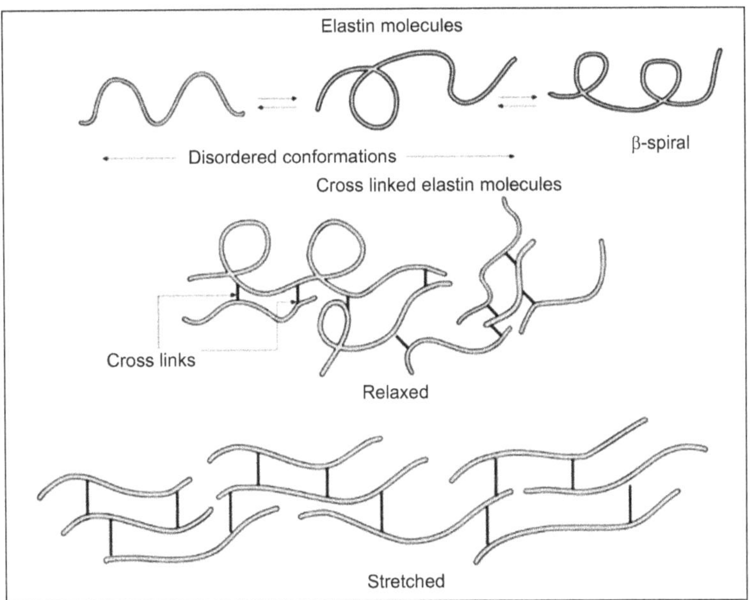

Figure 7.7: Elastin. *Elastin molecules have irregular conformation and sometimes take the form a β-spiral. Elastin molecules are joined together by cross links; elastin can stretch and relax due to straightening and folding of the elastin molecules.*

7.3.2 Integral Proteins

Integral proteins are present in the cell membrane and other membrane systems in the cell. The membranes are constituted by a bilipid layer in which proteins are integrated, either on the surface (peripheral), or passing through the entire bilipid layer (transmembrane). The integral proteins can help in transport of substances in and out of the cell, or act as receptors, to receive signals. The transport and receptor proteins are described below.

7.3.2(a) Transport Proteins

A cell requires exchange of large number of substances, from its inside to outside and vice versa. The exchange has to take place through the cell membranc. Even the cell organelles (e.g. mitochondria, chloroplasts, endoplasmic reticulum, Golgi complex), need to transport substances inside, or out, into the cytoplasm, through their membranes. The transport of even water molecules, across the lipid protein membranes of the cell, and its organelles, through simple diffusion, does not happen, at rates, that would suffice their needs. Glucose, which is the chief source of energy for cell metabolism, cannot enter a cell through its membrane, because of the polar nature of its molecules. (Polarity results, from the uneven partial charge distribution between various atoms in a compound). There are special proteins, that facilitate movement of substances across the membranes. They are called transport proteins.

Most transport proteins span the membrane and hence described as transmembrane. Each transport protein, is designed to transport a specific substance, or a group of very closely related substances, as needed. By helping movement of substances across membranes, transport proteins aid in causing nerve impulses and making cellular metabolism possible.

Transport proteins are basically of two types: channel proteins and carrier proteins. Channel proteins create a tunnel in the membrane, opening as a pore, on both sides of the membrane. Substances enter the pore and pass through the tunnel and thus transported. Some examples of channel proteins, are those, that facilitate movement of chloride, sodium, calcium, and potassium ions.

Ions such as sodium and potassium, sugars such as glucose, proteins and messenger molecules etc. are transported by transport proteins. Transport of substances is induced as per the need of the cell. The pore of a particular channel protein is open only when, the substance it transports needs to move in or out of the cell. Similarly, carrier proteins undergo conformational changes in shape to permit transport, only when the substance that it transports is required to be transported.

The transport proteins mediate transportation either in a passive or active mode. Passive mode involves movement of substances across a concentration gradient i.e., from a region of higher concentration, to a region of lower concentration. Such a movement is technically described as diffusion and since it is accelerated by the transport protein, this mechanism of transport of substances across the cell membranes, is termed 'facilitated diffusion'. In active mode of transport, substances move against a concentration gradient, with the help of transport proteins, through spending energy, in the form of ATP.

An example of active transport in humans, is the uptake during digestion of food in the ileum (small intestine). As the food is absorbed by the villi of the ileum, after some time, the concentration of food molecules inside villi, increases to a point, where transport cannot happen anymore, through diffusion.

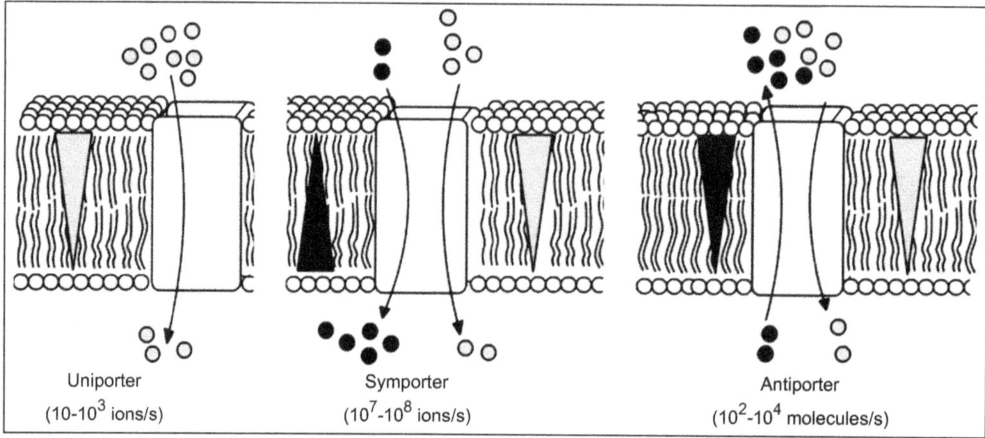

Figure 7.8: Transport mediated by different types of transmembrane transport proteins. *Uniporter. Transport in one direction only – down a concentration gradient; one type of ion/molecule moves from a region of higher to lower concentration.* **Symporter.** *Two types of ions/molecules move simultaneously (co-transport) in same direction.* **Antiporter.** *Two types of ions/molecules move simultaneously (co-transport) in opposite direction. During co-transport one of the ion/molecule, moves down a concentration gradient (higher to lower concentration) while the other moves in a direction opposite to the concentration gradient (lower to higher). The triangles in protein lipid cell membrane show the concentration gradient of the corresponding type of ions/ molecules (represented in the same colour as the ions: ash/black) on either side of the membrane: The triangle base, points to the region with the higher concentration of that particular ion/molecule. The bilipid membrane has the lipid molecules arranged with their hydrophilic heads facing outside and hydrophobic tails pointing inside.*

A transport protein can facilitate movement of substances or ions one-way or both ways. A transport protein, that helps in transport of one substance only, one way i.e., either from outside to inside or vice versa, of a cell or organelle is described as uniporter. Some transport proteins, mediate simultaneous transport of two substances, termed as coupled transport. When the two substances move in the same direction, the transport protein facilitating such movement is described as symporter, if they move in opposite direction it is described as antiporter (**Figure 7.8**).

7.3.2(a)1 Channel Proteins

Channel proteins act like pores in the membrane that let water molecules or small ions pass through, quickly. Therefore, they are also referred to as ion channel proteins. Channels are typically designed, so that, only one specific ion (or very closely related ions that are similar in size and charge), can pass

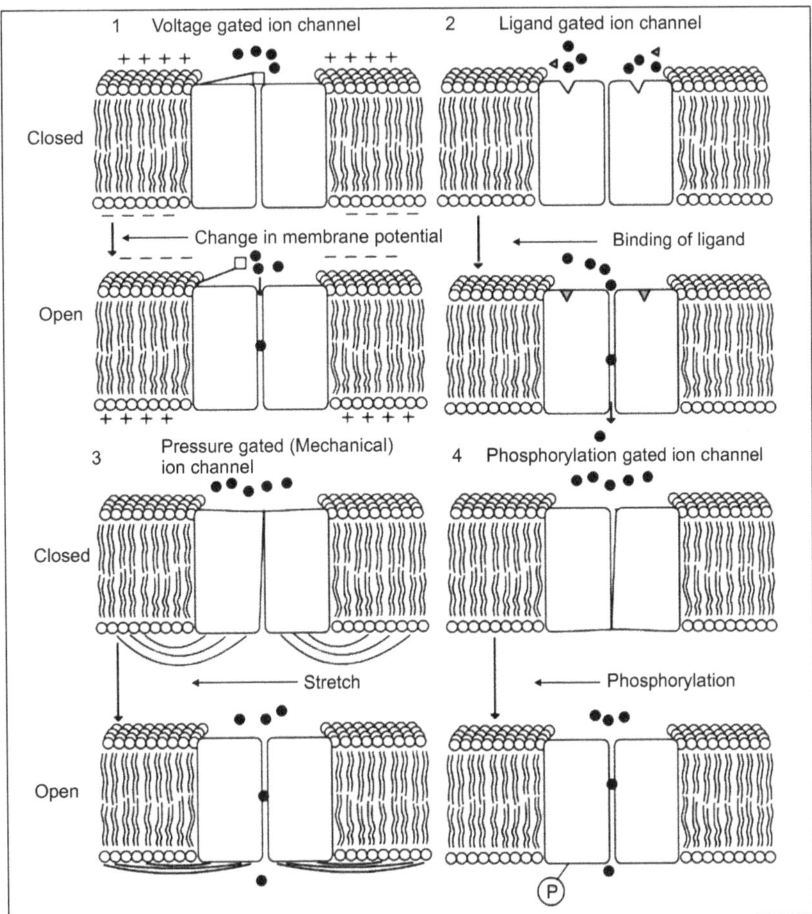

Figure 7.9: Ion channel transmembrane transport proteins – different types of gates. 1. *Voltage gated.* *Open and close due to changes in membrane potential leading to redistribution of charges on the membrane.* **2. *Ligand gated.*** *Open and close depending on the attachment/ detachment of a ligand.* **3. *Pressure gated.*** *Open and close due to stretching and contracting of cytoskeletal elements attached to the membrane.* **4. *Phosphorylation gated ion channel.*** *Opening and closing is due to phosphorylation/dephosphorylation mediated by ATP. The bilipid membrane has the lipid molecules arranged with their hydrophilic heads facing outside and hydrophobic tails pointing inside.*

through. A channel protein, serves as a tunnel across the membrane into the cell or organelle. The rate of transport of ions through these channel proteins is very high (10^6 ions or more per second), being transported downhill (i.e., from higher to lower concentration), by an electrochemical gradient (a function of ion concentration and membrane potential). Therefore, this movement is described as, facilitated diffusion. Examples of channel proteins include: chloride, sodium, calcium, and potassium ion channels.

A channel protein has its pore open for transport, only when the ion to which it is specific for, is required to be transported. Depending on the mechanism by which the pore is gated (kept open or closed), channel proteins are described as voltage gated (open/close in response to membrane potential), ligand gated (open in response to attachment of a specific ligand molecule to the receptor region of channel protein), mechanical gated (the pore being manipulated by the cytoskeletal elements attached to the pore), or due to activation e.g. phosphorylation by ATP (**Figure 7.9**).

Ion channels are assemblies of several identical or similar transmembrane proteins (sub units), closely arranged in a circular manner, around a water filled pore, through the plane of the lipid bilayer of the membrane. The Na^+, Ca^+ and K^+ ion channels, have four subunits (**Figure 7.10**). Some, have six subunits. Basically, an ion channel has three regions: the pore, channel and sensor (selectivity filter) (**Figure 7.10**).

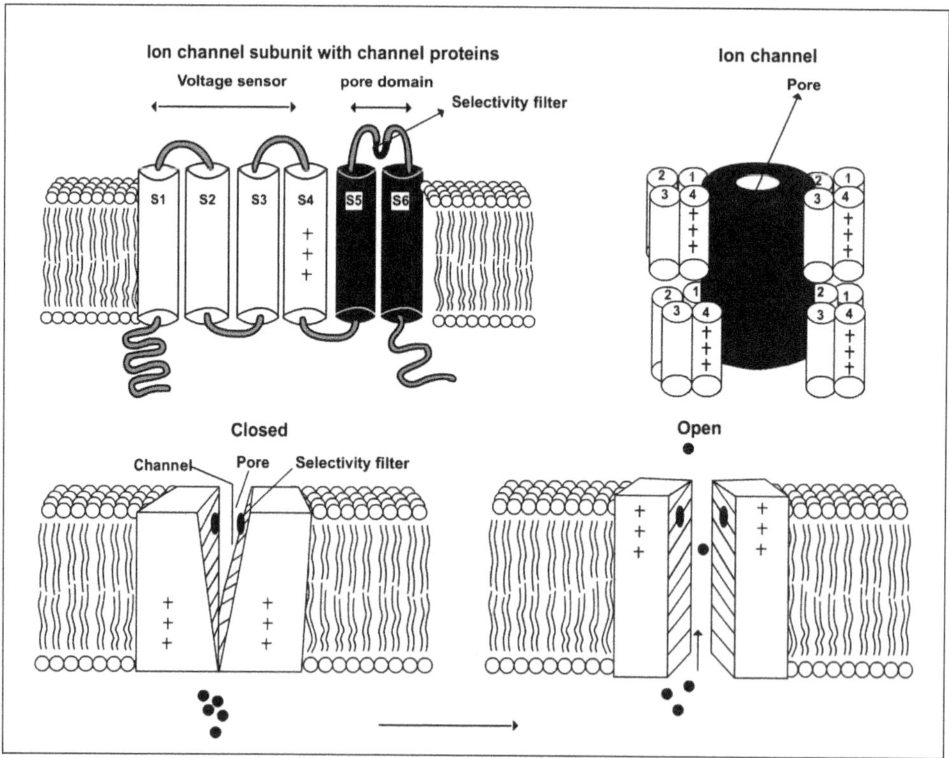

Figure 7.10: Transmembrane channel proteins organized into an ion channel. *Ion channel subunit. Six transmembrane channel proteins (S1 to S6) constitute a subunit of an ion channel; two of them (S5 and S6) form the lining of the pore and the remaining from the structural support around the pore and the channel.* **The ion channel.** *Made of four subunits. The ion channel has three regions – pore, channel and a selectivity filter (sensor). The pore opens or closes in response to change in membrane potential (redistribution of charge in the membrane) – an example of voltage gated ion channel. The bilipid membrane has the lipid molecules arranged with their hydrophilic heads facing outside and hydrophobic tails pointing inside.*

Several biological processes happen due to transport of ions through channel proteins e.g. conductance of nerve impulses, contraction of cardiac, skeletal and smooth muscles, release of insulin from pancreatic beta cells and hormones, controlling motility of migratory and growing cells etc.

There are gated ion channels in the hair cells of cochlea of the ear. They open by mechanical pressure due to swaying of the hairs on the cells to sound waves. Neurotransmitters are released from the cells through the open channels, which induce the nerves of the inner ear, to fire, enabling us to hear.

7.3.2(a)2 Carrier Proteins

Carrier proteins are transport proteins, that are only open to one side of the membrane, either inside or outside, at any given time. They transport substances against their concentration gradient, using energy, usually in the form of ATP and accomplish transport, through change of shape, when activated (**Figure 7.11**).

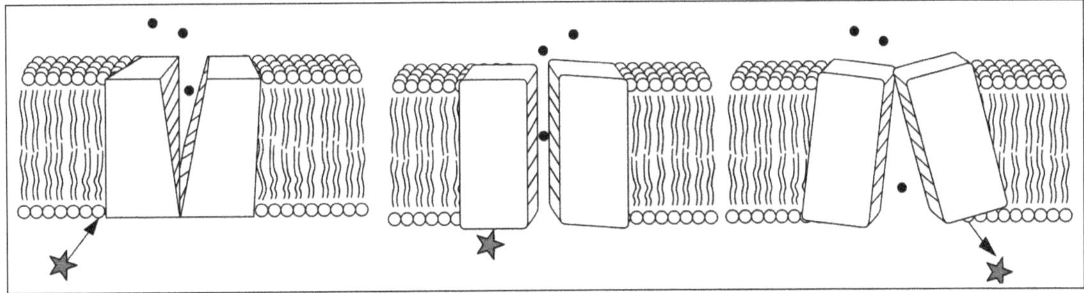

Figure 7.11: Transmembrane carrier protein – mechanism of active transport. *Conformational (shape) change in the protein (induced by a stimulus *) facilitates transport. The bilipid membrane has the lipid molecules arranged with their hydrophilic heads facing outside and hydrophobic tails pointing inside.*

The transport proteins that facilitate active transport are also called pumps. There are two main types of active transport: **1.** Primary (direct) active transport – metabolic energy (e.g. ATP hydrolysis) is used to mediate transport, **2.** Secondary (indirect) active transport – this is a two-way transport. The electrochemical gradient, created by an ion movement, that is moving down a concentration gradient, is made use of, by another molecule, to simultaneously move against a concentration gradient, without directly spending energy.

7.3.2(a)2.i Primary Active Transport through Sodium-Potassium Pump

The sodium-potassium pump was discovered in 1957, by the Danish scientist, Jens Christian Skou, who was awarded a Nobel Prize for it in 1997. This discovery helped in understanding how ions get into and out of cells. It has a particular significance in nerve cells, that depend on this pump, to respond to stimuli and transmit impulses.

Through the sodium-potassium pump, sodium is transported out, and, potassium into the cell, in a repeating cycle of conformational (shape) changes. Thus, an ion gradient is created with this pump, that results in maintenance of voltage across the membrane. Therefore, it is also called an electrogenic pump (**Figure 7.12**).

The sodium-potassium pump has binding sites for three sodium ions (Na^+) and two potassium ions (K^+). To begin, the Na^+ binding sites of the protein, face the inside of the cell. The Na^+ ions in the cell, bind to these sites because of their strong affinity to these sites. The protein then, binds to a molecule of

ATP and hydrolyses it into ADP and phosphate (PO_4). The PO_4 remains attached to the protein, while the ADP is released. Phosphorylation of the protein, results in a change of its shape, by which, the sodium binding site, now faces the extracellular side. The three sodium ions are released to the outside of the cell, while the K^+ on the outside of the cell bind to the two potassium-binding sites of the protein pump. The PO_4 attached to the pump is released and it reverts to its original shape. This facilitates release of K^+ into the cell (**Figure 7.12**).

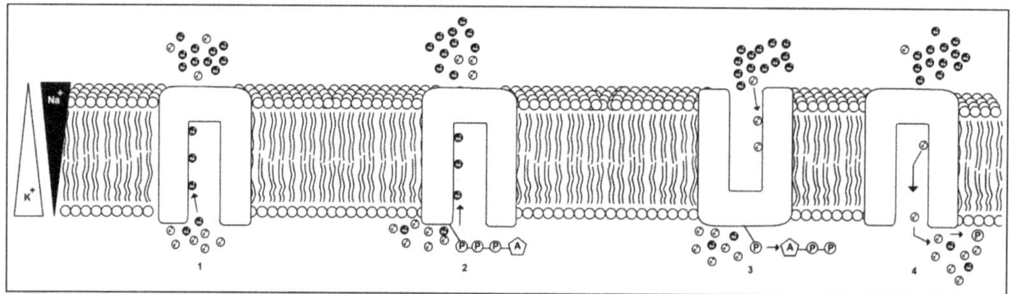

Figure 7.12: Transmembrane carrier protein mediated primary active transport – sodium-potassium pump (antiport). *The transport is achieved in four steps:* **1.** *Three sodium ions (Na^+) (black circles) attach to the carrier protein* **2.** *An ATP molecule binds to the carrier protein.* **3.** *ATP is hydrolysed to ADP and PO_4. The PO_4 remains attached to the carrier protein (ADP is released) which brings a change in its shape, resulting in the opening of the protein pump to outside of the cell. The Na^+ ions are released to the outside of the cell and two potassium ions (empty circles) from outside of the cell attach to the carrier protein.* **4.** *The PO_4 attached to the carrier protein is released bringing a conformational (shape) change in it resulting in the opening of the pump to the inside of the cell. The two potassium ions are released into the interior of the cell. For three sodium$^+$ ions released to the outside of the cell, two potassium$^+$ ions enter the cell. This creates an electrochemical gradient with more negative charge on the inside of the cell than the outside. The bilipid membrane has the lipid molecules arranged with hydrophilic heads facing outside and hydrophobic tails pointing inside.*

Thus, for each ATP molecule utilized by this pump, three positively charged Na^+ ions move out of the cell, while only two K^+ ions enter it. As a result, a strong concentration gradient is created with much more potassium inside the cell and much more sodium outside of it. Thus, electrochemical gradient is established, with the cell being more negatively charged inside compared to its outside.

7.3.2(a)2.ii Secondary Active Transport with Sodium-Glucose Transport Protein

The electrochemical gradients, set up by primary active transport, store energy which can be released, as the ions move back, down their gradients. Secondary active transport, uses the energy stored in these gradients, to move other substances, against their own gradients. Thus, co-transport of two different substances happens, through the symporter transport protein.

The sodium-glucose transport protein, uses secondary active transport, to move glucose into cells. It operates in intestinal cells and kidney cells, both of which, need to move glucose into the body's systems, against its concentration gradient. This requires energy, because, the cells have a higher concentration of glucose, than the extracellular fluid. The energy comes from the concentration gradient of sodium ions. Due to action of the sodium-potassium pump, there are much more Na^+ ions outside the cell, than its inside, with a strong concentration gradient, favouring their movement into the cell. This concentration gradient, serves as stored energy, to drive the movement of glucose, simultaneously with Na^+ ions. Binding of glucose to the channel proteins brings about a conformational change in them leading to the opening of the channel to the inside of the cell. Now both Na^+ ions and glucose molecules move into the cell (**Figure 7.13**).

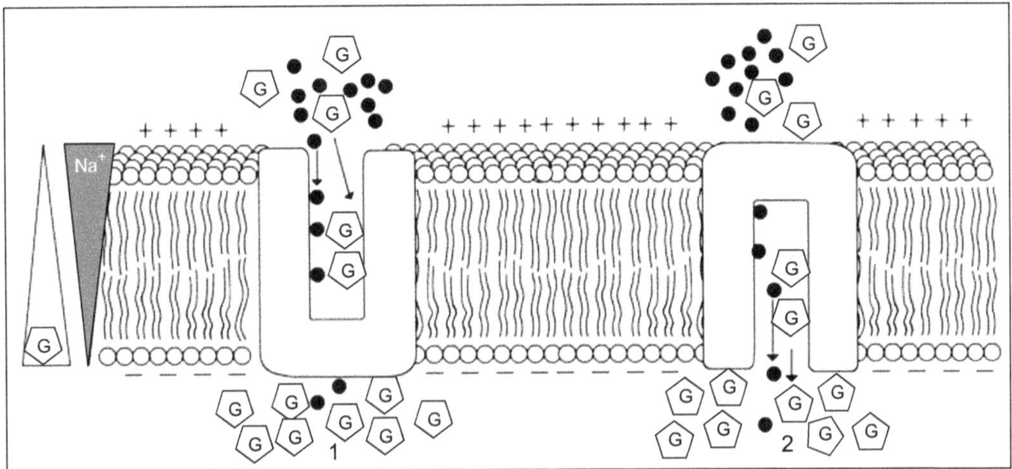

Figure 7.13: Transmembrane carrier protein mediated secondary active transport – sodium-glucose channel (symport). *Due to the previous activity of sodium-potassium pump, there is more sodium (black circles) on the outside of the cell, than its inside. The sodium flows back along its concentration gradient, through the carrier protein constituting the sodium glucose channel. The Na⁺ concentration gradient, serves as the energy, that drives simultaneously, glucose from the outside to the inside, against its concentration gradient (there is more glucose inside the cell than its outside). The attachment of the glucose molecules to the channel proteins brings about a conformational change in them resulting in the opening of the channel to the inside of the cell. Now, both Na⁺ ions and glucose molecules move into the cell. The bilipid membrane has the lipid molecules arranged with their hydrophilic heads facing outside and hydrophobic tails pointing inside. The triangles in the lipid bilayer indicate the concentration of the corresponding ion/molecule – the base represents higher concentration which gradually decreases to the tip of the triangle.*

7.3.2(a)3 Liposomes for Study of Molecules Transported by Different Transport Proteins

To detect the molecule(s) transported by different kinds of membrane transport proteins, experimental systems with a single type of membrane protein are devised. One such system is the liposome – a vesicle surrounded by pure phospholipid bilayer. A specific transport protein is extracted and purified; the purified protein then is incorporated into liposome membrane (**Figure 7.14**). The liposome is then dipped in a solution containing different ions/molecules to test which of them is accumulated in the liposome with that particular transport protein.

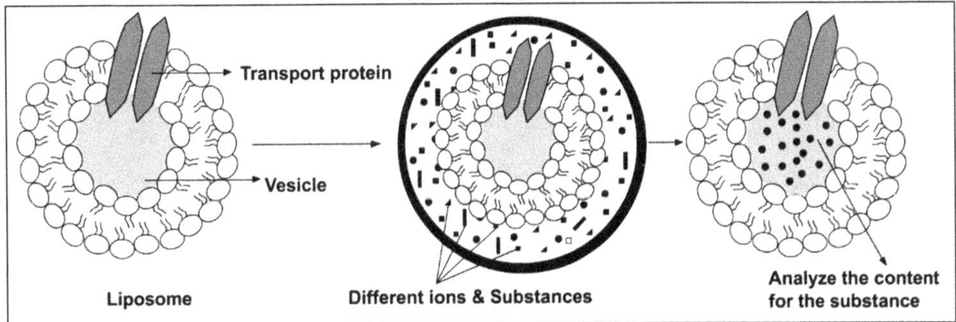

Figure 7.14: Liposomes to detect the molecules transported by a transport protein. *A transport protein is inserted in the lipid bilayer of the liposome. The liposome is immersed in a solution with different substances and ions. Later, the content in the vesicle of the liposome is analysed to find out the molecule/ion that has been transported by the transport protein inserted in the liposome membrane.*

Some transport proteins, like the haemoglobin, are not membrane bound, but present within the red blood cells, and transport oxygen to various parts of the body and carbon di oxide molecules back to the lungs.

7.3.2(b) Receptor Proteins

Communication is essential for life. The cell receives signals from its outside or inside and responds to them appropriately, for the smooth running of life processes. To accomplish this, the cells have receptors, which receive a signal and transduce it, most often, amplifying the signal during the transduction process, to elicit a suitable response.

Receptors are protein molecules, which may be present within the cell in its cytoplasm, or nucleus, or most often, they are present in the membranes of cells and organelles. The membrane receptor proteins, are integral proteins of the membrane, with their two ends, one facing the outside, and the other, inside. Hence, they are described as transmembrane. A membrane has several kinds of receptor proteins, usually clustered, rather than evenly distributed. Some of these receptors are cell specific. That is, different cell types, have different receptor proteins. The receptors, can be made up of, just only protein, or they may be glyco- (carbohydrate), or lipo-(lipid) proteins. The receptor proteins have a binding site for the signal. The signal is usually a molecule, which binds to its site, on the receptor. The signal is referred to as a ligand.

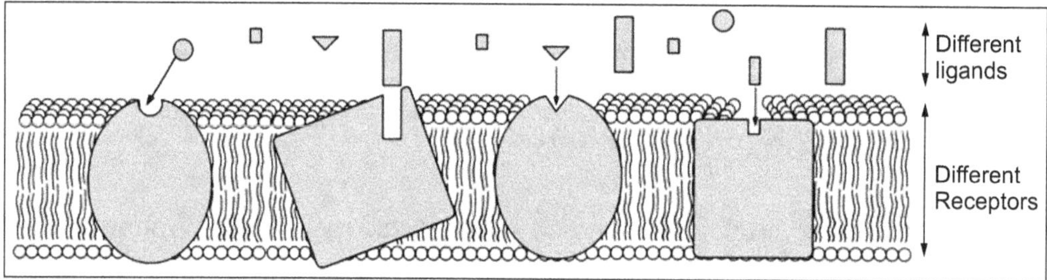

Figure 7.15: Different types of transmembrane receptor proteins each having a binding site for a specific ligand which fits like lock and key. *The bilipid membrane has the lipid molecules arranged with their hydrophilic heads facing outside and hydrophobic tails pointing inside.*

The relationship between a ligand (signal) and its binding site on the receptor, is one like a lock and key. Thus, every ligand (signal) has its own receptor (**Figure 7.15**). A ligand binding site can accommodate very closely related molecules. The ligand molecules can be peptides (like hormones), amino acids, organic molecules, steroids, gases like nitrous oxide, or pathogenic microbes, like bacteria and viruses.

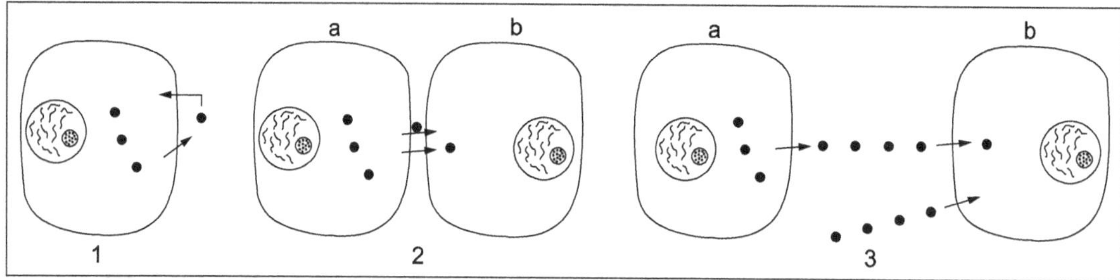

Figure 7.16: Signalling in cells. 1. *Autocrine.* *The signal is produced within the cell and released to the outside and perceived by the receptor on the membrane of the same cell.* **2 & 3. *Paracrine.* 2.** *The signal is produced by another close by cell.* **3.** *Signal is produced remotely by a distant cell, or received from the external environment, and travels long distance to reach the target cell.*

The ligands (signals) can be produced from within the cell (autocrine), or can come from its outside (paracrine), produced by a signalling cell, which might be close to the cell, or far away, or can come from outside environment (**Figure 7.16**).

7.3.2(b)i Ligand Types

Ligands can be broadly classified as of two types: hydrophilic (water loving) and hydrophobic (water hating). Hydrophilic ligands, bind to transmembrane receptor proteins, while the hydrophobic ligands (like steroids), can easily diffuse through the membranes and bind to intracellular receptor proteins. Testosterone, oestrogen, vitamin D (produced by skin cells on exposure to sunlight) are examples of hydrophobic steroids. Examples of hydrophilic ligands are: growth factors, hormones such as insulin and neurotransmitters. Hydrophilic ligands can be peptides and their size can range from, a few amino acids to more than hundred amino acids, or organic molecules, or standard or modified amino acids.

7.3.2(b)ii Types of Receptor Proteins

As discussed above receptor proteins are of two types: transmembrane and intracellular.

7.7.2(b)ii.1 Intracellular Receptor Proteins

Intracellular receptors are present in either the cytoplasm (type I), or the nucleus (type II). The intracellular receptor proteins have two binding sites, one for the ligand and the other, for a specific sequence of DNA. The receptors in the cytoplasm, when they bind to the hydrophobic ligands, that entered through the cell membrane, move through the nuclear membrane, and bind to the DNA in the chromatin of the nucleus, triggering gene activity, in response to the specific signal (ligand) (**Figure 7.17**). For the receptors in the

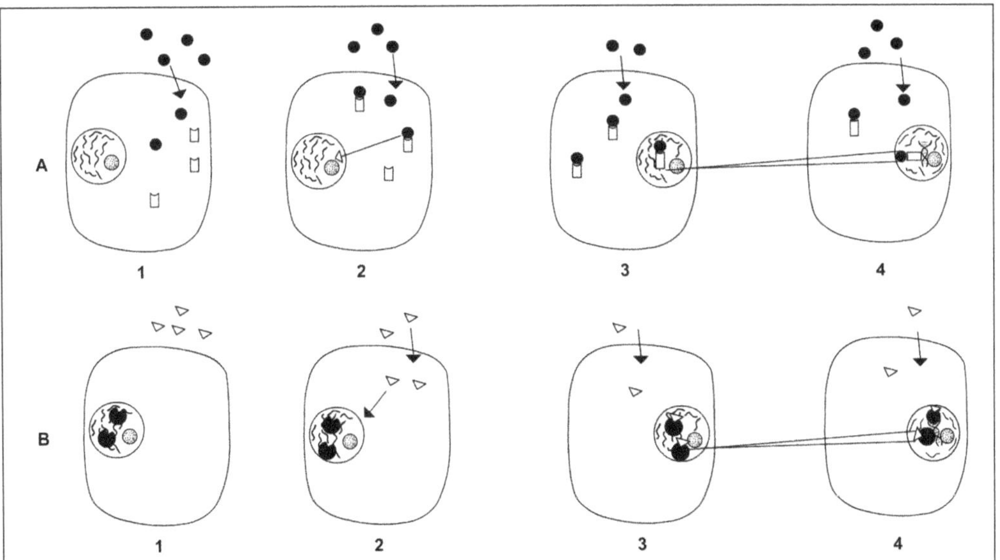

Figure 7.17: Intracellular receptors. *These receptors are either in the cytoplasm (A) or in the nucleus(B). The ligands (signal molecules) are hydrophobic.* **A. 1.** *The ligands (black circle) pass through the cell membrane and attach to the receptors in the cytoplasm.* **2&3.** *The intracellular receptors bound to the ligand, further travel through the nuclear membrane into the nucleus.* **B. 1&2.** *The ligands (triangle) pass through the cell and nuclear membranes.* **3.** *Ligands attach to the receptor in the nucleus.* **4.** *In both A and B, the ligand bound receptors, bind to a specific DNA sequence and induce gene expression as a response to the signal.*

nucleus, the ligands that entered the cell through the cell membrane, pass through the nuclear membrane as well, and bind to them (**Figure 7.17**). Since, the intracellular receptors bind to the DNA in the nucleus, to initiate response to a signal, they are also termed nuclear receptors. They act as transcription factors, that modulate gene expression, by initiating transcription of a gene, whose product is required, for response to the signal received. The response elicited by nuclear receptors is slow, since it involves synthesis of the response protein, on receiving the signal. They play a vital role in endocrine (thyroid, pancreas etc.) signalling and metabolic regulation and plant-pathogen interaction. Examples of nuclear receptors are: sex hormone receptors and cortisol receptor.

7.7.2(b)ii.2 Transmembrane Receptor Proteins

A transmembrane receptor protein has three regions – an external ligand binding region, a hydrophobic transmembrane region, and the intracellular region. Transmembrane receptor proteins are mainly of three kinds: ion channel-linked receptors, G-protein-linked/coupled receptors (GPCRs) and enzyme-linked receptors.

7.3.2(b)ii.2(a) Ion Channel-Linked Receptors (Ionophoric Receptors)

Ion channel-linked receptors are also called transport proteins since they help in movement of ions like Na^+, P^+, Ca^+, H^+ etc. They are also called ionotropic/ionophoric receptors since they aid in transport of ions. They are ligand gated transport proteins, which open, when the ligand binds to it, facilitating movement of ions, which triggers a response (**Figure 7.18**). Neurons (nerve cells) have ligand-gated channels and their ligands are termed neurotransmitters. Ionotropic receptors are involved in, generation and propagation of nerve impulses, muscle contraction, maintenance of balance of salts in the cells and release of hormones. Relaxants, anti-arrhythmics and anaesthetics used in medicine act by blocking ion channels.

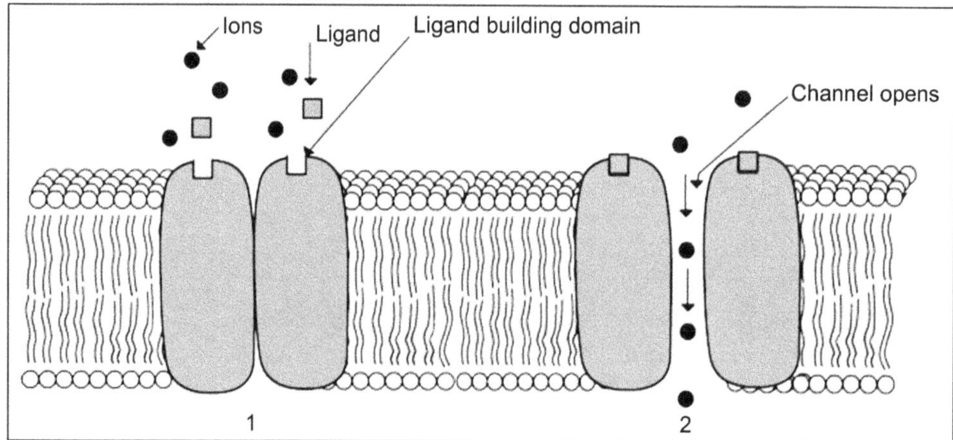

Figure 7.18: Ion channel-linked receptor protein (Ionophoric receptor). *The ion channel, opens when a ligand binds to the channel receptor protein, facilitating transport of ions into the cell which trigger a specific response. The bilipid membrane has the lipid molecules arranged with their hydrophilic heads facing outside and hydrophobic tails pointing inside.*

7.3.2(b)ii.2.(b) G-Protein-Linked/Coupled Receptors (Metabotropic Receptors)

G-protein-linked/coupled receptors (GPCRs), also termed metabotropic receptors are numerous. There are more than 1000 GPCRs in humans. They are a class of very unique membrane receptors and are the target of around 30 to 50% of all modern medicinal drugs. The GPCRs have seven alpha helices spanning the membrane, with the two extreme ends, one facing the outside, and the other, the inside of the cell or organelle (**Figure 7.19**). They are therefore often referred to as "7 transmembrane receptors." These receptors are bound to G protein in the domain facing the inside of the cell (**Figure 7.19**). The G proteins are, membrane resident heterotrimeric proteins, having three subunits – α, β and γ, with a guanosine diphosphate (GDP) attached to the α subunit. They are called G proteins, because they are linked to GDP.

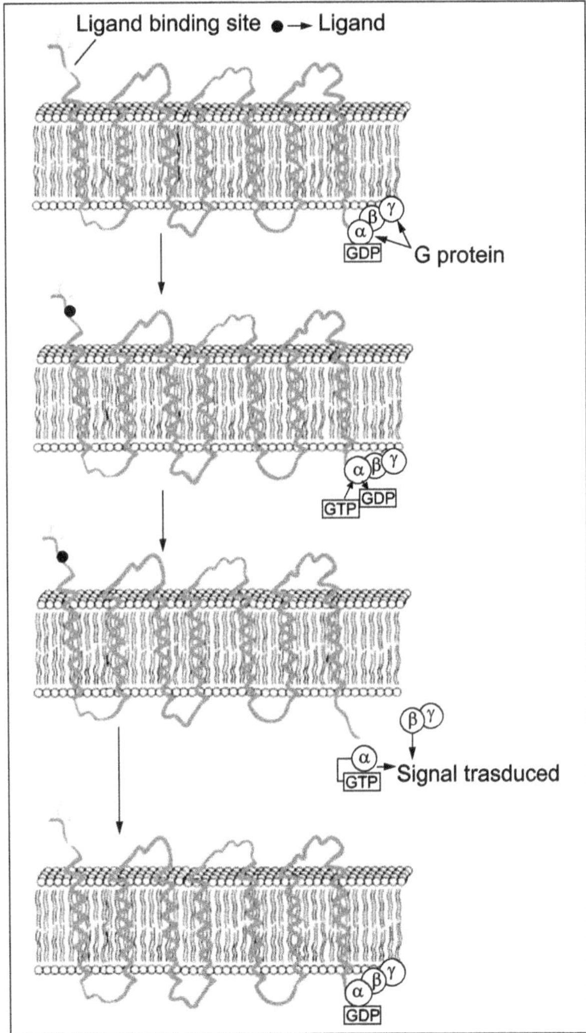

Figure 7.19: G-Protein-coupled receptor (GPCR) protein. *It is a transmembrane protein with seven α helices spanning the membrane (7 transmembrane protein). The ligand binding site on the coil faces the outside of the cell or organelle. To its end protruding into the cell or organelle, a G protein is attached. The G protein is a heterotrimeric protein, with α, β, γ subunits. To the α subunit, a GDP (Guanosine di phosphate) molecule is attached. When the ligand binds to the receptor, the α subunit exchanges its GDP with GTP (guanosine tri phosphate) and the G protein is separated from the GPCR. Next the GTP α subunit is separated from the β and γ subunits as a result of which the three subunits become activated. The activated subunits of the G protein, transduce the signal received from the ligand, by triggering a cascade of reactions which elicit cell response. Later, the GTP attached to α subunit is hydrolysed and it moves to reunite with the other two subunits which then link to GPCR. The bilipid membrane has the lipid molecules arranged with their hydrophilic heads facing outside and hydrophobic tails pointing inside.*

When a ligand binds to the receptor, the GDP in the α subunit of the trimeric G protein, is exchanged with GTP (guanosine triphosphate). As a result, the subunits of the G protein delink from the receptor and initiate transmission of a cascade of signals, that elicit a response by the cell. When the GTP in the G protein subunit is hydrolysed to GDP, all the subunits of the G protein come together and again bind to the GPCR (**Figure 7.19**).

The ligands that bind to GPCRs, range from, light sensitive compounds to odours, pheromones, hormones and even, neurotransmitters. GPCRs can regulate the immune system, growth, our sense of smell, taste, visual, behaviour and our mood. The GPCRs present in the cells of the inner nose help in sensing different smells (olfactory function). There are different GPCRs for different smells.

7.3.2(b)ii.2(c) Enzyme-Linked Receptors

Enzyme-linked receptors have, either an enzyme linked to the domain facing the inside of the cell, or, the intracellular domain of the receptor, can have enzyme function.

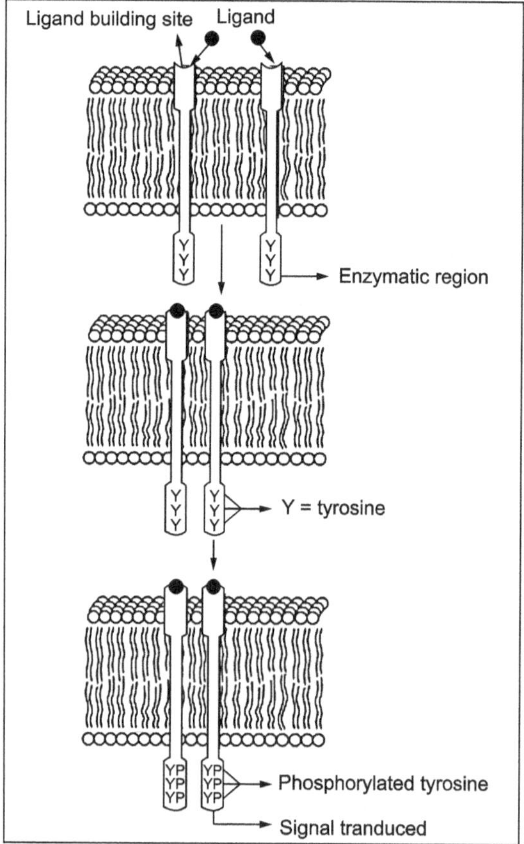

Figure 7.20: Enzyme-linked receptor protein – tyrosine kinase receptor. *It has a single transmembrane helix with the ligand binding site facing outside of the cell. The region intruding into the cell has kinase enzyme activity. When the ligand binds to the receptor, two receptor proteins nearby in the membrane come close together and dimerise. When they dimerise, the tyrosine (Y) amino acids in the enzymatic region of the receptor, auto phosphorylate. In this state it transduces the signal to effector molecules which elicit a response going through a cascade of events. The bilipid membrane has the lipid molecules arranged with their hydrophilic heads facing outside and hydrophobic tails pointing inside.*

Tyrosine receptor kinases (TRK) are examples of enzyme-linked receptors. Kinases attach phosphate groups which trigger responses. They have, a ligand domain on the region facing outside the membrane, a single αhelix transmembrane region, and their C end facing the inner membrane has enzyme activity (**Figure 7.20**).

Two TRKs, close by in the membrane, when bound to their specific ligand, dimerize. Several ligands that bind to TRKs, are themselves dimer molecules. Cytokines are one of the chief ligands of TRKs. In the dimerised form, the tyrosine amino acids in the enzyme domain of the receptor get auto phosphorylated. The phosphorylated receptor, interacts with other intracellular proteins, that initiates a signalling cascade, eliciting a response (**Figure 7.20**).

Receptor tyrosine kinases are crucial to many signalling processes in humans. For instance, they bind to growth factors, which are signalling molecules, that induce cell division, and, help in cell survival. Growth factors include, platelet-derived growth factor (PDGF) which plays a role in wound healing, nerve growth factor (NGF) which keeps the neurons alive. Overactive growth factor receptors, like polypeptide growth factors, are associated with some types of cancers. Protein-tyrosine receptors have therefore been particularly well studied, to understand the signalling mechanism involved in the control of animal cell growth and differentiation. About 50 receptor protein-tyrosine kinases have been identified to date. They are involved in growth, proliferation, differentiation or survival. They are slow in eliciting response since their signalling leads to switching on gene action.

Several diseases are associated with improper functioning of receptors: e.g. insulin resistant diabetes, night blindness, hypothyroidism, extreme obesity.

7.3.2(b)iii Receptors as Drug Targets

7.3.2(b)iii.1 Agonist and Antagonist Drug Molecules

There are several pharmaceutical drugs, targeted to receptor proteins. They interact with the specific receptor proteins, forming drug receptor complexes. Drugs can act as agonists or antagonists when bound to the receptor and change the response of the cell, resulting in control of disease symptoms (**Figure 7.21**).

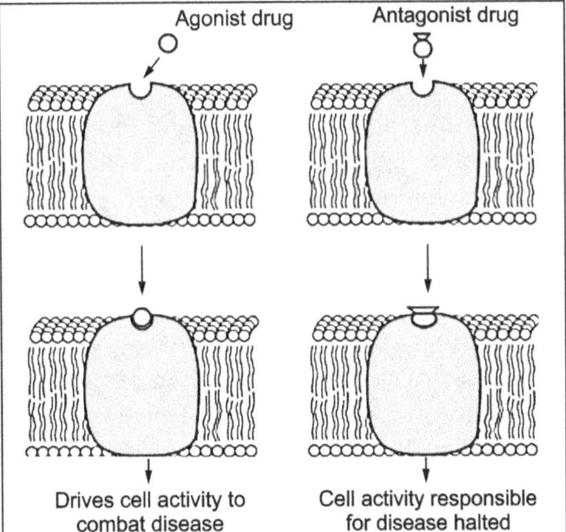

Figure 7.21: Receptor proteins as drug targets. *Agonist drugs induce a response that stops the symptoms of a disease.* ***Antagonist*** *drugs prevent the cell processes that are causing disease. The bilipid membrane has the lipid molecules arranged with their hydrophilic heads facing outside and hydrophobic tails pointing inside.*

7.3.2(b)iii.2 Targeted Drug Delivery

Theoretically, drugs can be targeted to specific cells. Drug delivery using polyhydroxyalkanoates (biodegradable substances) and ligands targeting a diseased cell specific receptor, is being conceptualized. Drug molecules are encapsulated in polyhydroxyalkanoate, to which, ligands that match the specific receptor on a diseased cell, are linked. For example, some types of cancer cells, are known to have novel receptor proteins, that are not present in normal cells. The drug, when targeted to these receptors, can reach the tumour cells, and, be delivered in them, through endocytosis (**Figure 7.22**). Thus, chemotherapy that targets only the malignant cells, is possible through this method, avoiding the side effects associated with the drug due to its effect on normal cells.

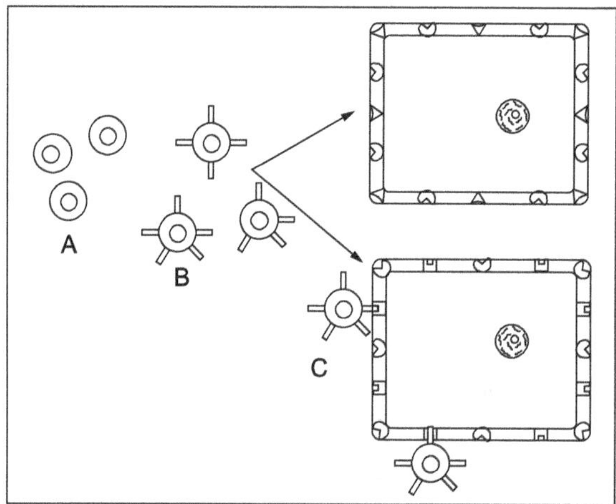

Figure 7.22: Receptor proteins on diseased cells targeted by drugs. A. *Drug molecules encapsulated in polyhydroxyalkanoate.* **B.** *Encapsulated drug molecules bound to ligands that match a specific receptor on a diseased cell.* **C.** *The ligand bound drug capsule binds only to the diseased cell with a matching receptor which is its target cell, and releases the drug into the cell. Top cell which is normal (not disease affected) does not have the receptor for the drug capsules. Thus, targeted drug delivery is achieved.*

7.3.2(b)iii.3 In-Silico Structure-Based Drug Design for Receptor Proteins

Potential drug molecules, to target specific receptor proteins, can now be identified using computer algorithms. This involves a bioinformatic approach and is termed, 'in-silico structure-based drug design'. Initially, the 3D structure of the ligand binding domain of the receptor protein is determined through methods such as, X ray crystallography and NMR spectroscopy. There are several ligand databases, with details of hundreds of molecules along with their structure. On a computer, several ligands from the data base, are docked into the 3D structure of the ligand binding site of the receptor protein using docking software. Ligand molecules that best fit the binding site on the receptor are thus, short listed for taking to preclinical testing (**Figure 7.23**). In-silico structure-based drug design, aids in speeding up of discovery of drugs and also decreasing costs involved, by reducing wet bench research to a great deal. It gives scope to screen a large number of potential molecules that can serve as a drug.

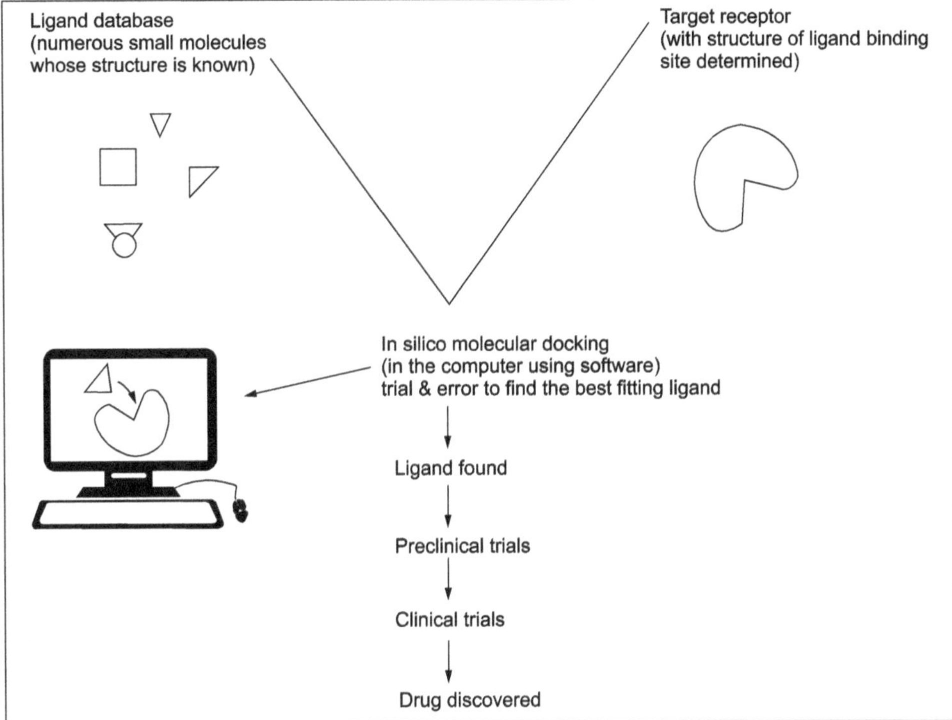

Figure 7.23. In-silico structure-based drug design through molecular docking. *Knowledge of the structure of the binding site in the receptor protein (drug target) is essential. The best fitting ligand molecule is identified from the repository of ligand molecules in the ligand databases through use of molecular docking software in the computer. The identified ligand is tested for its safety and therapeutic value in preclinical and clinical trials. Finally, a drug is discovered. Ligand identification through docking saves a lot of wet bench experiments (therefore costs) and is less time taking.*

7.4 ENZYMATIC PROTEINS

All biochemical reactions in a cell, take place at normal body temperature, at a rate that can meet an organism's needs, due to catalysis mediated by proteins, which are called enzymes. Enzymes are very specific, each catalysing a particular reaction. They make a biochemical reaction more likely to proceed, by lowering the activation energy of the reaction, thereby making these reactions proceed thousands or even millions of times faster than they would without a catalyst.

Enzymatic proteins are named based on the type of reaction they catalyse and the substrate of the reaction. For example, proteases break down proteins, and dehydrogenases oxidize a substrate by removing hydrogen atoms. As a general rule, the "-ase" suffix identifies a protein as an enzyme, whereas the first part of an enzyme's name refers to the reaction that it catalyses. For example an alcohol dehydrogenase enzyme, catalyses a dehydrogenation reaction, facilitating conversion of an alcohol into an aldehyde.

The enzyme has a binding site for the substrate molecule(s), The substrate molecules make a good fit into the binding site – the relation being similar to a lock and key (**Figure 7.24**). The binding site of the enzyme, when confronted with the substrate molecule(s), undergoes slight conformational change to make an exactly fitting site to accommodate the substrate – a phenomenon referred to as 'induced fit'. An enzyme-substrate complex is formed. Once the reaction is completed, the product molecule(s) leaves the enzyme, the enzyme remaining unchanged. Enzymatic proteins are dealt with in detail in chapter 5.

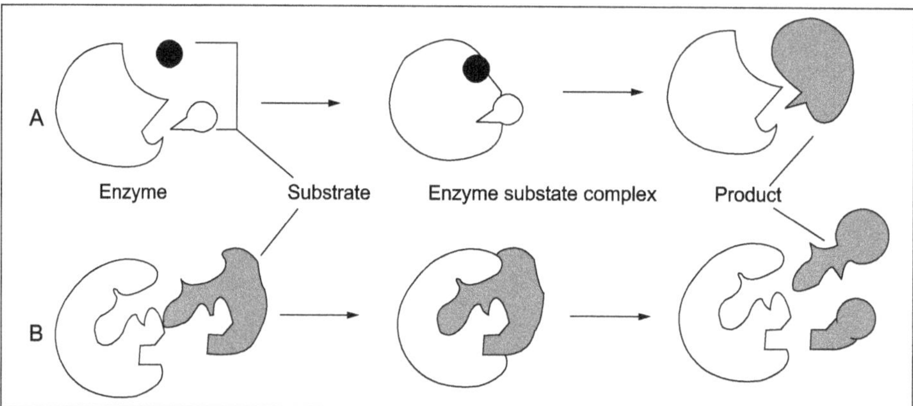

Figure 7.24: Enzyme mediated biochemical reactions. A. *combination and* **B.** *decomposition reactions. The enzyme has binding site(s) for substrate molecule(s). The binding site of the enzyme undergoes slight changes to perfectly fit in the substrate molecule (induced fit). In the enzyme substrate complex, the chemical reaction takes place at body temperature. The product(s) disassociates from the enzyme leaving the enzyme unchanged.*

7.5 SUMMARY

1. Proteins are a group of diverse macromolecules that constitute 50% of the dry mass of a cell. They are the constituents of structural components of the cells and drive biochemical reactions in the cell.

2. Proteins have a specific 3D structure. Different types of proteins have different structures. The function of a protein is governed by its structure. The structure of a protein is dependent on its amino acid sequence. Structure and therefore function, of a protein, can be predicted from the nucleotide sequence, of the gene from which it is transcribed and translated. Even a small change (mutation) in the sequence of a gene, can result in change in structure of the protein coded by it, which can profoundly affect its function.

3. For sake of basic understanding, protein structure is explained, based on a methodological reductionist approach. In this approach, protein structure is considered at four organizational levels – primary, secondary, tertiary, and quaternary. The amino acid sequence of a peptide (protein) constitutes its primary structure. The folding of the peptide leads to its secondary structure. The folding pattern depends on the type of amino acids and their sequence. The secondary structure consists of α helices, β sheets, loops and coils. The tertiary structure of the protein is attained by the folding of the polypeptide on itself, bringing distant amino acids closer together at specific points, giving a third dimension (3D) to the protein structure. Quaternary structure, refers to the non-covalent interactions of protein subunits and their spatial arrangement. The arrangement of different polypeptides that constitute subunits of a complex protein, with their acquired tertiary structure, gives the quaternary structure of a protein. Only oligo proteins (with more than one polypeptide chain) have a quaternary structure.

4. Proteins have a wide array of functions. They are the work horses of a cell. Based on function, proteins are classified into: structural, integral, enzymatic, hormone, transport, immunoglobulin and storage proteins.

5. Structural proteins give shape to the cell. They are part of the cartilage and bone in vertebrate animals. They constitute the hair, feathers, quills, nails and hoofs, and are important components

of the skin cells. The actin, α and β tubulin and few other structural proteins like keratin constitute the microfibrils, microtubules and intermediate filaments that comprise the cytoskeleton of a cell. The cytoskeleton gives shape to the cell and holds the organelles in place. Collagen and elastin are other structural proteins.

6. Integral proteins are in the membrane systems of the cells and cell organelles. They span the lipid bilayer with their two extreme ends extending into, and out of the cell or organelle. They are therefore called transmembrane proteins. Based on their function integral proteins are classified as transport proteins and receptor proteins.

7. Transport proteins can facilitate movement of small ions and molecules only one-way or both ways. A transport protein that mediates transport of one substance only one-way either into, or out of the cell is described as uniporter. When they aid in transport of two types of molecules/ions, simultaneously, in the same direction, they are described as symporter; and when in opposite direction they are termed antiporters.

8. Transport proteins are of two types: channel proteins and carrier proteins. Channel proteins act like pores in the membrane, that let water molecules, or small ions, pass through, quickly due to facilitated diffusion, a phenomenon that happens because of an electrochemical gradient. The pore created in the membrane due to channel proteins is open only when transport of particular ion/molecule is required. The opening and closing of the pore can be voltage gated, ligand gated, pressure gated, or, requires phosphorylation/dephosphorylation of the channel protein.

9. Carrier proteins are transport proteins, that are only open to one side of the membrane, either inside or outside, at any given time. They are also called pumps. They transport substances, against their concentration gradient, using energy, usually in the form of ATP. Activation of carrier proteins results in change in their shape which creates a passage for movement of molecules and ions. Some of the well studied carrier protein systems are the sodium-potassium pump and sodium-glucose pump.

10. Receptors are protein molecules, which may be present within the cell in its cytoplasm, or nucleus, or most often, they are present in the membranes of the cell or organelles. The receptor proteins have a binding site for the signal. The signal is referred to as a ligand. The relationship between a ligand (signal) and its binding site on the receptor, is one of a lock and key. Hydrophilic ligands bind to transmembrane receptor proteins, while hydrophobic ligands, move across the membrane and bind to intracellular receptors. The ligand bound receptors within the cell, bind to specific DNA sequences, to trigger a particular gene activity, as an appropriate response to the signal.

11. Transmembrane receptor proteins are three types: those linked to – ion channels, G proteins or enzymes. Those linked to ion channels facilitate movement of ions when bound by a ligand. Receptors linked to G proteins are called G protein coupled receptors (GPCRs). GPCR's are abundant. When bound by a ligand (signal), they initiate a cascade of reactions due to the activation of the G proteins linked to their end facing the inside of the cell or organelle. The activated G proteins initiate a cascade of reactions which culminates with the generation of an appropriate response. The enzyme-linked receptors have the ligand binding site on the end facing the outside of the cell or organelle and enzyme, to the end facing inside. Ligand binding to the receptor activates the enzyme which initiates a series of reactions that induce cell response.

12. Drugs can be targeted to specific receptor proteins, to either, activate or inactivate, certain functions in the cell, as a means of fighting (agonist) or preventing (antagonist) disease. Receptors can also be used for targeted drug delivery. Ligands that can act as drug molecules and fit the

binding site of a specific receptor on a cell, can be identified, through docking of ligands, using computer algorithms. This approach is called in-silico structure-based drug design. It saves time and lot of wet bench experiments thereby reducing costs of drug development.

13. Several proteins have enzymatic functions catalysing the array of biochemical reactions. Enzymes have a catalytic site (active site), to which the substrate molecule(s) bind, for the reaction to take place. The shape of the active site, fits the shape of the substrate molecule, so that an enzyme substrate complex is formed, for the reaction to proceed.

7.6 SAMPLE QUESTIONS

7.6(a) Subjective Questions

Q.1. How is the structure of the protein defined? Describe the four levels of protein structure. What is the significance of protein structure?

Q.2. Describe receptor proteins and their utility in disease treatment and targeted drug delivery.

Q.3. Describe the cytoskeletal proteins.

Q.4. What are carrier proteins? How do they operate?

Q.5. What are ion channels proteins? How do they function?

Q.6. Describe G protein coupled receptors and the mechanism by which they function.

7.6(b) Objective Questions

Q.1. The structure of a protein is dependent on

(*a*) The sequence of peptides.   (*b*) The sequence of amino acids in the peptide.

(*c*) The number of amino acids in the polypeptide.

(*d*) The proportion of purine and pyramidine nucleotides in the gene coding for it.

Q.2. Chaperone proteins help in

(*a*) Preventing improper folding of a peptide.

(*b*) Help in proper folding of a peptide.

(*c*) Facilitate cross link formations between folds of a peptide.

(*d*) Prevent the peptide from degradation.

Q.3. Receptor proteins

(*a*) Are present only in the membranes.   (*b*) Only in the cell.

(*c*) Only in the nucleus.   (*d*) In all the above.

Q.4. Carrier proteins

(*a*) Create a pore on the membrane.

(*b*) Facilitate transport through conformational change.

(*c*) Transport materials in a passive way.

(*d*) Transport material across a electrochemical gradient.

Q.5. Ion channel proteins aid in

(*a*) Active transport.   (*b*) Passive transport.

(*c*) Movement of organic molecules and ions.

(*d*) Transport of hormones.

Q.6. Structural proteins that constitute all components of cytoskeleton

(*a*) Actin, tubulin, collagen.

(*b*) Actin, α tubulin, β tubulin and keratin.

(*c*) Tubulin, elastin, actin.

(*d*) α Tubulin, β tubulin, Actin, keratin or other structural proteins.

Q.7. Receptor proteins bind to ligands due to

(*a*) Affinity.

(*b*) Charge balance.

(*c*) Matching shape to the binding site.

(*d*) Enzymatic reaction.

Q.8. Ligand gated carrier proteins

(*a*) Are both receptor and transport proteins.

(*b*) Are ion channels.

(*c*) Are sodium pumps.

(*d*) Act as drug carriers.

ANSWERS

1. (*b*)　　**2.** (*a*)　　**3.** (*d*)　　**4.** (*b*)　　**5.** (*b*)　　**6.** (*d*)　　**7.** (*c*)　　**8.** (*a*)

8 Laws of Thermodynamics at Play in Metabolism

> *A theory is the more impressive the greater the simplicity of its premises is, the more different kinds of things it relates and the more extended is its area of applicability. Therefore, the deep impression which classical thermodynamics made upon me. It is the only physical theory of universal content concerning which I am convinced that within the framework of the applicability of its basic concepts it will never be overthrown.*
> ~ **J.W. Gibbs**

8.1 INTRODUCTION

Thermodynamics is a branch of physical science, that deals with the relation between heat and other forms of energy (mechanical, electrical, or chemical). In sum, it explains the relationship between all forms of energy. 'Thermo' stands for heat (in broader sense – energy) and 'dynamics' indicates change. Essentially, it is the study of energy, and its conversion from one form to another. Thermodynamics is important because, it helps us understand, how small sub-microscopic atoms are connected to the large things, we see every day.

Thermodynamics was studied earlier to learn how to make the steam engines move faster. Now, it is being pursued in many fields, from devising better engines to understanding black holes. Thermodynamics is also used in chemistry, to explain which reactions will work and which will not. All activities in living organisms involve, flow of energy and conversion of one form of energy to another. Thus, thermodynamics applies to biological systems as well. Like the cell theory, concept of evolution and homeostasis, principles of thermodynamics also form the foundation for understanding biological systems.

Energy is the ability to do work. It is basically of two types: kinetic and potential. Kinetic energy is energy of motion. Potential energy is stored energy. Objects that have a capacity to move, but are not moving, possess potential energy. Energy is in many forms (**Figure 8.1**). Because energy can exist in so many forms, it can be measured in many ways. The most convenient measure, is in terms of heat, because, all other forms of energy can be converted into heat. The common unit of heat is kilo calorie. One kilo calorie = 1000 calories. One calorie is equal to the amount of heat required to raise the temperature of one gram of water by one degree Celsius.

All organisms require energy to stay alive. They are energy transformers. They take in energy and transform it to various forms to do different kinds of work (**Table 8.1**). All chemical reactions in a cell are energy transactions. Thermodynamics is about storage, transformation and dissipation of energy. Cells store energy; they transform it and dissipate it to carry forward unfavourable reactions. Thus, all physical, chemical and biological phenomena are regulated by the laws of thermodynamics. In living cells, thermodynamic changes are essential for biological functions.

Life can exist only where molecules and cells remain organized. Cells need energy to do various kinds of work: to generate and maintain structure, for various movements, to create electrical gradients across cell membranes, to maintain body temperature etc. The study of the energy transactions of a cell is called bioenergetics. Living beings take in energy and transduce it to various forms to do different kinds of work. The flow of energy maintains order and life. The matter flowing into a living organism is of high energy potential and that which flows out of it is of lower energy. The energy changes that occur between these two events are used to do physical and chemical work in the living systems.

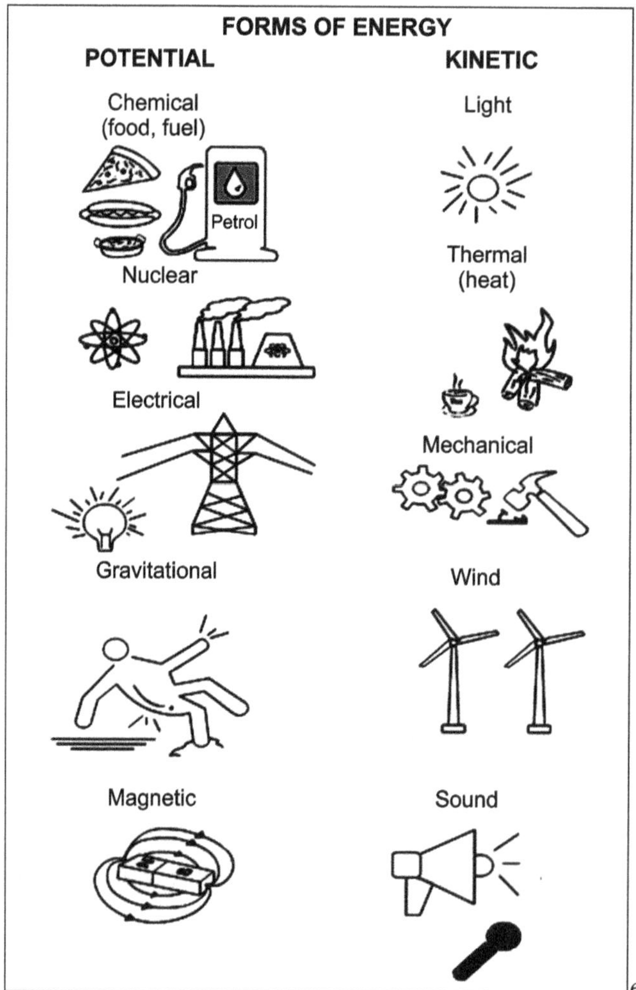

Figure 8.1: Different forms of energy. *Inherent energy is called potential energy. Energy that is being put to use for work is kinetic energy.*

Table 8.1: Examples of some energy transactions in living beings

Energy transformation	Organ– Process–Result
Light to chemical	Plant leaf – Photosynthesis – Sugar – Starch $6CO_2 + 6H_2O \longrightarrow C_6H_{12}O_6 + 6O_2$

Energy transformation	Organ– Process–Result
Light to electrical	Eye – Vision – Image
Chemical to mechanical	Muscle – Movement – Work
Chemical to electrical	Nerve – Neurotransmission – Impulse
Latent chemical to ready-to-use chemical energy – Adenosine Tri Phosphate((ATP))	Cytoplasm & Mitochondria – Respiration – Energy $C_6H_{12}O_6 + 6O_2 \longrightarrow 6CO_2 + 6H_2O + 36ATP$

8.2 THERMODYNAMICS AS APPLIED TO BIOLOGICAL SYSTEMS

The behaviour of energy has been expressed in terms of reliable observations known as the laws of thermodynamics. Thermodynamics deals with systems and its surrounding environment. A system is the portion of the universe that one is considering – a rock, a living organism or an ecosystem. A system is an assemblage of matter that interacts with energy. The decision of what constitutes a system is arbitrary (up to the observer) and depends on the purpose of study. The system is separated from its surroundings by a boundary. The system and the surroundings together make up the universe.

Energy and matter can be exchanged between the system and its surrounding environment. Based on the nature of exchange, systems are of three kinds: isolated (no exchange of matter or energy), open (exchange of both matter and energy) and closed (no exchange of matter but exchange of energy) (**Figure 8.2**). By this definition, the 'universe' is a closed system because there is nothing outside of it.

There are three states of energy within a system: initial state (energy in the system before the process took place), final state (energy in the system after the process) and equilibrium state (no further exchange of energy between the system and its surrounding takes place). Biological systems are open systems. For instance, eating a snack, or lifting a bag of laundry onto a table, or simply breathing – involve exchange of energy and matter with surrounding environment.

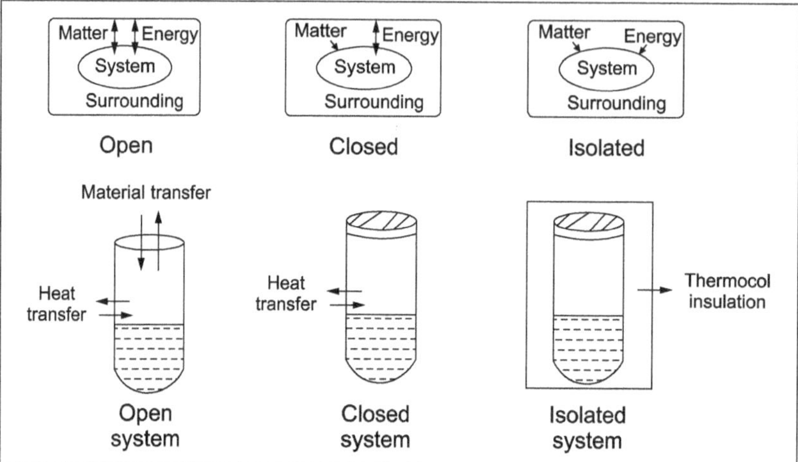

Figure 8.2: Different types of thermodynamic systems. *Open*. *Both matter and energy are exchanged by the system with the surrounding. **Closed**. Matter is not exchanged between system and its surrounding but energy is exchanged. **Isolated**. neither matter nor energy are exchanged between the system and its surrounding.*

All energy exchanges that take place in a living being during various metabolic reactions and between the living organism and its surrounding can be described by the same laws of physics as energy exchanges between hot and cold objects, or gas molecules. How the physical laws of thermodynamics – the first and second laws apply to biological systems is described below.

8.2.1 First Law of Thermodynamics

The first law of thermodynamics, also known as the law of conservation of energy, states that, 'energy can neither be created nor destroyed'. It may change from one form to another, but the energy in a isolated system remains constant. Stated in another way, energy can be exchanged between the system and its surroundings but the total energy of the system + the surrounding is constant. i.e., the energy of the universe, remains constant. Some examples of the energy transformations in biological systems are given below.

Plants execute the most important biological energy transformation on earth. Through photosynthesis, they trap the radiant energy from sunlight and convert it into chemical energy, stored within the resultant glucose. The glucose can polymerize to form starch and other complex carbohydrates like cellulose that is necessary to build their mass. The energy stored in glucose can be released through cellular respiration. Cellular respiration allows plants and animals to access energy stored in biomolecules like carbohydrates, lipids etc.

The chemical energy stored in the food taken by a human, is transformed to kinetic energy when he breathes, walks, does work with his hands etc. In man-made machines, energy is transferred or used as heat. In biological systems, heat is not a useful form of energy to do work. Heat can do work only if there is a temperature differential through which it can act i.e., heat can pass, from a body with a higher temperature, to one with a lower temperature. Small temperature differences do not exist within a cell. Living organisms are isothermal. Therefore, heat transfers cannot take place within a cell.

The living organisms need to continuously obtain energy from their surroundings, in forms, that they can transfer, or transform into energy, usable to do work. They accomplish this task, by transforming the chemical energy stored in organic molecules like carbohydrates, fats etc. through a string of chemical reactions, into the chemical energy of ATP (Adenosine tri phosphate). Energy in ATP is readily accessible to do various kinds of work like, movement, transport of materials, contraction of muscles etc.

However, none of the energy transfers that take place during the sequence of metabolic reactions in the cell are totally efficient. During transfer, some of the energy is lost as thermal energy (heat). Every step of biochemical reactions and every movement of a living system, involves some loss of energy as heat. Therefore, more energy, than required to do work, must be taken by a living system. Thus, while energy can neither be created nor destroyed, it can transform from a useful form to an unusable form (heat). Heat being unusable, causes randomness within the system. The degree of randomness or disorderliness within a system is termed entropy.

8.2.2 Second Law of Thermodynamics

It is called law of degradation of energy or law of entropy. It states that, 'every energy transfer that takes place, will increase the entropy of the system and reduce the amount of usable energy available to do work'. In other words, any process, such as a chemical reaction, or set of connected reactions, will proceed in a direction, that increases the overall entropy of the system. Entropy increases as energy is transferred. Thus, in a closed system, the amount of available energy decreases and its entropy increases constantly.

The second law of thermodynamics can be stated in another way as: 'a system and its surrounding, always proceed to a state of maximum disorder, or maximum entropy'. In an isolated system, entropy or disorderliness increases.

Biological systems are open systems, where material and energy are exchanged with the surroundings. They remain highly organized – the cells have a well-organized internal structure. The cells are structured into tissues and tissues into organs. How do living beings fight entropy or randomness? Through continuous uptake of energy rich food, the living systems maintain order in their body. To offset entropy in living organisms, energy input should exceed the energy used for doing work and the energy lost. Similarly, a car or house must be constantly maintained with work (expenditure of energy), in order to keep them in an orderly state. Left alone, the entropy of the house or car gradually increases through dust, degradation, rust etc.

If one considers the energy exchanges that take place in an animal for example while walking: as the muscles of the legs are contracted to move the body forward, the chemical energy in the biomolecules like glucose is converted to kinetic energy. However, this energy transformation process is of very low energy efficiency with a lot of energy being transformed to heat in the process. Some of this heat is useful for maintaining the body temperature, but most of it is dissipated into the surrounding. This increases the entropy of the surrounding. Living things are highly ordered and maintain a state of low entropy. However, the entropy of the universe in total is constantly increasing, due to the loss of unusable energy with each energy transfer that occurs in biological systems.

8.2.3 Consequence of Energy Loss During Biochemical Reactions of Living Systems in a Food Chain

Photosynthetic and chemosynthetic organisms are the only living beings which can synthesize their own food. Therefore, they are called 'producers'. All other organisms depend on plants directly or indirectly for their food requirements. They are referred to as consumers. The consumers are mainly of two kinds – herbivores that feed on plants (consumer of first order) e.g. rabbit, grasshopper deer, elephant and carnivores that consume herbivores or other animals. There are different orders among carnivores, e.g. a frog that eats a grasshopper which is a herbivore is a consumer of second order, a snake that eats a frog is a consumer of third order and eagle that feeds on a snake is the top consumer. A top consumer is one which is not killed and eaten by any other animal. The plant, grasshopper, frog, snake and the eagle each represents a trophic level constituting a food chain. Due to utilization of energy to carry on life processes (metabolism), and a great loss of energy during energy transformations, the amount of energy available to the next trophic level in a food chain decreases from one trophic level to the next (**Figure 8.3**).

The lower the available energy at a trophic level, the lesser the number of organisms it can support. Thus, there are more producers than consumers, and among the consumers, more herbivores than the carnivores in an ecosystem. This drives one to a crucial point about human nutrition. Eating producers (plant-based products – vegetarianism, rather veganism) which are at the bottom of the food chain, is the most efficient way to acquire energy for living.

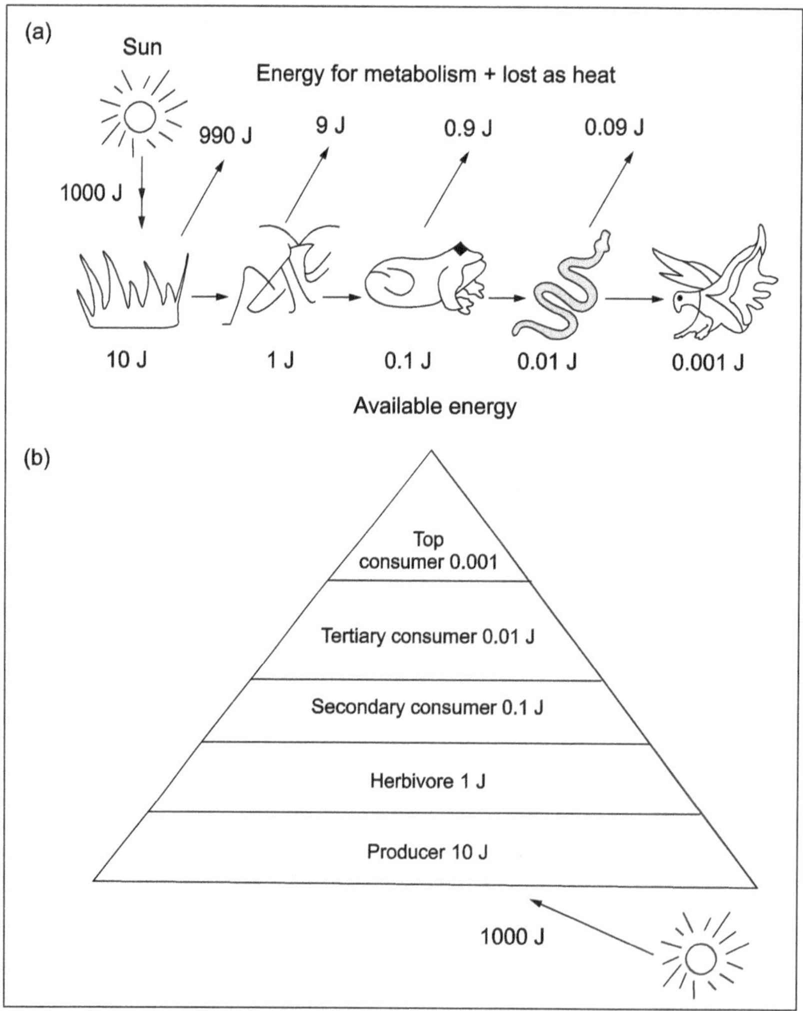

Figure 8.3: Energy transfer in a food chain in an ecosystem. *The energy absorbed by an organism is used for its metabolism and most of it is lost as heat, with just ~10%of it, remaining and available to the organism that feeds on it.* **a**. *An example of a food chain showing energy available at different trophic levels.* **b**. *Energy pyramid of an ecosystem depicting the energy at different trophic levels.*

8.3 EXOTHERMIC AND ENDOTHERMIC VERSUS EXERGONIC AND ENDERGONIC REACTIONS

8.3.1 Exothermic and Endothermic Reactions

Exothermic reactions are those in which, heat is released due to the reaction. In these reactions, the products of the reaction have less energy than the reactants; some of the energy of the reactants is lost as heat. The reaction vessel of an exothermic reaction gets hot. Any combustion reaction is an exothermic reaction. When the energy of the reactants of a reaction is lesser than that of its products, for the reaction to happen, energy in the form of heat is absorbed. Such types of reactions are described as endothermic. Melting of ice is an example of endothermic reaction. The reaction vessel of an endothermic reaction becomes cool. For any chemical reaction to occur, the energy of the reacting molecules is increased. This energy required to induce the reactant molecules to participate in a chemical reaction is termed the activation energy of the reaction. The activation energy of an exothermic reaction is less than that of an endothermic reaction; since in endothermic reactions, the energy of the reactants is lower than its products (**Figure 8.4**).

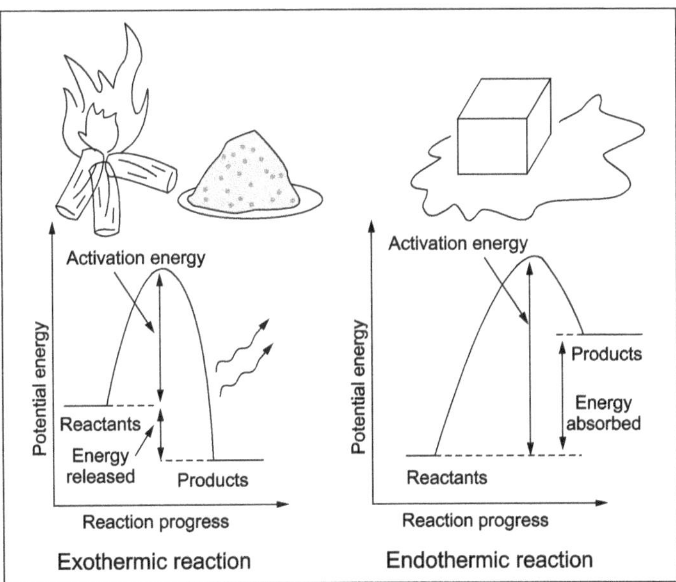

Figure 8.4: Exothermic and endothermic reactions. *Heat is released during exothermic reaction, energy in the form of heat is absorbed during endothermic reaction. The energy of the reactants of a chemical reaction is more than that of the products in exothermic reaction while the energy of the products is more than that of the reactants in endothermic reaction. Therefore, higher activation energy is required to drive an endothermic reaction.*

Change in phase (solid, liquid, gas), that results in an increase in the distance between molecules, like: conversion of solid to liquid (melting), liquid to gas (evaporation) and solid to gas (sublimation), require an input of energy in the form of heat and therefore, considered endothermic reactions. Phase changes that bring molecules closer together, decreasing distance between them, like: condensation (gas to liquid), freezing (liquid to solid), deposition (gas to solid) result in release of energy as heat and are thus exothermic.

8.3.2 Enthalpy, Entropy, Free Energy

The energy changes in chemical and biological reactions are described by three quantities: enthalpy, entropy and free energy.

Enthalpy is the total heat content of a system. It depends on the number of chemical bonds in a substance. When a reaction occurs at constant pressure, enthalpy tells how much heat and work was added or removed from the substance. A negative change in enthalpy means that heat is released, while a positive change in enthalpy, indicates that heat is absorbed. Enthalpy is denoted as H and change in enthalpy is denoted as ΔH (delta H). Thus, ΔH is positive for endothermic reactions (more than zero) and negative (less than zero) for exothermic reactions. Thus:

Exothermic reaction ΔH (Enthalpy) < 0 i.e., it is – ve (negative); heat released

Endothermic reaction ΔH (Enthalpy) > 0 i.e., it is + ve (positive); heat absorbed

ΔH of a chemical reaction is given by:

Sum of enthalpies of the products – sum of enthalpies of the reactants.

If the sum of the enthalpies of the reactants is greater than the products, ΔH will be – ve and the reaction will be exothermic. The reaction is endothermic when the products side has a larger enthalpy.

Entropy (S), is the level of disorder or randomness in a system. Entropy represents energy dispersion. An organized or ordered state, is a low energy state while a disordered state is high energy state. Entropy indicates the amount of energy that is unavailable during energy transformations.

Free energy refers to the amount of energy available in a thermodynamic system during a chemical reaction to do cellular work i.e., the internal energy minus any energy that is unavailable to perform work. If the reactions take place under constant pressure and temperature, which is the case with open systems like the biological systems, free energy is referred to as Gibbs free energy (G). It is given by:

$$G = H - TS \quad \text{and} \quad \Delta G = H - T\Delta S$$

Where G is Gibbs free energy, ΔG is change in free energy of a system, H is enthalpy, T is temperature in Kelvin (K), where K = 273 + °C, S is entropy and ΔS is change in entropy.

Thus, Gibbs free energy combines enthalpy and entropy into a single value. Gibbs free energy indicates whether a reaction is favoured or not. A reaction is favoured when the free energy of the system decreases. If the free energy of the system increases, a reaction is not favoured. ΔG is the difference in the energy between reactants and products. For a reaction taking place at constant temperature and pressure, ΔG represents, the portion of the total energy change, that is available (free) to do work. For example, if ΔG is −270 KJ for a reaction, it means 270KJ is available to do work from the reaction. If on the other hand, for a reaction, ΔG is + 270KJ, it means 270KJ of energy is required to make the reaction happen.

8.3.3 Exergonic and Endergonic Reactions

An exergonic (comes from the word ergonomic which means convenient to do work) reaction releases work energy. An endergonic reaction takes in work energy. Work energy means energy available to do work. It is termed the free energy. In an exergonic reaction the energy of the reactants is more than that of the products and the reverse is true for an endergonic reaction. Exergonic and endergonic reactions do not indicate how hot or cold a system and its surroundings become. It is about whether the reaction is feasible.

Exo/endo- thermic, indicate the relative change in heat/enthalpy in a system, while exer/ender- gonic, denote the relative change in the free energy of a system. Every chemical reaction involves a change in free energy(ΔG). Since endergonic reactions require an input of energy, the ΔG for that reaction will be a positive value. Exergonic reactions release free energy; the ΔG for that reaction will be a negative value. The differences between exer- and ender- gonic reactions are tabulated in **Table 8.2.**

Table 8.2. Differences between exergonic and endergonic reactions

Exergonic	Endergonic
Catabolic	Anabolic
Coupled to ATP formation	Coupled to ATP utilization
Spontaneous	Non-spontaneous
Δ G is negative	Δ G is positive
Δ H is negative	Δ H is positive

8.4 CONCEPT OF K_{eq} AND ITS RELATION TO STANDARD FREE ENERGY

If not interfered with, all chemical reactions come to an equilibrium. Chemical equilibrium is a state in which the concentration of reactants and products remain constant with no further change over time. That is, the forward and reverse reactions occur at equal rates. This state of equilibrium can be described by the equilibrium constant, K_{eq}. For a chemical reaction:

$$aA + bB \longrightarrow cC + dD$$

Equilibrium constant $K_{eq} = [C]^c[D]^d/[A]^a[B]^b$

Here the products are in the numerator and the reactants in the denominator. The value between [ ] represents the molar concentrations of A, B (reactants), C and D (products); the lower-case letters a, b, c, d represent the numbers before the reactants and products in the chemical equation inserted to balance the reaction. If the reactants and products are gases then:

Equilibrium constant $K_{eq} =[PC]^c[PD]^d/[PA]^a[PB]^b$

P represents the partial pressure in atmospheres.

Gibb's free energy ΔG is free energy of a reaction under constant temperature and pressure. $\Delta G°$ is the standard free energy for the reaction (as carried out with 1M concentration of each reactant, at 298K (temperature in Kelvin) and at 1 atm. pressure). The change in Gibb's free energy ΔG under non-standard conditions can be determined from the standard change in Gibbs free energy represented by $\Delta G°$ as follows:

$$\Delta G = \Delta G° + RT \ln Q$$

R is the ideal gas constant (8.314 J/mol K), Q is the reaction quotient, and T is the temperature in Kelvin. Reaction quotient Q for solids is given by $[C]^c[D]^d/[A]^a[B]^b$ and gases by $[PC]^c[PD]^d/[PA]^a[PB]^b$ (similar to equilibrium constant). Reaction quotient is an estimate of the relative amount of products to reactants at a particular point in time of the reaction while equilibrium constant is reaction quotient when the reaction has reached an equilibrium. Under standard conditions, if the reactant/products are gases, the pressure of the gases is 1 and if the reactant/ product are in the form of solution, the concentrations are 1 M, therefore Q is equal to 1. Taking the natural logarithm simplifies the equation to:

$$\Delta G = \Delta G° \text{ (under standard conditions)}$$

Under nonstandard conditions, Q must be calculated (in a manner similar to the calculation for an equilibrium constant). For gases, the concentrations are expressed as partial pressures in the units of either atmospheres or bars, and solutes in the units of molarity.

The equilibrium constant K_{eq} of a reaction has a direct relationship to the Gibbs free energy change under standard conditions $\Delta G°$. This relationship can be established by substituting the values for ΔG and Q for a chemical reaction at equilibrium. At equilibrium state, $\Delta G = 0$ (since there is no more change in Gibbs free energy), and reaction quotient becomes equal to equilibrium constant; i.e., $Q = K_{eq}$. Substituting these values into the equation for determining free energy change under nonstandard conditions, we can draw an equation of relation of $\Delta G°$ to K_{eq} as follows:

Equation for free energy change ΔG under non-standard conditions is:

$$\Delta G = \Delta G° + RT \ln Q$$

At equilibrium state of a chemical reaction, $\Delta G°$ is 0 and reaction quotient Q = equilibrium constant K_{eq}. Therefore, the equation becomes:

$$0 = \Delta G° + RT \ln K_{eq}$$
$$\Delta G° = - RT \ln K_{eq}$$

This equation helps to relate the equilibrium constant directly to changes in enthalpy and entropy.

8.5 SPONTANEITY OF A REACTION

A spontaneous reaction is one that occurs without the addition of external energy. Once started, it continues on its own without input of energy. Therefore, a spontaneous reaction is termed a natural process. A non-spontaneous process needs a continuous input of energy. A reaction will always move spontaneously towards equilibrium and never spontaneously moves away from equilibrium. Spontaneity of a reaction is not related to the speed of the reaction; it can be either fast or slow. For example, diamonds left to themselves decay to graphite. It is a spontaneous reaction but a very slow reaction and a few human generations are not sufficient to observe the change of diamond to graphite – it takes millions of years. Spontaneous reactions are often exothermic but not always. An endothermic reaction can also be spontaneous under certain conditions. That means spontaneity of a reaction is not related to change in enthalpy of a process. Spontaneity of a reaction is related to change in entropy. Entropy change measures the dispersal of energy – how much energy is released and how widely it is spread out in a particular process at a specific temperature. A spontaneous change always results in the dispersal (spreading out) of energy and also matter.

The second law of thermodynamics is related to entropy. It states that change in free energy of a system $\Delta G = \Delta H - T\Delta S$. From the value of ΔG, it can be predicted whether the reaction is spontaneous or not (**Figure 8.5**).

	$\Delta H < 0$ Exothermic	$\Delta H > 0$ Endothermic
$\Delta S < 0$ endergonic	Spontaneous at low temperature when ΔTS is small	$\Delta G > 0$, Non-spontaneous at all temperatures
$\Delta S > 0$ exergonic	$\Delta G < 0$, Spontaneous at all temperatures	Spontaneous at high temperatures when ΔTS is large

Figure 8.5: Spontaneity of a reaction as related to change in enthalpy, entropy and free energy of a system. ΔG, ΔH and ΔS are change in free energy, enthalpy and entropy respectively.

In the equation above, value of ΔG depends on three factors: H which is enthalpy, T which is temperature in Kelvin unit $(= 273 + °C)$ and ΔS which is change in entropy. Temperature is always positive because it is in Kelvin units. Therefore, $T\Delta S$ will have the same symbol +/– as ΔS.

In an exothermic reaction, ΔH will be < 0 (because heat is lost from the system), if entropy S increases i.e., $\Delta S > 0$, the value of ΔG will be negative. That means the reaction is spontaneous.

When the reaction is endothermic ΔH will be > 0, if the entropy of the system decreases i.e., $\Delta S < 0$, the value of ΔG will be positive. That means the reaction is not spontaneous (**Figure 8.5**).

Exothermic reactions ($\Delta H < 0$) in which entropy S decreases ($\Delta S < 0$), are spontaneous at low temperatures (**Figure 8.5**).

Endothermic reactions ($\Delta H > 0$) in which entropy S increase ($\Delta S > 0$), are spontaneous at high temperatures (**Figure 8.5**).

Thus, ΔG values are more indicative of the spontaneity of a reaction. When ΔG is negative i.e., $\Delta G < 0$, the reaction can start spontaneously in the forward direction to form more of the product. When ΔG is positive i.e., $\Delta G > 0$, the reaction cannot start spontaneously in forward direction but the backward direction is spontaneous making more of the reactants. Thus, a spontaneous reaction is one that moves a chemical reaction to equilibrium. A non-spontaneous reaction is one which moves the system away from equilibrium. When ΔG is 0, the reaction is at equilibrium and the concentration of the reactants and the products remains constant.

8.5.1 Activation Energy of a Reaction

Even a spontaneous reaction needs a little push to take off. This is called activation energy of the reaction. For reaction to take place, the energy of the reactants must be raised so that they reach an unstable transition state. For example, in the simple sugar glucose which is the chief energy source of cells, the energy within the bonds of the atoms of glucose is released, as it breaks down to CO_2 and water. Since, a single large molecule of glucose is broken down, to large number of smaller molecules (six molecules of CO_2 and six molecules of water), entropy of the system increases. That means, it is an exergonic reaction and energetically favourable to start spontaneously. But, if glucose is kept in a dish, it will not spontaneously break down to CO_2 and water, unless heated. After initially applying heat, the sugar will continue to burn, even when the heat source is removed. This initial energy, that is supplied for a reaction to take off is termed activation energy. Activation energy is required for both exergonic and endergonic reactions. More activation energy is required for endergonic than exergonic reactions just as the case with exothermic vs endothermic reactions (**Figure 8.4**). In living systems, the reactions take place at a reasonably fast rate because the activation energy of a reaction is lowered by the enzymes which catalyse the reaction (**Figure 8.6**).

8.6 ATP AS AN ENERGY CURRENCY

Cells need energy for all life activities. Energy is the capacity to do work. Cells perform several kinds of work: chemical work to build, rearrange or tear apart compounds; mechanical work to move cilia or flagella or flex a muscle; electrochemical work like nerve impulses etc. The original source of energy for most life on earth is from the sun, which is converted to chemical energy by plants, and is stored in the bonds of the glucose molecules that are formed. The glucose is converted to other complex carbohydrates and fatty acids (components of fats) and amino acids (components of proteins). The living beings obtain energy from their food. Food is broken down during cellular respiration and energy contained in the bonds of the biomolecules of the food is released. The covalent bonds in the food molecules are of low energy and not very useful to do most of kinds of work in the cells. Therefore, the energy released from the breaking of the chemical bonds in food is stored in a nucleotide called the Adenosine tri phosphate (ATP). All life processes that require energy are driven utilizing the energy stored in ATP. Therefore, ATP is referred to as the energy currency of the cell.

All organisms, from the simplest bacteria to humans, use ATP as their primary energy currency. Even viruses utilize ATP. It powers every activity of a cell and an organism. ATP is thus, a critically important macromolecule next in importance only to DNA in a cell. It is the most widely distributed high-energy compound within an organism. A human cell at any given time is estimated to contain one billion ATP molecules. An adult human is made up of 100 trillion cells.

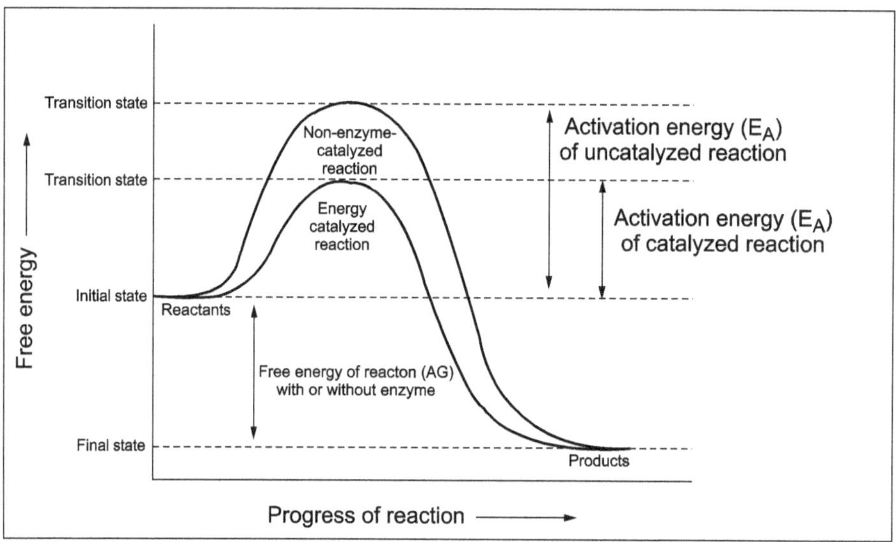

Figure 8.6: Activation energy of an enzyme catalysed and an uncatalyzed reaction. *Activation energy is reduced in the presence of enzyme and therefore the reaction occurs faster.*

ATP is a complex nanomachine, with a level of mechanical organization, equivalent to, a research microscope, or a standard television. Each ATP molecule is over 500 atomic mass units (500 AMUs). It consists of a sugar (ribose), a nitrogen base (adenine) and three phosphate groups (**Figure 8.7**). The bonds between the phosphate groups in ATP are high-energy bonds and therefore unstable. The energy contained in these bonds is readily released when ATP is hydrolysed in cellular reactions. When the terminal bond between the second and third phosphate group in ATP is broken with the addition of water (hydrolysis) in the presence of the enzyme ATPase, 7.3 Kcal/Mole of energy is released and ADP (Adenine di phosphate) and inorganic phosphate (Pi) are formed (**Figure 8.7**). ATP hydrolysis is spontaneous exergonic favourable reaction. The energy released due to hydrolysis is just the right amount for most biological reactions and no energy is wasted. If more energy is required, the next phosphate bond (between the second and the first) is broken, to meet the energy requirement and one phosphate group (Pi) is released. ATP is resynthesized again in a reverse reaction with the help of the enzyme ATP synthetase (**Figure 8.7**). ATP is continuously used by a cell to do work; it is estimated that in each ATP molecule, the terminal phosphate is added and removed three times in a minute. It is estimated that, in a human adult, the energy used per day, requires the hydrolysis of 100 to 150 moles of ATP. Hooking and unhooking that last phosphate on ATP, is what keeps the whole living world operating. Thus, ATP is not a storehouse of energy that is set aside for future use. It is produced by one set of reactions and is almost immediately consumed by another. It is rather an energy-coupling agent.

ATP is used in many cell functions: transport of substances across membranes, mechanical work like muscle contraction of the skeletal (for body movement or movement of cilia and flagella) and heart muscles (for blood circulation) and spindle fibres during cell division (for movement of chromosomes).

The chief role of ATP however is chemical work i.e., supplying energy for the synthesis of several thousands of macromolecules that different cells need. Thus, ATP is a perfectly-designed intricate molecule that plays a critical role in providing right sized energy packet for scores of thousands of reactions that occur in cells. The ATP energy supply system is quick, highly efficient with a rapid turnover rate enabling the cells to efficiently respond to energy demands.

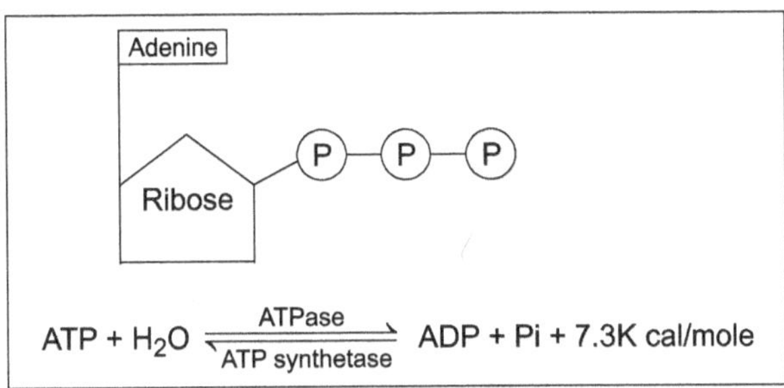

Figure 8.7: Adenosine tri phosphate (ATP). *Consists of ribose sugar, a nitrogen base, adenine and three phosphate groups. Reaction showing hydrolysis and synthesis of ATP. Hydrolysis of ATP results in the release of energy.*

8.7 ENERGY YIELDING AND ENERGY CONSUMING REACTIONS

Cells make energy in two ways – photosynthesis and respiration. Photosynthesis is an energy consuming reaction carried out by chlorophyll containing organisms. The solar energy is utilized to synthesize carbohydrate (glucose) which serves as the energy source for both plants and animals. Animals obtain energy through their food which is obtained either directly or indirectly from plants. Food has energy in the chemical bonds in its molecules. This energy is released when the chemical bonds are broken. This energy yielding process is termed respiration. In both energy consuming and yielding reactions, ATP is involved.

8.7.1 Photosynthesis: Energy Consuming Reactions

The general equation of photosynthesis is written as follows:

$$6CO_2 + 6H_2O \xrightarrow[\text{Chlorophyll}]{\text{Sunlight}} C_6H_{12}O_6 + 6O_2$$

Photosynthesis occurs in the chloroplasts. It involves two sets of reactions – the light reactions and dark reactions. Light energy (from the sun), is trapped by the chlorophyll molecule and converted to chemical energy, in a series of steps. Some of the energy is used to split water and oxygen is released. The hydrogen ions released due to splitting of water, reduce NADP (nicotinamide adenine diphosphate) to NADPH. As a result of light reactions, ATP and NADPH are formed and oxygen is released (**Figure 8.8**).

$$12H_2O + 12NADP^+ + 18ADP + 18P_i \xrightarrow[\text{Chlorophyll}]{} 6O_2 + 12NADPH + 18ATP$$

The ATP and NADPH formed in the light reaction are utilized during the dark reaction to convert carbon di oxide to glucose in a cyclic series of several reactions called the Calvin's cycle.

$$6CO_2 + 12NADPH \xrightarrow{\text{Energy from 18ATP}} C_6H_{12}O_6 + 12NADP + 18Pi + 18ADP$$

The dark reactions are so called because these do not require solar energy – they use the energy of the ATP produced during light reaction (**Figure 8.8**).

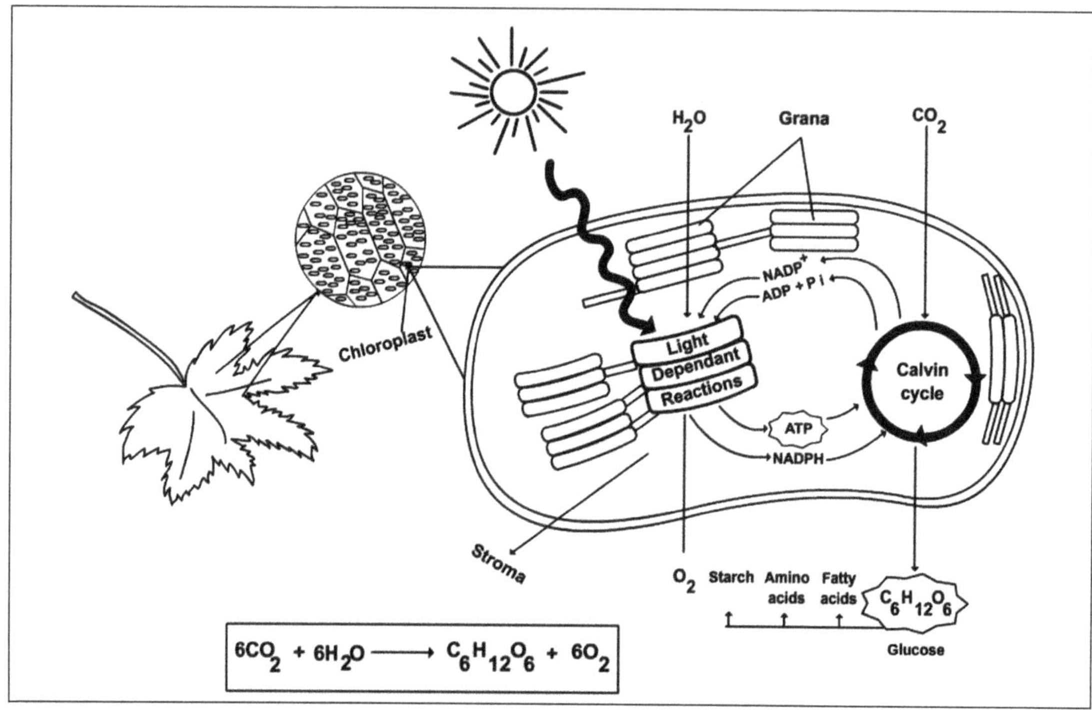

Figure 8.8: Energy consuming reactions of photosynthesis. *Light energy from the sun is utilized by the plant to convert to chemical energy in the bonds of glucose and the other products made from it like starch, fatty acids, amino acids etc. Both the light dependent and independent (dark – Calvin cycle) reactions occur in the chloroplast. Light reactions take place in grana (lamellae stacked as discs) that contains chlorophyll which traps sunlight. Dark reactions take place in the stroma (the space within the chloroplast). The light is trapped by chlorophyll (in the grana) and utilized to convert the CO_2 taken up by the plant to glucose. Oxygen is released.*

8.7.2 Respiration: Energy Yielding Reactions

Respiration is the process of breaking down food molecules to release energy. There are three steps in this process: Glycolysis, Krebs cycle and electron transport chain. The first step occurs in the cytoplasm of the cell and the next two steps happen in the mitochondria (**Figure 8.9**). During all the three steps ATP is formed.

The coenzyme NADP participates in some of the steps of glycolysis and Krebs cycle and is converted to NADPH. During glycolysis, glucose is converted to pyruvate. The pyruvate moves from the cytoplasm into the matrix of the mitochondrion. Here it is converted to acetyl coenzyme A. Acetyl CoA, then undergoes a cyclic set of reactions, termed the Kreb's cycle, during which CO_2 is released. The NADPH formed during glycolysis and Kreb's cycle is stripped of H during the electron transport chain to form NADP. The electron transport chain occurs in protein complexes in the inner membrane of the mitochondrion (**Figure 8.9**). Through the electro chemical gradient established due to electron transport chain, the hydrogen ions from NADPH, reach the site of ATP synthetase (in the form of tennis rackets) present in the inner membrane of the mitochondrion, and are used for synthesis of ATP. This movement of hydrogen ions due to electrochemical gradient is termed chemiosmosis. The process of electron transport chain and chemiosmosis together which result in formation of ATP is termed oxidative phosphorylation. The hydrogen stripped from NADPH finally reacts with oxygen to form water. A total of 38 ATP molecules are formed from the oxidation of one glucose molecule during cellular respiration (**Figure 8.9**). The ATP as discussed earlier is the ready energy currency to do various kinds of work.

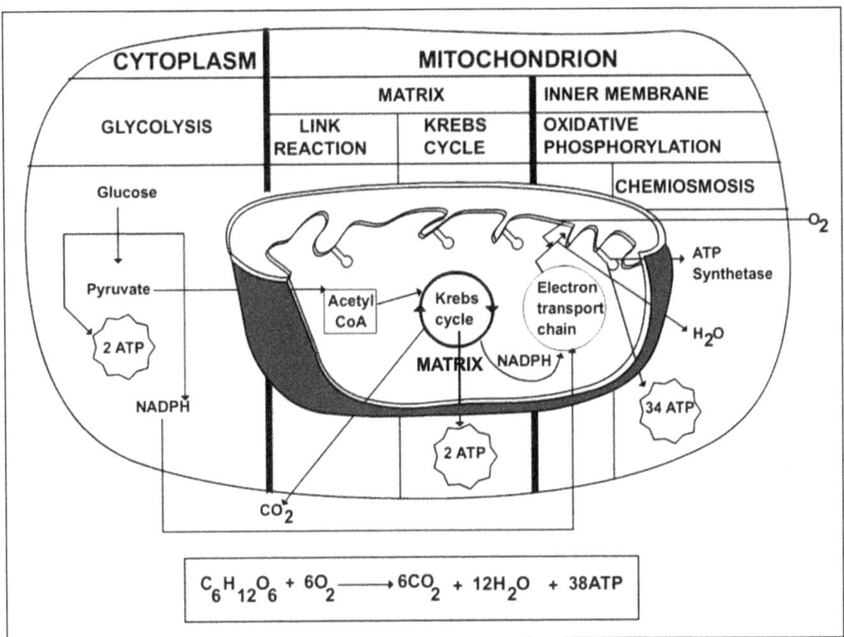

$$C_6H_{12}O_6 + 6O_2 \longrightarrow 6CO_2 + 12H_2O + 38ATP$$

Figure 8.9: Energy releasing reactions of cellular respiration. *The chemical energy contained in the bonds in glucose is released as usable form of energy – ATP (38 molecules per each glucose molecule) during respiration. The glycolysis part of respiration occurs in cytoplasm, the Krebs cycle reactions take place in the matrix of the mitochondrion. The NADPH generated in glycolysis and Krebs cycle passes its electrons through membrane proteins in the inner membrane of the mitochondrion – the electron transport chain which finally convert ADP to ATP with the help of ATP synthetase (a large enzyme molecule in the form of a tennis racket on the inner membrane of the mitochondrion). The synthesis of ATP due to the electron transport chain is termed oxidative phosphorylation. Oxygen is taken up during this process to form water. Carbon di oxide is released during the reactions of Krebs cycle.*

8.8 CONCEPT OF ENERGY CHANGE

As per the first law of thermodynamics – the law of conservation of energy, energy can neither be created nor destroyed. Energy is defined as the capacity to do work. All activity of living beings and man-made machines involves doing work. Energy needs to be spent to do work. If continuous work involves continuous expenditure of energy, then how does the law of conservation of energy stand? As we have seen earlier in this chapter, energy is of two kinds – potential and kinetic, both of which can be in many forms: light, heat, sound, chemical, mechanical, electrical etc. Work involves not expenditure of energy, but it happens due to change or transformation of energy, from one form to another. Some examples are given below to illustrate the point.

8.8.1 Energy Changes During Production of Electricity in a Hydroelectric Dam

In a hydroelectric dam, the water in the reservoir has potential energy (**Figure 8.10**). When the retention gate of the reservoir is opened, water begins to flow through the penstock, and its energy is converted to kinetic energy. The water moving down the pen stock, when it reaches the turbine, it starts to move its shaft, with the result that, its kinetic energy, is converted to mechanical energy. The mechanical energy moving the turbine shafts, is converted to electrical energy in the generator attached to the turbine. Finally, the electrical energy is sent to a transformer, for voltage adjustments, and from there supplied through wires to various electrical sub stations (**Figure 8.10**).

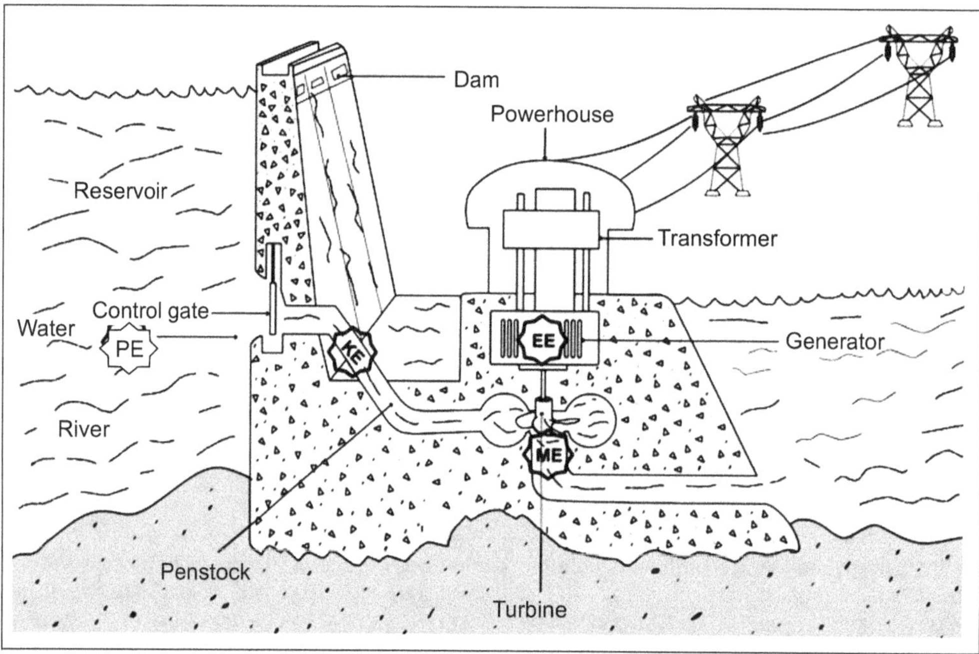

Figure 8.10: Energy transformations in a hydroelectric project. *The potential energy (PE) of water in the reservoir is converted to kinetic energy (KE) when the control gates are opened and it passes through the penstock. The kinetic energy of the water moves the turbine blades, thereby transforming to mechanical energy (ME). The mechanical energy created by the turbine is converted to electrical energy (EE) in the generator connected to it. The voltage of the electrical energy is adjusted in the transformer and it is sent through power lines.*

8.8.2 Energy Changes During Production of Electricity in a Thermal Power Project

In the thermal power project, the energy that turns the turbines to generate electricity, is derived from, the steam evaporating from water, that is heated up by coal. The chemical energy in coal, which is a fossil fuel, is converted to heat energy, that heats the water. The water boils as a result of heating, and the kinetic energy in the resulting steam, moves the shafts of the turbines getting converted to mechanical energy, which is then transformed to electrical energy, in the generator connected to the turbine.

8.8.3 Energy Changes During Synthesis of Animal Food and Formation of Fossil Fuels

The first energy conversion begins in the sun, where, due to nuclear fusion i.e., hydrogen atoms coming together to form helium, a tremendous amount of energy is formed, and neutrons are released. During this reaction, matter is converted to energy called the electromagnetic radiation. It consists of visible light, x rays, ultraviolet light, infrared, gamma rays, microwaves, and radio waves. This energy pours out of the sun into space, some of which falls on earth where it is called solar energy (**Figure 8.11**).

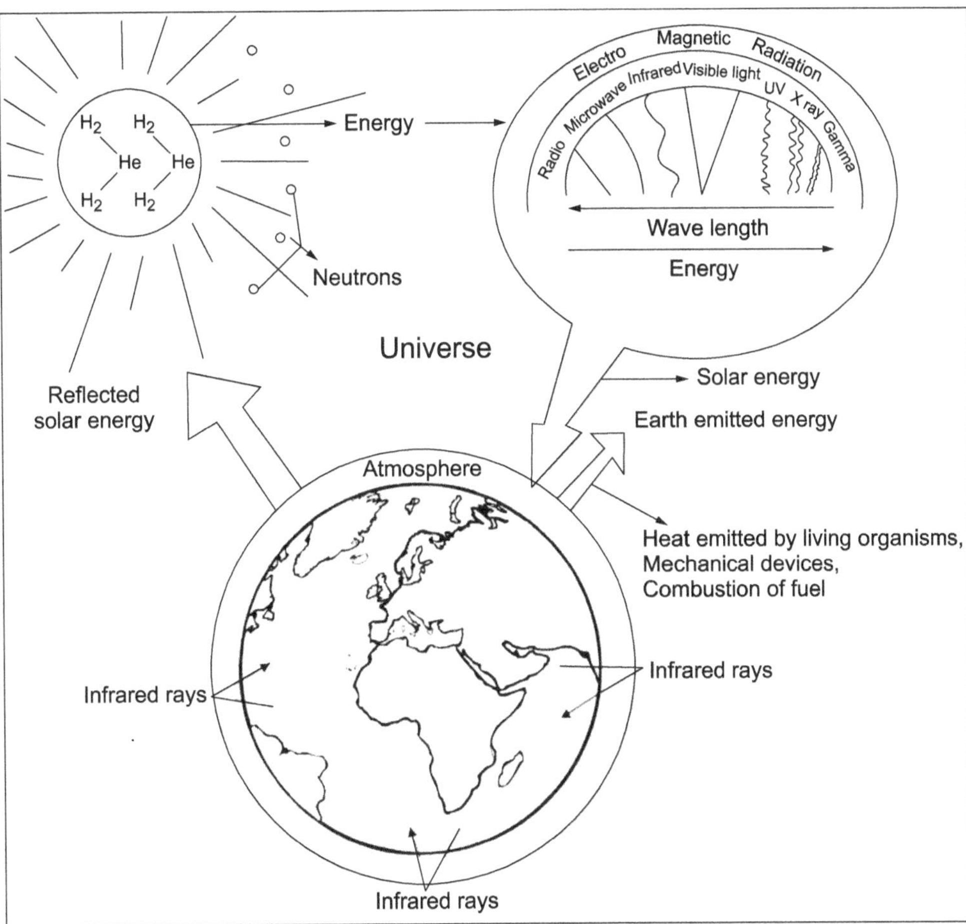

Figure 8.11: Electromagnetic radiation from sun and its utilization on earth. *In the sun, hydrogen atoms fuse to form helium. These nuclear fusions release huge amount of radiation called the electromagnetic radiation and neutrons. The electromagnetic energy consists of several kinds of rays of a wide range of wave length and energy. The electromagnetic radiation from the sun reaches the earth. The light energy is converted by chlorophyll containing organisms to chemical energy through photosynthesis. The plants and animals make use of this energy for doing work. Energy transformations during the activities of living organisms and mechanical devises release heat which to some extent is absorbed by earth and some of it is radiated to the universe. Some of the infra-red rays of the sun are reflected back to the universe. If the infra-red rays received from the sun and heat generated from activities on earth are prevented from being transmitted to the universe, the earth gets warmed up – a phenomenon termed "Global warming". Global warming has adverse effects on climate which are detrimental to life on earth.*

The light energy from the sun is utilized by plants during photosynthesis to synthesize its food and grow. The light energy is converted to chemical energy stored in the chemical bonds of the various biomolecules in the plant body. Animals feed directly or indirectly on plants, and thus the chemical energy in the food, is distributed in their body cells as well. The chemical energy is transformed to various other kinds of energy for the plants and animals to carry out their life activities.

Eventually, when the plants and animals die and naturally get buried for eons of time, they are converted to coal and petroleum. Coal is extracted from the coal mines. Crude petroleum is pumped out of the earth, and purified in an oil refinery, to various forms of automobile fuel like, petrol (gasoline), diesel, jet fuel, kerosene etc. Thus, light energy from the sun is stored as chemical energy in coal and petroleum.

8.8.4 Energy Changes During Movement of an Automobile

The energy transformations that result in movement of an automobile be it a car, truck etc. are detailed below (**Figure 8.12**).

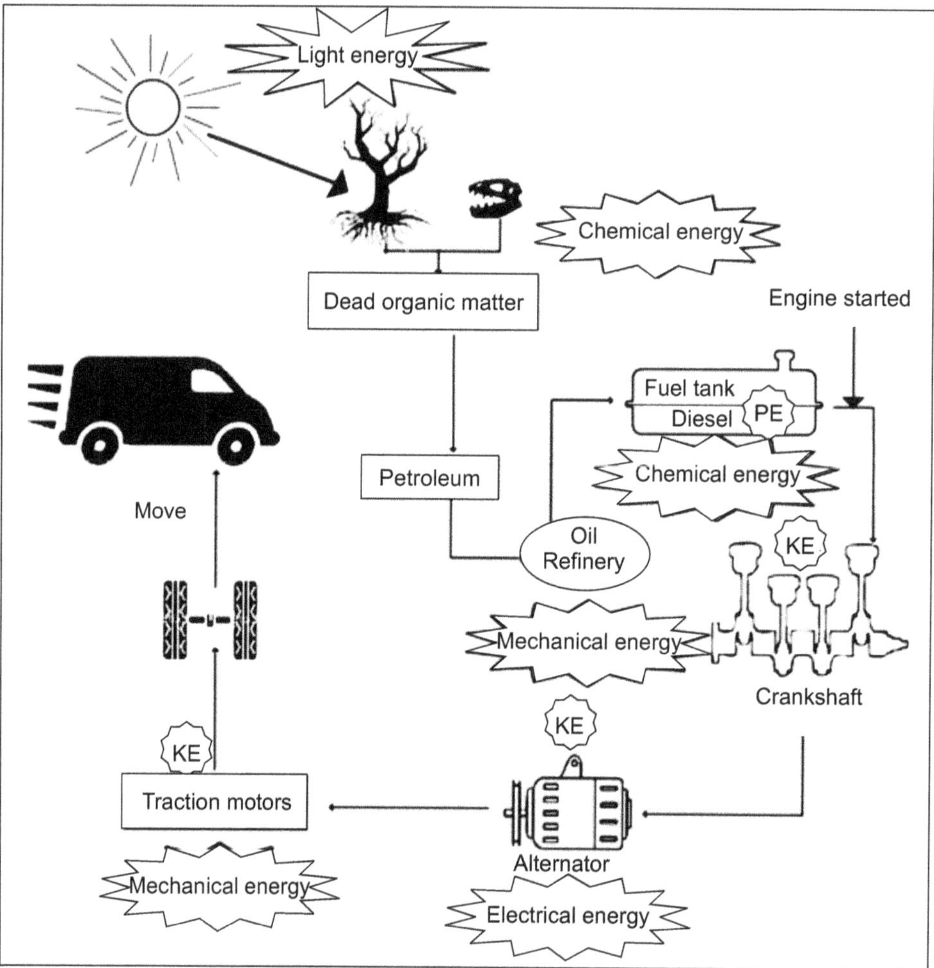

Figure 8.12: Energy transformations during the running of an automobile. *The light energy from the sun is converted to chemical energy by chlorophyll containing organisms on which all living organisms directly or indirectly depend for their energy needs. The dead organisms get buried under the soil or sea floor for millions of years. The pressure on the dead living matter converts them to coal and petroleum (fossil fuels). The crude petroleum has chemical energy. After refining, the diesel/petrol is used by automobiles. The chemical energy in the diesel is in the form of potential energy (PE) in the fuel tank of the automobile. When the engine is started, the diesel moves to the crank shaft where its energy is converted to mechanical energy. The movement of the crank shaft due to the mechanical energy causes it to be converted to electrical energy in the alternator connected to the crankshaft. The electrical energy is received by the traction motors attached to the wheels of the automobile. The traction motors convert electrical energy to mechanical energy resulting in the movement of the wheels and thus the automobile. Mechanical and electrical energy are a form of kinetic energy (KE).*

The chemical energy in the automobile fuel like gasoline (petrol) or diesel, is in the form of potential energy, as long as it is the fuel tank of the automobile. When the automobile engine is started, the chemical energy that remained as potential energy in the fuel, is converted to the kinetic form of mechanical energy, in the crankshaft. The moving crankshaft, transfers its mechanical energy to the alternator attached to it, which converts the mechanical energy to electric energy. The electrical energy goes to the giant traction motors, which convert the electrical energy, back to mechanical energy in the rotating mechanical torque. This mechanical energy is a form of kinetic energy that turns the wheels. The turning wheels move the automobile (**Figure 8.12**).

The best locomotive engines can transform only some proportion of the fuel energy into useful mechanical work. For example, in diesel engine run automobiles about 44% of the energy of the diesel they burn is converted into useful mechanical work of moving forward the automobile, nearly 33% goes out as hot exhaust gas and the remaining ~ 20% goes into the cooling water and engine oil. All of the fuel energy, even the part that's turned into useful work, eventually ends up heating the air which is radiated out into space. This explains the second law of thermodynamics – no energy transfer is efficient. During energy transformations a lot of energy is lost as heat.

8.8.5 Energy Transformations During Walking of an Animal

Most animals move from one place in another. The vertebrate animals move through muscle movement of the limbs. Let us examine the energy transformations during this process. An animal gets its food directly or indirectly from plants. The energy in the food is in the form potential chemical energy. The chemical energy in the food, is the result of transformation of light energy by plants, through photosynthesis. In the muscle of the animal, the chemical energy derived from food (the energy in the food is transferred to ATP to be in ready-to-use form of chemical energy during cellular respiration), is converted to mechanical energy and heat. The mechanical energy, being kinetic energy, causes contraction of the fibres in the muscle, and the legs move forward, resulting in walking. Some of the mechanical energy is converted to heat. This heat energy warms up the body. If walking is done for a considerable period of time, the body heats up too much, resulting in sweating in humans (**Figure 8.13**).

The working of any mechanical devise for e.g. a toaster, heater, grinder, motor, iron, gas stove, battery, involves energy transformations. Similarly, all living processes are carried out through energy transformations. No energy is lost but changed into different forms and distributed to different spaces.

Heat is a by-product in all energy transformations. In an animal, the heat generated due to work is used to maintain its normal body temperature. When excess work is done, and too much heat is generated, it is dissipated either through, sweat (humans), panting (dogs), fanning ears (elephants), etc. to cool down. All heat from mechanical devises or living organisms flows into the surrounding air heating it up. Eventually it is radiated back to space. The energy flow into earth is continuous from the sun to compensate for the energy radiated back to space. Thus, energy content of the total universe remains constant. Heat warms the atmosphere on earth. If the infra-red radiation from sun and the heat generated on earth is hindered from radiating into the universe (due to build-up of green-house gases in the earth's atmosphere), the earth gets heated up, a phenomenon we term as global warming. Global warming results in climate changes that make life difficult, or, even the earth uninhabitable ,if it is left unchecked (**Figure 8.11**).

In the word thermodynamics 'thermo' refers to heat or energy, 'dynamics' indicates movement or change. Thermodynamics talks about energy changes. Energy is interesting because it changes and during this process, gets work done.

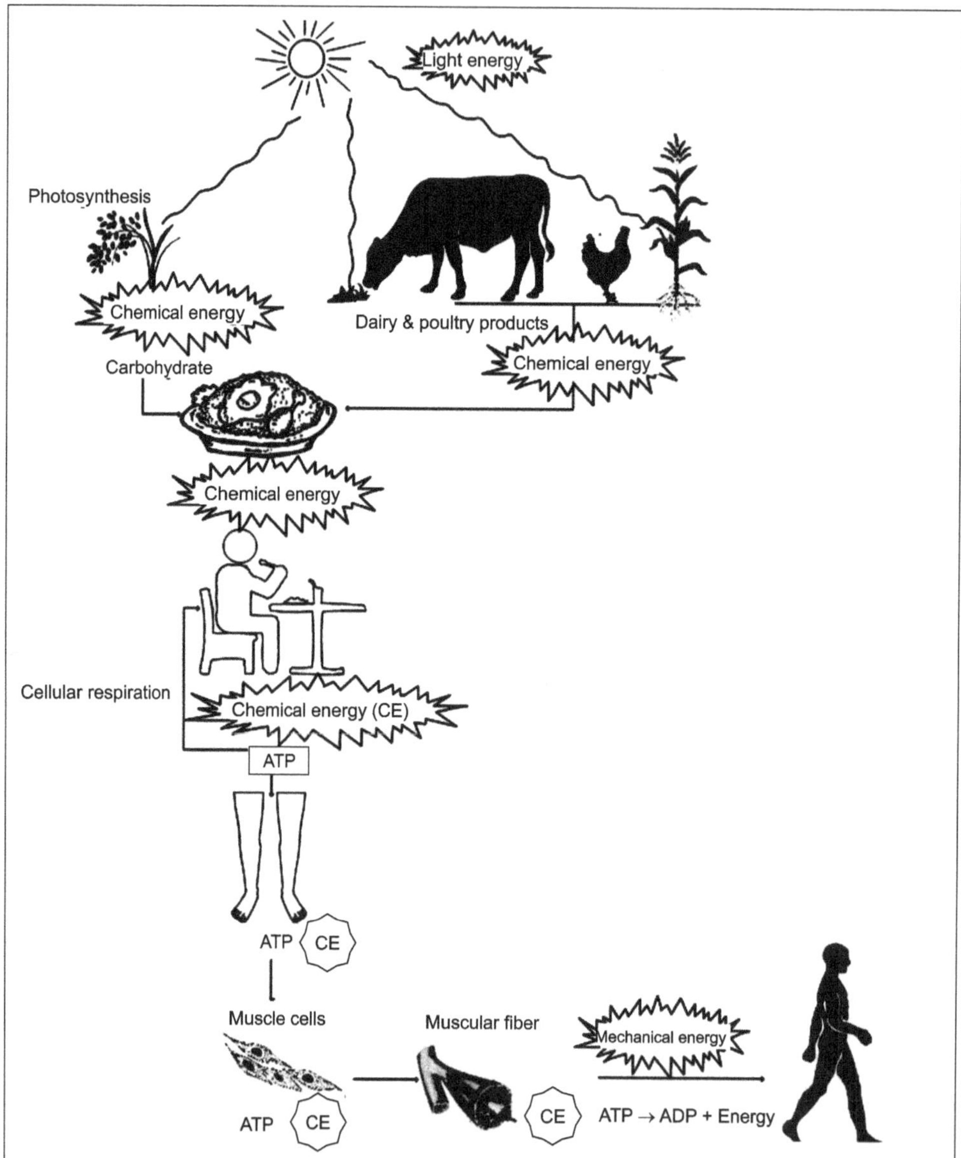

Figure 8.13: Energy changes involved during walking. *The light energy from sun is converted to chemical energy by plants. The animals derive food directly or indirectly from plants. The chemical energy in the food is transferred to ATP during cellular respiration; energy in ATP is in ready-to-use form. The chemical energy in the leg muscle cells is converted to mechanical energy in the fibres present in the muscle cells resulting in the contraction of the leg muscle and walking.*

8.9 SUMMARY

1. Thermodynamics is a branch of physical science that explains the relationship between all forms of energy: mechanical, electrical and chemical energy. Essentially, it is the study of energy, and its conversion from one form, to another. Principles of thermodynamics form the foundation

for understanding biological systems. Thermodynamics is about storage, transformation and dissipation of energy. Cells store energy; they transform it and dissipate it to carry forward unfavourable reactions. Thus, all physical, chemical and biological phenomena are regulated by the laws of thermodynamics. In living cells, thermodynamic changes are essential for biological functions.

2. All energy exchanges that take place in a living being during various metabolic reactions, and between the living organism and its surrounding, can be described, by the same laws of physics as energy exchanges between hot and cold objects, or gas molecules. The first law of thermodynamics, also known as the law of conservation of energy, states that, 'energy can neither be created nor destroyed'. It may change from one form to another. Plants execute the most important biological energy transformation on earth. Through photosynthesis, they trap the radiant energy from sunlight and convert it into chemical energy, stored within the resultant glucose. The energy stored in glucose can be released through cellular respiration. Cellular respiration allows plants and animals to access energy stored in biomolecules like carbohydrates, lipids etc. However, none of the energy transfers that take place during the sequence of metabolic reactions in the cell are totally efficient. During transfer, some of the energy is lost as thermal energy (heat).

3. The second law of thermodynamics states that, 'every energy transfer that takes place, will increase the entropy of the system and reduce the amount of usable energy available to do work'. Biological systems are open systems, where material and energy are exchanged with the surroundings. They remain highly organized – the cells have a well-organized internal structure. Through continuous uptake of energy rich food, the living systems maintain order in their body. To offset entropy in living organisms, energy input should exceed the energy used for doing work and the energy lost.

4. Among the living organisms, only the photo- and chemo- synthetic organisms can produce their own food. All other organisms depend on them directly or indirectly for their food (energy requirements). They are called the consumers. Consumers are of different trophic levels: herbivores, 1st order consumers, 2nd order consumers and finally top consumers. Due to utilization of energy to carry on life processes (metabolism), and a great loss of energy during energy transformations, the amount of energy available to the next trophic level in a food chain decreases from one trophic level to the next.

5. Exothermic reactions are those in which, heat is released due to the reaction. In these reactions, the products of the reaction have less energy than the reactants. When the energy of the reactants of a reaction is lesser than that of its products, for the reaction to happen, energy in the form of heat is absorbed. Such types of reactions are described as endothermic. Therefore, activation energy (energy required for a reaction to proceed) of an exothermic reaction is less than that of an endothermic reaction.

6. The energy changes in chemical and biological reactions are described by three quantities: enthalpy, entropy and free energy. Enthalpy is the total heat content of a system. It depends on the number of chemical bonds in a substance. Enthalpy is denoted as H and change in enthalpy is denoted as ΔH (delta H). Thus, ΔH is positive for endothermic reactions (more than zero) and negative (less than zero) for exothermic reactions. Entropy (S), is the level of disorder or randomness in a system. Entropy indicates the amount of energy that is unavailable during energy transformations. Free energy refers to the amount of energy available in a thermodynamic system during a chemical reaction to do cellular work. If the reactions take place under constant

pressure and temperature, which is the case with open systems like the biological systems, free energy is referred to as Gibbs free energy (G). Thus, Gibbs free energy combines enthalpy and entropy into a single value. Gibbs free energy indicates whether a reaction is favoured or not.

7. An exergonic reaction releases work energy. An endergonic reaction takes in work energy. Exergonic reactions are spontaneous, catabolic and coupled to ATP formation while endergonic reactions are non-spontaneous, anabolic and coupled to ATP hydrolysis. Exo/endo- thermic, indicate the relative change in heat/enthalpy in a system, while exer/ender- gonic, denote the relative change in the free energy of a system

8. If not interfered with, all chemical reactions come to an equilibrium. This state of equilibrium can be described by the equilibrium constant, K_{eq}. The relation between equilibrium constant directly to changes in enthalpy and entropy is given by the equation $\Delta G° = - RT \ln K_{eq}$ where $\Delta G°$ is Gibbs free energy R is the ideal gas constant: 8.314 J/mol K, and T is the temperature in Kelvin.

9. A spontaneous reaction is one that occurs without the addition of external energy. Once started, it continues on its own without input of energy. A non-spontaneous process needs a continual input of energy. From the value of ΔG (change in free energy of a system), it can be predicted whether the reaction is spontaneous or not. When ΔG is negative i.e., $\Delta G < 0$ the reaction can start spontaneously in the forward direction to form more of the product. When ΔG is positive i.e., $\Delta G > 0$, the reaction cannot start spontaneously in forward direction but the backward direction is spontaneous making more of the reactants. Thus, a spontaneous reaction is one that moves a chemical reaction to equilibrium. A non-spontaneous reaction is one which moves the system away from equilibrium. When ΔG is 0, the reaction is at equilibrium and the concentration of the reactants and the products remain constant.

10. For a reaction to take place, the energy of the reactants must be raised so that they reach an unstable transition state. This initial energy, that is supplied for a reaction to take off, is termed the activation energy. Activation energy is required for both exergonic and endergonic reactions. More activation energy is required for endergonic than exergonic reactions just as the case with exothermic vs endothermic reactions. In living systems, the reactions take place at a reasonably fast rate because the activation energy of a reaction is lowered by the enzymes which catalyse the reaction.

11. The covalent bonds in the food molecules are of low energy and not very useful to do most of kinds of work in the cells. Therefore, the energy released from the breaking of the chemical bonds in food is stored in a nucleotide called the Adenosine tri phosphate (ATP). All life processes that require energy are driven utilizing the energy stored in ATP. Therefore, ATP is referred to as the energy currency of the cell. The bonds between the phosphate groups in ATP are high-energy bonds and therefore unstable. The energy contained in these bonds is readily released when ATP is hydrolysed in cellular reactions.

12. Metabolic reactions in living organisms can be energy consuming or energy yielding. Photosynthesis is energy consuming, utilizing solar energy and converting it into chemical energy. Photosynthesis occurs in chlorophyll containing organisms. It involves two sets of reactions – the light reactions and dark reactions. During this process the solar (light) energy is fixed in the chemical bonds of glucose that is synthesized from CO_2 and water.

13. Respiration is the process of breaking down food molecules (glucose) to release energy. There are three steps in this process: Glycolysis, Krebs cycle and electron transport chain. The first step

occurs in the cytoplasm of the cell and the next two steps happen in the mitochondria. A total of 38 ATP molecules are formed from the oxidation of one glucose molecule during cellular respiration.

14. Work involves not expenditure of energy, but it happens due to change or transformation of energy, from one form to another. The change of energy to different forms can be visualised in several processes like hydroelectric and thermoelectric production, running of an automobile and walking of an animal.

8.10 SAMPLE QUESTIONS

8.10(a) Subjective Questions

Q.1. State and explain the laws of thermodynamics. Describe how they apply to biological systems.

Q.2. What is a spontaneous reaction? How can it be explained based on free energy, entropy and enthalpy?

Q.3. With two examples explain the concept of transformation of energy. How do they relate to law of thermodynamics?

Q.4. Describe briefly the reactions of energy consuming and energy releasing reactions in living beings. Add a critical note on ATP.

8.10(b) Objective Questions

Q.1. Electromagnetic radiation is
   (*a*) Radiation due to magnetic properties of earth.
   (*b*) Constitutes the energy from the sun containing waves and light of different wavelengths.
   (*c*) Are produced due to cosmic rays.
   (*d*) None of the above.

Q.2. Dark reactions of photosynthesis
   (*a*) Occurs during the night.   (*b*) Are light independent.
   (*c*) Occurs in the grana of the chloroplast. (*d*) Involve capture of solar energy.

Q.3. Activation energy is
   (*a*) High for endergonic reactions.   (*b*) Lower in endergonic reaction.
   (*c*) High in exergonic reaction.
   (*a*) Is the same for exergonic and endergonic reactions.

Q.4. Spontaneous reactions
   (*a*) Happen at high speed.   (*b*) Are mostly exothermic.
   (*c*) Are mostly endothermic.   (*d*) Always move toward equilibrium.

Q.5. Exergonic reactions are
   (*a*) Catabolic.   (*b*) Anabolic.
   (*c*) Non-spontaneous.   (*d*) Coupled to ATP utilization.

Q.6. Free energy of a reaction
   (*a*) Indicates whether a reaction is favoured.
   (*b*) Indicates if heat is released during a reaction.

    (*c*) Is energy released during a reaction.

    (*d*) Is decreased in an enzyme catalysed reaction.

Q.7. Exothermic reactions

    (*a*) The products have lower energy than the reactants.

    (*b*) Require higher activation energy.

    (*c*) The reactants have lower energy than the products.

    (*d*) Energy is absorbed.

Q.8. Organisms at higher trophic level

    (*a*) Have more available energy than the preceding trophic level in the food chain.

    (*b*) Have less available energy than the preceding trophic level in the food chain.

    (*c*) Support more organisms than the preceding trophic level in the food chain.

    (*d*) Are always herbivores.

ANSWERS

1. (*b*)　　**2.** (*b*)　　**3.** (*a*)　　**4.** (*d*)　　**5.** (*a*)　　**6.** (*a*)　　**7.** (*a*)　　**8.** (*b*)

9 Microbiology

I believe that thirty million of these animalcules together would not take up as much room, or be as big, as a coarse grain of sand. **~ Antonie Van Leeuwenhoek**

"When Orson Welles said "We're born alone, we live alone, we die alone", he was mistaken. Even when we are alone, we are never alone. We exist in symbiosis – a wonderful term that refers to different organisms living together. Some animals are colonised by microbes while they are still unfertilized eggs; others pick up their first partners at the moment of birth. We then proceed through our lives in their presence. When we eat, so do they. When we travel, they come along. When we die, they consume us. Every one of us is a zoo in our own right – a colony enclosed within a single body. A multi-species collective. An entire world." **~ Ed Yong**

9.1 INTRODUCTION

Microbiology is a science that involves the study of microbes – organisms that are too small to be seen individually by the naked eye. Though invisible, their numbers far exceed the gregarious organisms of the living world. They are however, less diverse than the macro living world. Microbes are characterized by very rapid multiplication rates. They do not lead a solitary life, but, live as a population, in the form of colonies or communities. The entire prokaryotic world and protists are microbes. A sub group of the sac fungi consisting of yeasts, are microbes. They inhabit any and all places on earth: soil, water and air, including places of extreme environment like hot springs, ocean vents, frozen lands (Arctic zone and Antarctica) and very salty water bodies (dead sea).

Microbes are thus, abundant and ubiquitous. Their effect on environment and other living organisms is profound. They are the engines of mineral recycling on earth (biogeochemical cycles). They are the sinks of carbon di oxide in the oceans and water bodies. They constitute the first-tier primary producers that support all life in these aquatic ecosystems. Curdling of milk, making of cheese, bread, and beverages, be it, beer, brandy or wine, infectious disease epidemics and pandemics on crops, animals or humans are all because of the microbes. They are source of drugs and are used as live factories, to produce several drugs and enzymes, through fermentation technologies.

Based on fossil evidence, it is clear that the prokaryotes are the first living beings on earth, appearing 3.5 to 3.8 billion years ago. They are estimated to constitute, more than 50% of the total protoplasmic biomass, on earth. It is estimated that, there are 100,000,000 (hundred million/10 crore) times, more microbial cells on the earth, than there are stars in the observable universe. This is astounding, since they are invisible to the naked eye. Their existence became evident only after the discovery of the microscope by

Anton Von Leeuwenhoek. An adult human is made up of approximately 10 trillion cells. It is estimated, he harbours ten times that number, i.e., 100 trillion of microbes.

9.2 CONCEPT OF SINGLE CELLED ORGANISMS

Cells are called the building blocks of life because plants and animals are made up of thousands of cells. However, most microbes are just one celled. They are the unicellular or single celled organisms. These organisms do perform many of the complex activities carried out by large multicellular organisms. They respire, feed, digest, excrete, and reproduce like a typical multicellular animal. Unicelled plant-like protists like microalgae, diatoms and some prokaryotes, synthesize their own food through photo- or, chemo- synthesis. Unicellular organisms are, 'smoking gun evidence' of the cell theory, which states that, cells are the basic functional units in the living world. Being single celled, they are microscopic. The viruses, among the microbes, are visible only under the electron microscope. The viruses are distinctive in that, they cannot be described by conventional parameters, as living organisms. They have no cell metabolism. Viruses reproduce only within a living host, using the metabolic machinery of the host cell.

9.3 CLASSIFICATION OF MICROORGANISMS

Among the three domains of the living world, Bacteria and Archaea are exclusively microbial. The third domain of Eukarya, consists of four kingdoms: Protista, Fungi, Plantae and Animalia. Protista is predominantly constituted by unicellular microbes, with the exception of macroscopic algal groups. A few members of fungi are microbes. Kingdoms of Animalia and Plantae do not have unicellular microbes.

9.3.1 Classification of Single Celled Prokaryotes: Bacteria (Eubacteria) and Archaebacteria

A reliable classification of prokaryotes based on external characters is difficult, because they have simple structures. Before the advent of molecular phylogeny (classification based on DNA sequences), bacterial classification was based on results of Gram staining. Named after its discoverer, Gram staining involves, staining bacteria with crystal violet, adding iodine, washing with ethanol and counter staining with safranine or fuchsin. Some species of bacteria stain violet (Gram positive), while some species lose the violet colour during ethanol treatment, and take the red colour of safranine or pink of fuchsin stain (Gram negative) (**Figure 9.1**). Berge's manual of "Determinative Microbiology (9th edition,1994)", follows the classification of bacteria based on Gram staining. It was essentially designed to help identification of prokaryotic microbes. In this classification, the prokaryotes are classified into four major categories and 35 groups (**Table 9.1**).

Table 9.1: Berge's classification of eu- and archaea- bacteria

Category	No of groups and examples
Gram negative bacteria (stain pink/red with Gram's stain)	16 groups (1 to 16) e.g. sulphate & sulphur reducing bacteria, *Rickettsia*, *Chlamydia*, anoxygenic and oxygenic phototrophic bacteria (cyanobacteria, also called blue green algae).
Gram positive bacteria (stain violet with Gram's stain)	13 groups (17 to 29) e.g. endospore forming rods and cocci, non-spore forming rods, Mycobacteria, Actinomycetes, Streptomycetes.
Gram variable bacteria lacking cell wall (mix of pink/red and violet cells)	One group (30) e.g. Mycobacterium, Mycoplasma.

Archaebacteria (uneven Gram stain)	Five groups (31 to 35) e.g. Extremophiles like extreme halophytes, thermophilic and hyper thermophilic sulphur reducing bacteria, methanogens.

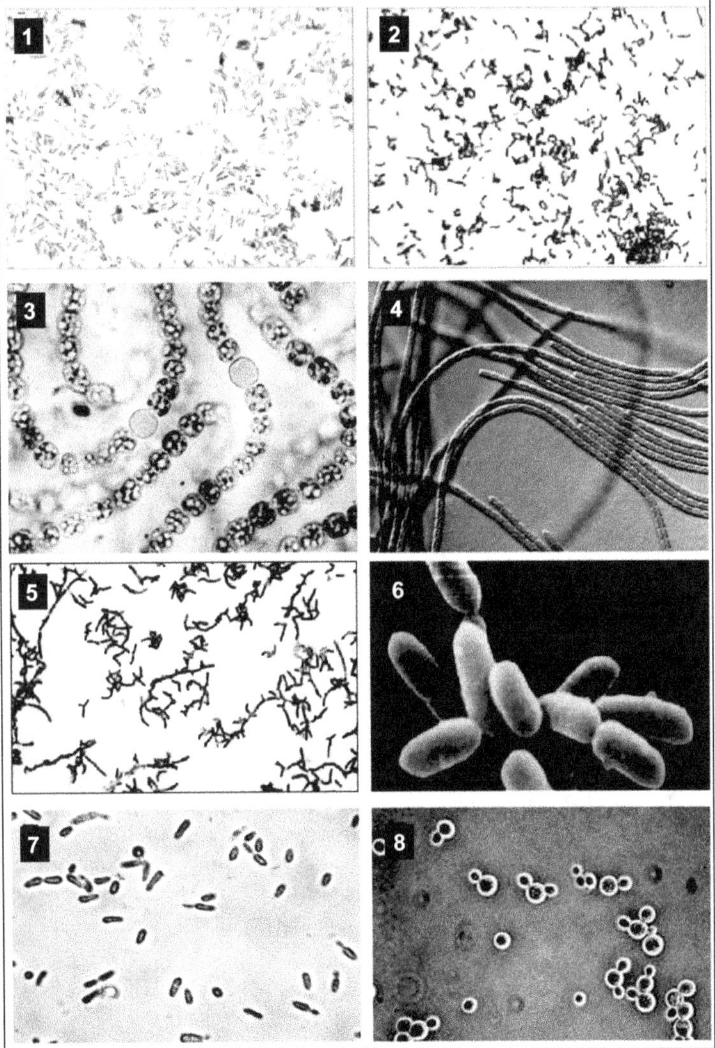

Figure 9.1: Eu-& archae- bacteria and eukaryotic microbes. 1&2. *Gram negative and Gram positive bacteria respectively.* **3&4.** *Photosynthetic cyanobacteria, Nostoc and Oscillatoria respectively.* **5.** *Actinomyces spp.* **6.** *Archaebacteria: Halobacterium spp.* **7&8.** *Eukaryotic microbes: fission and budding yeast (Saccharomyces spp.) respectively.* Picture credits: Figures 1&2. https://ib.bioninja.com.au/options/untitled/b1-microbiology-organisms/gram-staining.html; Figure 3. https://www.pinterest.com/pin/523754631645291204/; Figure 4. http://cfb.unh.edu/phycokey; Figure 5. https://microbe-canvas.com/uploads/image/bacterien/actinomyces-graevenitzii/actinomyces-graevenitzii_an-74-75_cult-3-2_f-350x220.jpg; Figure 6. https://karaskingdoms.weebly.com/archaebacteria.html; Figures 7&8. https://wineserver.ucdavis.edu/sites/g/files/dgvnsk2676/files/inline-images/s.bayanus40x.jpg. *Magnifications are not given since the pictures were resized to fit the figure plate. The source images may be referred for magnification.*

The archaebacteria are considered prokaryotes, as they have no membrane system within the cell and differ from the eubacteria in many features. A unique feature of archaebacteria is, most of them inhabit places with extreme environments like, very high/low temperature, high salinity, high acidity etc. They are extremophilic. Photomicrographs of some eubacteria and archaebacteria are shown in **Figure 9.1**.

Since bacteria are microscopic, and do not have many physical differences, to distinguish from each other, there is uncertainty in the classification based on external and biochemical features alone. To overcome this difficulty, classification that is presently followed is based on molecular (DNA) features. In 1987, Carl Woese divided Eubacteria, into 11 divisions, based on the gene sequence coding for 16S ribosomal RNA of ribosomal SSU – small subunit of the ribosome.

9.3.2 Classification of Microbes in Protista

Eukaryotic organisms which cannot be placed in any of the three groups of Eukaryota – fungi, plantae or animalia are placed in a separate group called Protista. It is thus an artificial group. Not all members of Protista are microbes. The heterotrophic animal like and fungi like protists are unicellular. In the autotrophic plant like protists, some phyla are exclusively microbial, some have a few microbial members, while some phyla have only multicellular organisms. The classification of the microbial members of Protista is given in **Table 9.2**. Some of the members of this group are shown in **Figure 9.2**.

9.3.3 Classification of Microbes Among Fungi

Among the fungi, the yeasts are unicellular, e.g. *Saccharomyces spp.* (Baker's yeast) (**Figure 9.1**) and *Candida* (pathogenic yeast). They belong to the subclass Hemiascomycomycetidae of the class Ascomycetes (sac fungi).

The viruses are pseudo living, as they do not have the conventional properties of a living organism in the physical world. They reproduce, only when they inhabit the cells of a living host. Hence, they are not discussed here.

Table 9.2: Microbes in Protista

Type	Phylum

Plant like (autotrophic) (ALGAE)	Phyla having exclusively unicellular organisms
	1. Euglenophyta: *Euglena*
	2. Chrysophyta (diatoms)
	3. Pyrrophyta (dinoflagellates)
	Phyla with unicellular and multicellular forms
	1. Chlorophyta
	  Orders Volvocales and Chlorococcales are mostly constituted by unicellular or colonial phytoplankton. e.g. *Chlamydomonas, Chlorella, Volvox, Pediastrum.*
	2. Rhodophyta
	  Orders Cyanidialies (e.g. *Cyanidioschyzon merolae* and Rhodellales (e.g. *Rhodella*) have a few unicellular organisms.

Animal like (heterotrophic) (PROTOZOA)	1. Sarcodina: amoeba 2. Ciliophora: *Paramecium* 3. Zoomastigophora: *Trypanosoma* 4. Sporozoa: *Plasmodium*
Fungi like (heterotrophic)	Myxomycota: slime moulds (can live as independent single cells and also aggregate to form colonies). Oomycota: watermolds (basically unicellular – the body is filamentous with numerous nucleic with no cell membrane separating them – described as coenocytic).

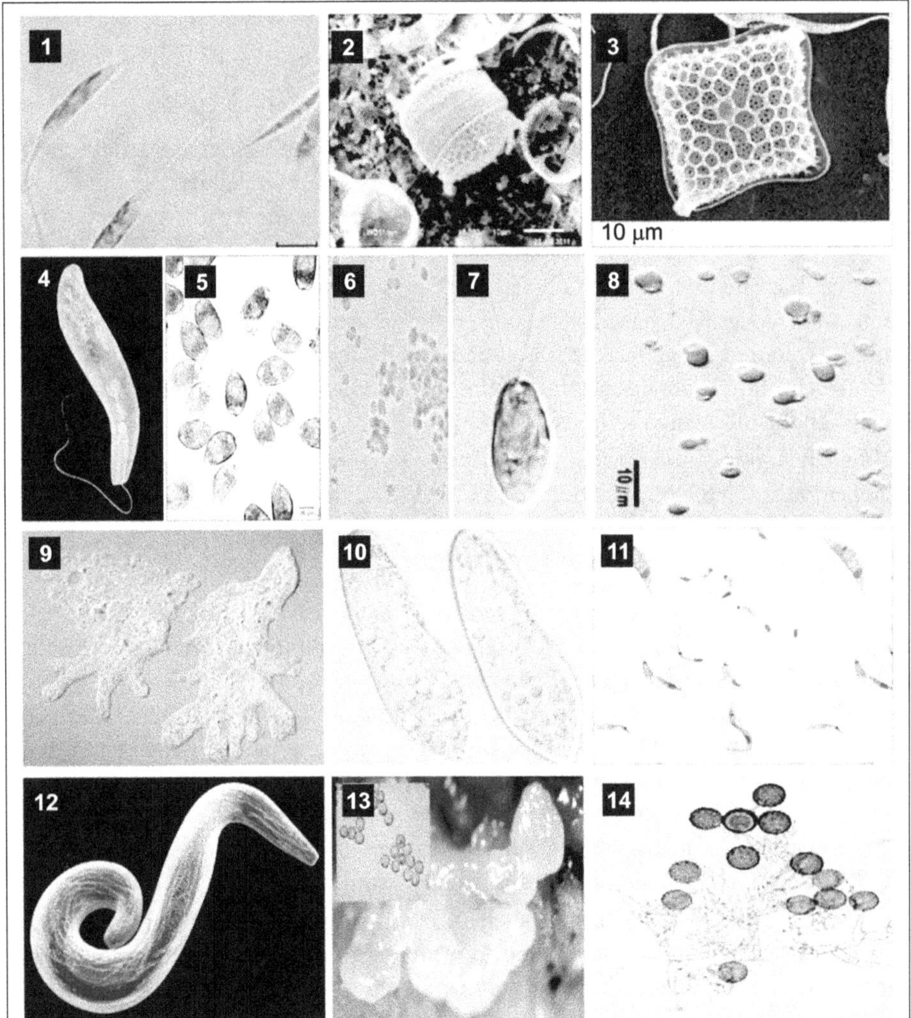

Figure 9.2: Single celled Protists. 1 to 3. Diatoms: **1.** *Cylindrotheca*. **2&3.** *Two views of SEM (Scanning electron microscope) images of Odontella.* **4.** *Euglenophyta – Euglena.* **5.** *Dinoflagellate– Procentrum.* **6&7.** *Chlorococcales – Chlorella* and *Chlamydomonas respectively.* **8.** *Rhodophyte – Cyanidioschyzon.* **9 to 12.** *animal like Protists*: **9.** *Amoeba.* **10.** *Paramecium.* **11.** *Trypanosoma.* **12.** *Plasmodium.* **13 &14.** *Fungi like Protists.* **13.** *Slime mold with cells under microscope in the inset.* **14.** *Water mold.* Picture credits: Figs 1,2,3,6 & 7 from the author's lab; Figure 4. https://aquariumstoredepot.com/blogs. Figure 5. Procentrum not known. Figure 8. https://alchetron.com/Cyanidioschyzon-merolae. Figure 9. https://www.sciencenewsforstudents.org/article/scientists-say-amoeba. Figure 10. https://pulpbits.net/8-paramecium-images/paramecium-cell/. Figure 11. https://www.pinterest.com/pin/506232814335553200/. Figure 12. https://thenativeantigencompany.com/products/plasmodium-falciparum-msp1-protein/. Figure 13. https://www.youtube.com/watch?v=Nx3Uu1hfl6Q. Inset https://www.discoverlife.org/. Figure 14. http://website.nbm-mnb.ca/mycologywebpages/Moulds/Examination.html. Magnification of some of the images is not given as the original photomicrographs were resized to fit the picture plate. Sizes may be referred from the original image.

9.4 IDENTIFICATION OF MICROBES

9.4.1 Identification of Bacteria

Identification of bacteria, is to determine to which group within a classification scheme, a bacterium of interest belongs, and what genus/species/strain it is. Compared to multicellular organisms, identification of bacteria is a challenging task because of their small size and less diversity with respect to morphological features.

Culture of the microbe in different media are set up as a first step for identification. The microbe in culture, is sometimes recognized, based on certain end products of its metabolic reactions, either because they are originally coloured, or, an indicator is added in the culture medium, that stains the end product. From culture results, a hint is obtained about of the type of microbe. An authentic identification is then carried out through the following methods:

1. Traditional: phenotypic and biochemical, **2.** Immunochemical: serological, **3.** Genetic: single strand DNA-DNA hybridization (SS DNA hybridization), % G+C content, 16S rRNA gene sequence.

9.4.1(a) Traditional Methods for Identification

Phenotypic characterization is done both at macroscopic and microscopic level. At macroscopic level, cultures on solid medium are characterized for their colony morphology – size, shape, margin (regular or irregular), surface (even or elevated), colour and consistency. In liquid medium, following characteristics are examined: degree of growth (absence, scanty, moderate, abundant etc.), nature of growth on the surface of the liquid medium, odour, presence of turbidity and its nature, presence of deposit and its character. The physical conditions required for growth in culture medium, such as, requirement of oxygen (aerobic or anaerobic), optimum temperature, pH, nutritional requirements, are also helpful in phenotyping the microbe. At microscopic level, the features considered are, size and shape of the cell, cellular composition, sporulation, Gram staining and cell motility.

Preliminary phenotypic characterization, such as Gram staining and growth characteristics, give hint, as to the panel of biochemical tests that need to be done for identification. For biochemical characterization, a large number of tests are available. They usually reveal the enzymatic properties of the microbe. Not more than 15 biochemical tests are required to identify a bacterium up to species level.

Phenotypic-biochemical tests are at three levels: **a.** universal, **b.** differential and **c.** specific. Universal tests are preliminarily done for every microbe to be identified. The results of these tests, serve as a guide, to choose the set of appropriate tests, that are required to be done next, to get a reliable identification. The differential tests are a common set of tests done in combination to identify a microbe up to species level. A single differential test is not sufficient; usually, a set of tests are done for authentic identification. Specific tests are biochemical tests that are specific for a particular species or sub-types within the species. A single specific test would be sufficient to identify the organism. Several developments to improve classical biochemical tests have been devised and automated equipment is available to conduct the tests.

Phenotypic methods are most widely used for identification, due to their relatively lower cost, compared to genotypic tests. However, the difficulty with phenotypic tests is that, the characters examined in these tests, may not always be consistent, being dependant on the type of culture medium used, and growth conditions. Genotypic characteristics are not influenced by these factors. It is also not possible, to artificially culture all types of microbes.

9.4.1(b) Immunochemical Methods for Identification

The immunochemical tests involve, analysing the antibody elicited by the microbe, which acts like an antigen. For these tests, the serum is assayed. Therefore, they are also called serological tests. The immunoassays used are: precipitation and agglutination tests, enzyme immunoassays (ELISA- Enzyme linked Immuno Sorbent Assay), direct and indirect fluorescence immune assays (FIA) and flow cytometry.

9.4.1(c) Genotypic Methods for Identification

Genotypic methods have become indispensable for reliable identification of microbes. These are the only methods for identification of microbes, which do not lend to artificial culture, and, those which grow extremely slow in culture. The methods used are: estimation of %G+C content, single stranded (SS) DNA-DNA hybridization method, and sequencing of 16S rRNA gene.

9.4.1(c)i Estimation of G+C Content of DNA

This was the first nucleic acid technology applied to prokaryote systematics, and initially proved to be a useful and routine way, of distinguishing between phenotypically similar, but genomically different, strains. It is one of the usual genome characteristics recommended for the description of species and genera. The G+C content is calculated as a percentage: $\%G+C = [G+C/ G+C+A+T] \times 100$. The G+C content in DNA is also reflected in its melting (becoming single stranded on heating) point (T^M) – the higher the G+C content, the higher is the T^M (because in DNA, there are three bonds between G and C while only two bonds between A and T).

9.4.1(c)ii (SS) DNA-DNA Hybridization Method

This method is an indirect measure of genomic (total DNA) similarity between organisms compared. Single stranded (SS) DNA from two different microbes is mixed in solution and the percentage that has hybridized and become double stranded is measured. A high hybridization percentage indicates, that the two microbes share similar nucleotide sequences. This is a standard technique for species delineation, because, it was found to correlate well with phenotypic similarity in numerous studies. The DNA is made single stranded by denaturing it, through heating to above 100°C and immediately placing on ice, or, treating it with a strong acid or alkali.

9.4.1(c)iii 16S rRNA Gene Sequencing

The 16S rRNA approach, is one of the most widely used standard techniques in microbial taxonomy. A comprehensive set of several lakhs of 16S rRNA gene sequences is available in widely accessible databases. These are helpful for in comparison to know the relatedness of a bacterium to other known bacteria. Bioinformatic tools are used to compare the gene sequences and group the related sequences (organisms)together. The 16S rRNA gene of the microbe to be identified is sequenced to compare with the sequences in the rRNA sequence databases.

9.4.2 Identification of Eukaryotic Microbes

9.4.2(a) Identification of Microbial Protists

Identification of eukaryotic microbial species is comparatively much easier than prokaryotes, because of the larger cell size. Examination of the cell morphology under the microscope, is sufficient to identify up to species level. The phylogenetic relationship between the different species and strains within a species, is determined from comparison of sequences of marker genes. The most common marker gene used is the 18S rRNA gene.

9.5 CONCEPT OF SPECIES AND STRAINS

Species is the basic unit in biological classification. A species is defined as a group of organisms capable of interbreeding and producing fertile progeny. Prokaryotes multiply asexually and there is no typical sexual reproduction. So, in prokaryotes, defining species boundaries based on reproductive compatibility (ability to breed) is not possible. The most commonly accepted definition of a species in bacteria, is the polyphasic species definition, which takes into account as many parameters as possible – phenotypic, biochemical and genetic to circumscribe a species. In prokaryote taxonomy, a species is considered to be *'a group of strains that show a high degree of overall similarity and differ considerably from related strain groups with respect to many independent characteristics'.*

Prokaryotes constitute less than 1% of the total described species in the living world. This is an unbelievably small number, given their long evolutionary history, the fact that they are the first living organisms to have originated on earth (3.5 billion years ago), and, the great genetic diversity with which they are endowed, which enables them to reside in all ecosystems on earth. The small proportion of bacteria in the described species of the biological world is due to two reasons: 1. To recognize a prokaryote as a new species requires that, it be cultured as a pure culture, which in many instances is quite difficult, or, sometimes even impossible, 2. Extensive exchange of genetic material between species occurs in prokaryotes through lateral or horizontal gene transfer (from one species/strain to another) as against the vertical (from parent to offspring). This blurs the distinctions between species.

A strain is a genetic variant, a sub-type, or a culture variant within a biological species, which can be distinguished from other organisms within the species, based on certain characteristics. A strain comprises of descendants from a single spore/cell culture. Different strains with close resemblance together form a species. They can be differing in morphological features (morphovars), or biochemical characters (biovars), or antigenic properties (serovars), or genotypically.

To determine which strains, belong to a species, a "type strain" is used for reference. A type strain is a live strain, that serves as the representative of a species. The type strain of a species, is often the strain which was first discovered and it may not be very "typical" for a given species. The pure culture (constituted by descendants from a single cell) of a type strain is deposited in a culture collection and is available to all scientists. There are several type culture collections: e.g., American type culture collection (ATCC), USA, Microbial type culture collection (MTCC) maintained by IMTEC of Chandigarh, India. Culture inoculum of a type strain can be obtained from the culture collection by a scientist, for identification of bacterial strains. Strains which are very similar to the type strain are all grouped in one species. In addition to the morphological, biochemical and antigenic properties, genetic relatedness is considered, to determine the similarity of a microbe to a type strain and to include it in a particular species. A threshold value of similarity is set for each genotypic feature, and a strain which has similarity with the type strain above the threshold value for all the characters considered, is included in that species. For SS DNA-DNA hybridization, a similarity of more than 70% with the type strains qualifies the microbe to belong to that species. Among the prokaryotes, G+C contents vary between 20 and 80 mol%. A 5 mol% is the common range of difference in G+C content found within a species, which means a difference of less than 5% in G+C content (or 5 to 7°C difference in DNA melting point), from the type strain, would mean that the microbe is a strain of that species. A similarity of above 97% in 16S rRNA sequence to the type strain is required, for a microbe to be placed within the species of the type strain.

There is general agreement among the microbial taxonomists that, the currently adopted species concept is useful, pragmatic and universally applicable within the prokaryotic world.

9.6 MICROSCOPY

A microscope is a must, to observe microbes, as they are too small to be seen by naked eyes. It is a devise, which magnifies the image of a specimen and enables to visualize it, at a good resolution. The image can be seen with our eyes, or is projected as a digital image on a computer screen. Pictures of the images can be captured with a camera and these are called photomicrographs.

There are basically two types of microscopes: optical and electron microscopes. Optical microscope, also called light microscope, uses light (photons) and has glass lenses for magnification. Electron microscope uses fast moving electrons and has electro-magnetic lenses. The magnifying power of an electron microscope, is much higher than the optical microscope, because, the electrons have much shorter wave length, than photons. The resolving power of the microscope, depends on the wave length of the illuminating particles. The images formed in the electron microscope are visualized on a screen. Stains are used to enhance the clarity of the images in light microscopy while they are not of use in electron microscopy. Elaborate methods are used to prepare the specimen to examine under an electron microscope. Both the microscope and specimen preparation are much more expensive in electron microscopy compared to light microscope. Two types of electron microscopes are commonly used. The transmission electron microscope (TEM) and scanning electron microscope (SEM) which gives images of the surface topology of structures.

9.6.1 Optical Microscopy

Several kinds of optical microscopes are available. The most commonly used optical microscopes are compound microscopes which have two lenses – an objective lens with a very short focal length and an ocular lens (eyepiece), with a longer focal length, both mounted in the same tube. The compound microscope has three components:

Magnifying parts — Used to enlarge the specimen (eye piece and the objective).

Illuminating parts — Provide light (source of light, condenser with iris diaphragm, mirror to focus light).

Mechanical parts — Provide support and adjust the above two parts (stage for slide, knobs to adjust the resolution of the image, knobs to adjust the height of the condenser tube to hold the eye piece and objectives, nose piece holding several objective lenses with different magnification, foot for the support of the microscope to stay erect).

The number of times an object is magnified depends on the power of the lenses used and it is a product of the power of the ocular and objective lenses used. A magnification of 1000 times is possible with optical microscopes

Several kinds of light microscopes are available. They differ in the optics and type of light (visible or ultraviolet) used. Depending on the purpose of study and the level of resolution required, different microscopes are used.

The different types of optical and electron microscopes used in the study of microbes with their features and the images obtained with them are given in **Table 9.3.**

Table 9.3: Different types of optical and electron microscopes used in the study of microbes – their features and the images obtained with them

Type of Microscopy	Features
Brightfield	Most commonly used microscope. Produces dark images on a bright background. Stains (photochromatic dyes) are used to enhance the images.
	Image with brightfield microscopy *Staphylococcus aureus (https://www.memorangapp.com/)*
Darkfield	An opaque disc is placed between the illuminator and the condenser of a brightfield microscope. No staining required and therefore, live specimens can be examined. High resolution, high contrast, bright images are formed against a dark background.
	Image with darkfield microscopy *Lactobacilli (https://twitter.com/zeiss_micro/)*
Phase contrast microscopy	Images are created by altering the wavelengths of light rays passing through the specimen. This is achieved through use of an annular stop in the condenser and a phase plate with a phase ring in the objective. This causes interference and refraction of light, thereby highlighting the minute differences in thickness between different structures in the specimen. High contrast, high resolution images are produced. No staining is required. Very useful for observing live cells.

Type of Microscopy	Features
	Image with phase contrast microscopy 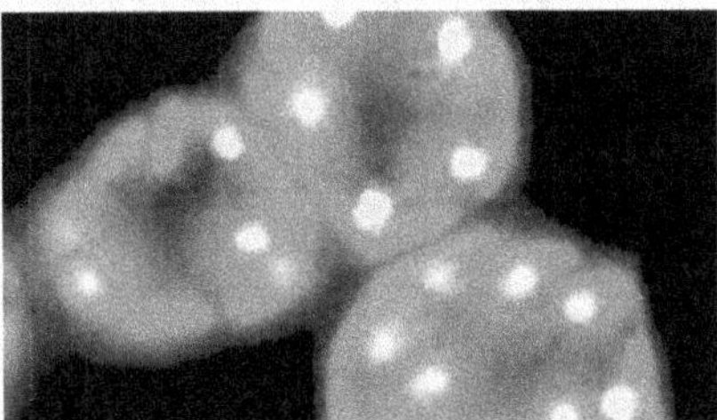 Amoeba (https://www.canadiannaturephotographer.com/)
Fluorescence	Fluorescent dyes called fluorochromes are used. Or else, natural fluorescence of the organism (e.g. chlorophyll) becomes visible. Short wave length UV light is used. The chromophores absorb this light and emit longer wavelength light in the visible spectrum. The UV light emerging from the specimen is then filtered out (since it is harmful to the eyes) and only visible light passes through the eyepiece lens. A bright colour image against a dark background is formed. Fluorochromes are vital dyes, so live cells can be stained. It can be used to distinguish living from dead cells, or to locate particular molecules within a cell. It is also used in immunofluorescence techniques for identification of microbes.
	Image with fluorescence microscopy 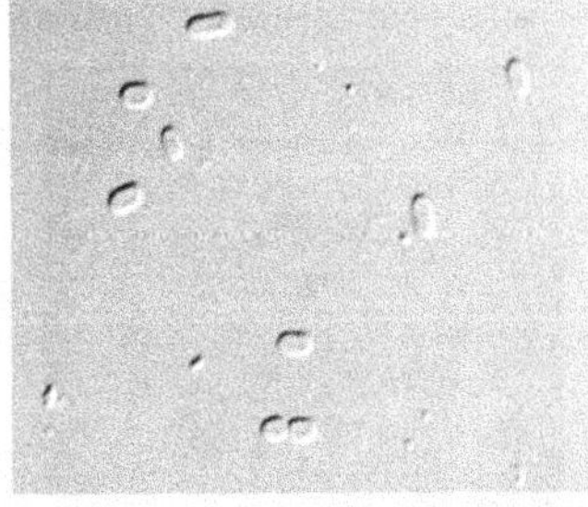 Cells of a marine protist with foreign genes included (seen as green dots but in this B & W print as white dots) (https://www.sciencemag.org/)
Differential interference contrast (DIC) (Nomarski optics)	Similar to phase contrast microscope optics, but two beams of light are used for illuminating the specimen. No staining is required. Images of live cells of high contrast and resolution with 3D appearance are formed.
	Image with differential interference contrast microscopy *E. coli* (https://www.asmscience.org/VisualLibrary)

Type of Microscopy	Features
Confocal	A laser is used to scan the specimen in multiple planes capturing multiple 2D pictures at different depth of the specimen (optical sectioning) to create a 3D image. The resolution and contrast of the image is increased in this optical imaging technique, with the help of a spatial pinhole, that blocks out-of-focus light in image formation.
	Image with confocal laser microscopy 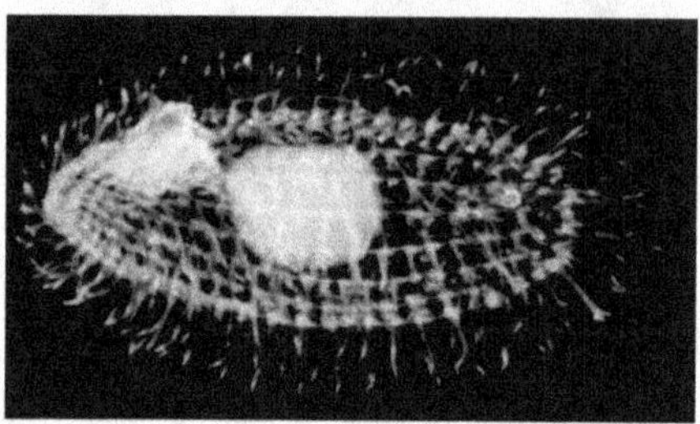 *Tetrahymena thermophilia*, a ciliated protozoan – a model organism for research (https://manoa.hawaii.edu/ Image courtesy Richard Robinson, PLOS Biology)
Transmission electron microscopy (TEM)	Provides for detailed study of the internal ultrastructure of cells. A beam of electrons is transmitted through the specimen for a 2D view. Image resolution is increased 1,000-fold over brightfield microscope and magnification is increased up to 1,00,000x. Electron microscopy can resolve subcellular structures.
	Image with transmission electron microscopy 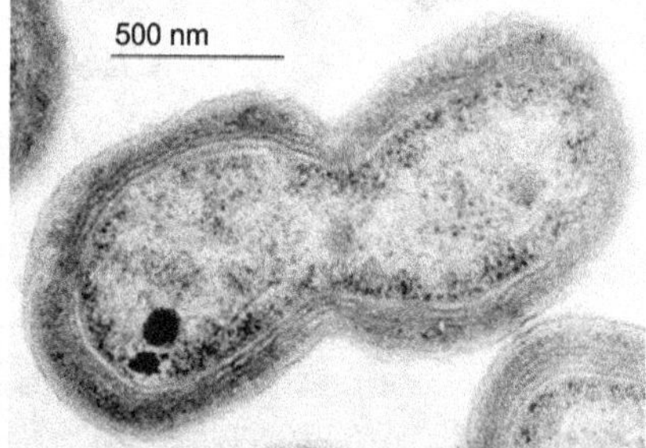 *Prochlorococcus marinus* , a cyanobacterium (dividing cell). A very small prokaryote which would appear as a tiny speck of dust under a light microscope. (https://www.quora.com/What-are-examples-of-single-cell-organisms)
Scanning electron microscopy (SEM)	Electron beams are used to scan the surface of the specimen. The electrons that are knocked off the specimen, create a detailed picture of the surface, which appears like a 3D image. The image is visualized on a digital screen.
	Image with scanning electron microscopy 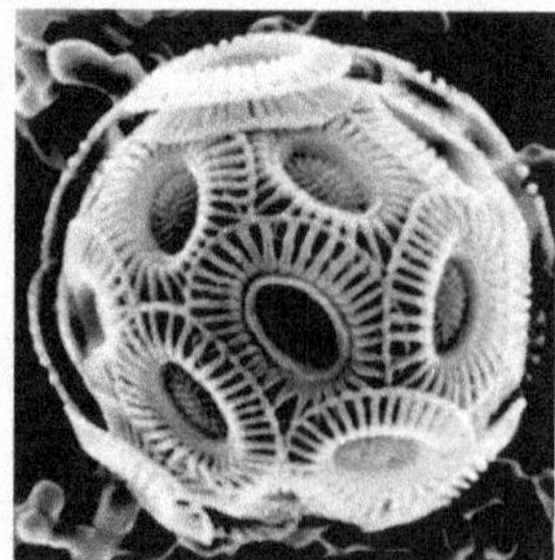 A single celled alga *Emiliana huxleyi.* (https://web.uri.edu/gsc/image-gallery/)

9.7 ECOLOGICAL ASPECTS OF SINGLE CELLED ORGANISMS

Microbes live and thrive everywhere, on every square inch of earth, including regions that are hostile and inhabitable for other living organisms. The tremendous diversity in microbes endows them with this competence. The microbes that inhabit a particular ecological niche have metabolic properties that suit that environment. *Thermophilus acquaticus*, a bacterium present in the hot springs, for example, can reproduce and thrive in the extremely hot temperature of the natural geysers, because, the enzyme that helps replicate its genetic material – the DNA polymerase is tolerant to very high temperatures; it does not get denatured even at 100°C. Because microbes are ubiquitous, they profoundly affect the biosphere.

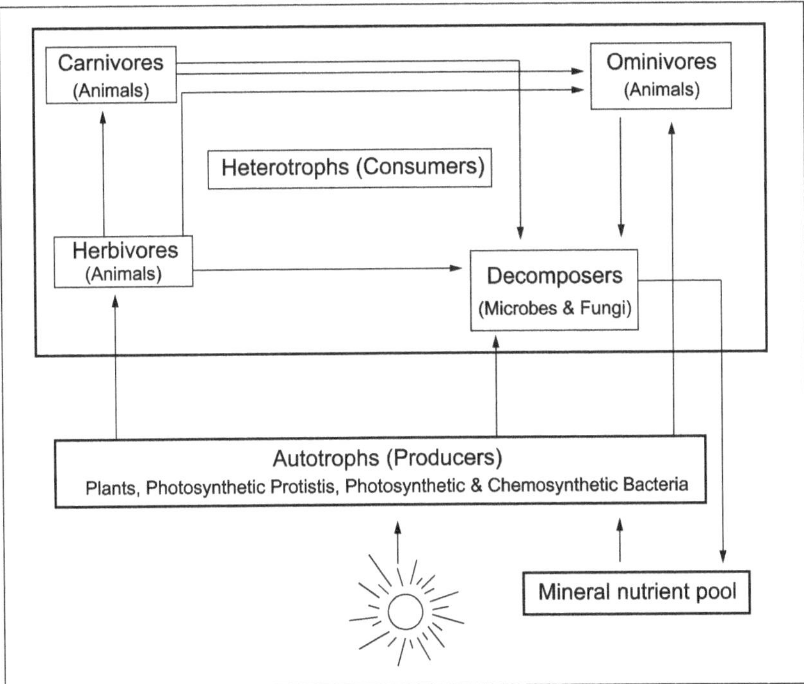

Figure 9.3: Microbes in the food web. *Microbes are producers in some ecosystems, but in all ecosystems, they decompose the dead organic matter and recycle the nutrients trapped in them.*

Microbes are an important, rather essential, component of all ecosystems on earth. They play a vital role in the complex nutrient exchange system in an ecosystem. An ecosystem consists of producers (autotrophs), consumers (heterotrophs) and decomposers (exclusively composed of microbes and fungi) (**Figure 9.3**). Microbial ecology involves the study of the numerous interrelationships of the microbes with the living and non-living components of the environment around them.

9.7.1 Microbes In Different Environments

9.7.1(a) Aquatic

Microbes constitute the primary producers in water bodies including the oceans. The phytoplankton (constituted by cyanobacteria, microalgae and diatoms), are the primary producer in these aquatic systems, which support the huge biotic component in their ecosystem. The microalgae are very efficient in photosynthesis. *Synechococcus*, a cyanobacterium is responsible for nearly 25% of the primary production in oceans. The phytoplankton account for 50% of the total photosynthesis on earth. Chemosynthetic bacteria occur in deep parts of the ocean, where light is not available. They synthesize sugars from CO_2, using energy from inorganic substances, a process termed chemosynthesis. The autotrophic microbes were the first inhabitants of earth and they enriched the atmosphere with oxygen because of their photosynthesis. Fifty percent of the oxygen in the atmosphere, is contributed by the photosynthetic activity of microbes. The bacteria and fungi in aquatic bodies constitute the decomposers in this ecosystem.

9.7.1(b) Soil

Microbes in soil, thrive on dead organisms, decomposing them in the process. They are responsible for recycling of the nutrients trapped in the bodies of dead organisms. There is no known naturally occurring organic molecule, that cannot be decomposed by some microbe. The inorganic nutrients in the dead organisms are recycled by the microbes. They are engines for the biogeochemical cycling of important elements like carbon, nitrogen and sulphur. A biogeochemical cycle is the rotation of elements between the biotic (living) and abiotic (non-living) components in the ecosystem. Nitrogen cycling by microbes is depicted in **Figure 9.4**.

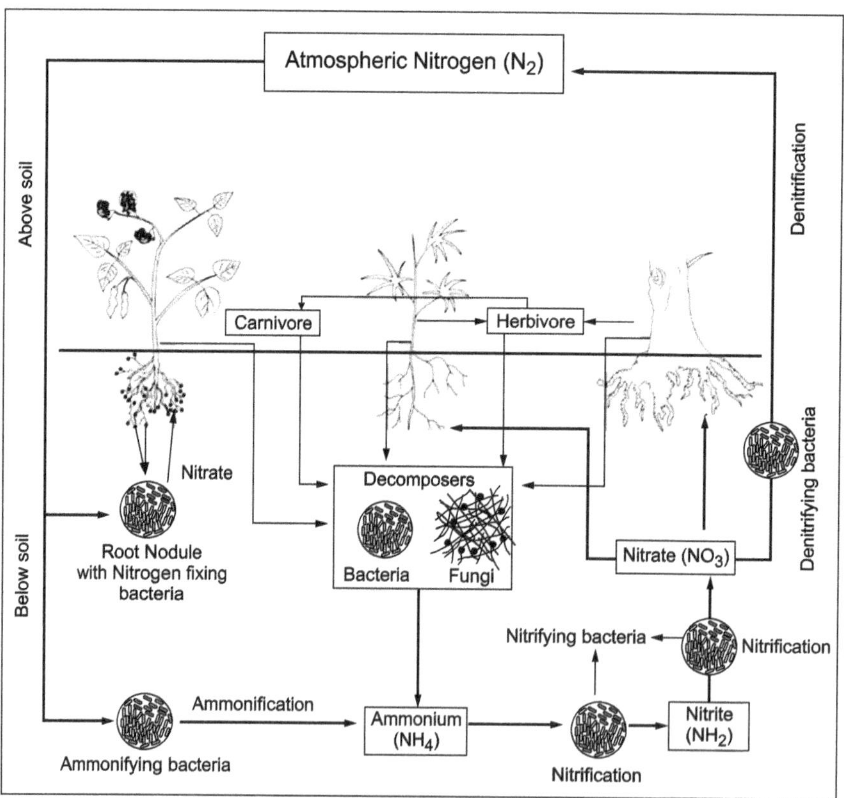

Figure 9.4: Microbes in biogeochemical cycling of nitrogen and fixing atmospheric nitrogen in plant absorbable form as nitrate.

Nitrogen is an essential element of all living organisms. Microbes are the only living organisms that can fix atmospheric nitrogen in the soil and water. Some plants, harbour nitrogen fixing bacteria in their roots, due to which they become nodulated. Some soil dwelling microbes, secrete enzymes and vitamins that promote plant growth. They are termed plant growth promoting bacteria (PGPB).

9.7.1(c) Arid Regions

These include the deserts and lands covered with snow like the Tundra (Arctic circle) and Taiga (sub-arctic region), which have the snow forests (boreal forests), that constitute the largest land biome on earth. Cactus vegetation is possible in the deserts, solely because of the soil dwelling nitrogen fixing bacteria. In the snow-covered regions, lichens, constitute an important component of the vegetation. In Tundra, the lichens dominate the vegetation as the primary producers. Lichens are associations of microalgae or cyanobacteria with fungi. The cyanobacteria in lichens, fix atmospheric nitrogen, which leaches into the soil, and supports the dominant vegetation constituted by huge trees (conifers), in the snow forests in Taiga. It is estimated that 6% of earth's land surface is covered by lichens.

9.7.2 Microbes in Associations with Other Living Organisms

Microbes make several types of associations with plants, animals and other microbes. Based on the relationship between the partners in the association, it is described as: mutualistic (both partners are benefited), commensalistic (one partner is benefited while the other is neither benefited nor loses), neutralistic (neither partners is benefited or harmed), or parasitic (one partner exploits the other and harms it).

9.7.2(a) Microbes in Associations with Animals

Microbes are present both on the surface and inside of many living organisms. Some examples are described.

9.7.2(a)i Gut Bacteria (Mutualism)

Some types of bacteria, live in the guts of some insects and vertebrate animals including humans. The gut microbiota is constituted by a diversity of bacterial species. The types of microbes in different parts of the gastrointestinal tract are different. These microbial associations are mutualistic. The microbe obtains nutrients and shelter from the host, while they ferment the dietary fibre into short chain fatty acids. Gut microbes produce vitamins (k and biotin), growth factors and hormones useful to their host. The microbes in the rumen (special stomach present in ruminants that regurgitate their food) of cattle, are cellulolytic and help in digestion of the cellulose in the plant material. The termite guts, are inhabited by a protozoan, which breaks down cellulose and releases acetate, which can be absorbed and used by the termite. Neither of them can live without this mutual relationship.

9.7.2(a)ii Association of the Bacterium *Buchnera Aphidicola* and Aphids – A Type of Insects (Obligate Mutualism)

The bacterium resides in the cytoplasm of special cells – the bacteriocyte cells of the aphid. *Buchnera*, the bacterium lost its ability to live freely and is an obligate endosymbiont. It over synthesises some ten different amino acids which the aphid utilizes. The plant sap diet of the aphid lacks these amino acids; the aphid thus cannot survive without the microbial partner.

9.7.2(a)iii Association of Dinoflagellates and Marine Invertebrates (Facultative Mutualism)

Zooxanthellae, a group of dinoflagellates lead a symbiotic life with marine invertebrates like reef building corals, sponges, jelly fish, some bivalve molluscs and flat worms. Being photosynthetic, they produce carbohydrates and cater to nearly 90% of the energy needs of the marine invertebrate. The dinoflagellates in return get nutrients and carbon di oxide and an access to the elevated region within the ocean ecosystem for availability of sunlight. The corals cannot survive without the dinoflagellate symbiont and bleach off if the dinoflagellate is lost. The dinoflagellate can survive as a motile cell (with flagella), but usually lives in a cysted form in marine invertebrates.

9.7.2(a)iv Association of Chemosynthetic Bacteria with Giant Tube Worms in Hydrothermal and Cold Vents on the Ocean Floor (Obligate Mutualism)

The hydrothermal and cold vents on the ocean floor (up to 5000ft below the ocean surface), are mineral rich and have high concentrations of hydrogen sulphide, iron, methane etc. They are one of the most hostile and uninhabitable environments on earth, with utter darkness and high pressure, besides toxic concentration of inorganic compounds. But, these vents on ocean floor are like oasis in a desert. They support animal life, thanks to the chemosynthetic bacteria. One of the dominating animals in this ecosystem, are the giant tube worms, the fastest growing marine invertebrates (growing two feet per year). They thrive in this ecosystem, reaching eight feet in length and 1.5 inches in width surrounded by a chitin shell. The giant tube worms have no mouth, stomach or digestive tract. They are fed by the sugars synthesized by endosymbiotic bacteria that fix carbon dioxide using hydrogen sulphide and oxygen supplied from the surrounding environment by the tube worm. Neither the bacteria, nor the tube worm, can live by themselves, in this lethal environment. The chemosynthetic bacteria, cannot live in the harsh physical conditions in these ocean vents. The tube worm provides a safe habitat for the chemosynthetic bacteria, which live inside its body in a region called trophosome. The tube worm is entirely dependent on food on the chemosynthetic bacteria. The tube worms are predated by deep sea crabs and shrimps, giant clams and large brown mussels.

9.7.2(b) Association of Bacteria with Plants (Mutualism)

Methane rich peat bogs are inhabited by the peat moss *Sphagnum*. In these bogs, *Sphagnum* is submerged by water and has no availability of CO_2. The methanotrophic bacteria harboured in *Sphagnum,* oxidise methane and make available CO_2 to the moss for photosynthesis. In turn, the methanotrophic bacteria receive oxygen released by the moss during photosynthesis. The plant environment provides an elevated concentration of oxygen than the outside environment in the peat bog. The methanogenic bacteria, thus reduce emissions of the greenhouse gas – methane from the peat bogs and cycle the carbon in-situ.

Nitrogen fixing bacteria in root nodules of plants is an example of yet another association of mutualism.

9.7.2(c) Association of Microbes with Fungi

Cyanobacteria and microalgae associate with fungi to form lichens. The lichen is a composite organism, wherein neither of the partners of the association, appear as they do when they live separately. Lichens are in many forms (**Figure 9.5**) and ubiquitous, being present in most ecosystems. There are over 14,000 named species of lichens. The relationship is one of mutualism, commensalism or even parasitism. While the alga, the photobiont partner of the lichen, can live independently, its ecological distribution is widened, because of the physical protection afforded by the fungus. Lichens constitute important part of the vegetation in frozen forests (Taiga) and Tundra regions. The fungus is dependent on the food synthesized by the alga.

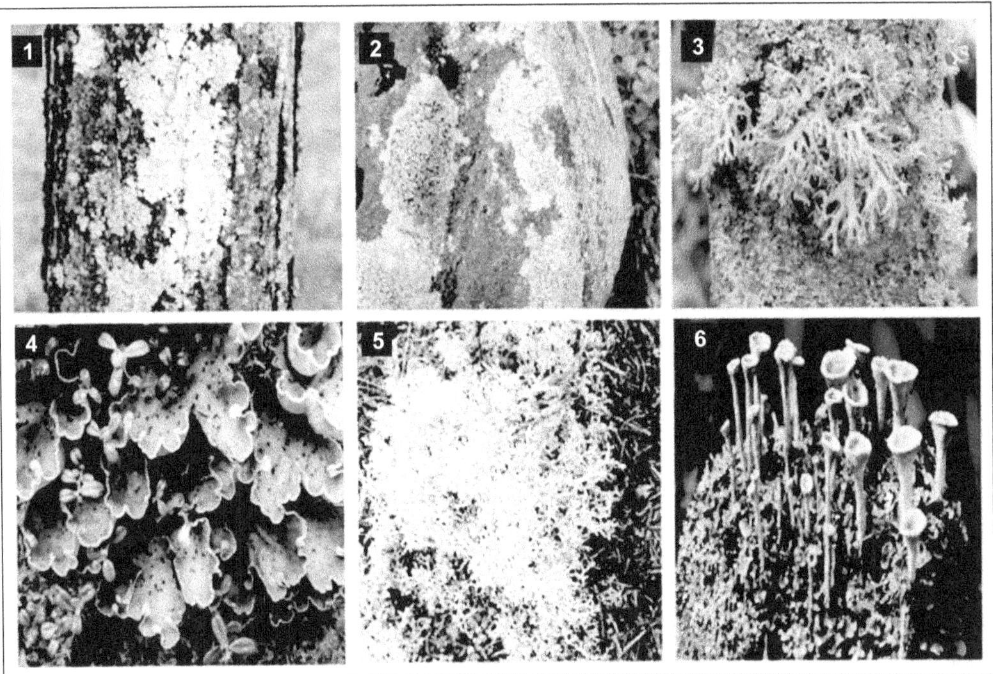

Figure 9.5: Lichens (symbiotic relationship between microalgae or cyanobacteria (photobiont) and a fungus (mycobiont). **1&2.** *Crustose lichens on tree trunks and rock.* **3.** *Foliose lichen.* **4.** *Foliose lichen of Tundra.* **5.** *Reindeer lichen of Tundra and Taiga region.* **6.** *Pixie cup lichen of boreal forests of Taiga.* Picture credit for 1. https://njaes.rutgers.edu/fs1205/Tree. 2. https://www.flickr.com/photos/brewbooks/2924427336. 3. https://www.nytimes.com/2016/07/22/science/lichen-symbiotic-relationship.html. 4. https://www.nps.gov/articles/aps-16-1-13.html. 5. https://botit.botany.wisc.edu/toms_fungi/dec2000.html. 6. https://www.mtpr.org/post/exploring-landscape-pixie-cup-lichen

9.7.2(d) Association Between Microbes

7.7.2(d)i Microbes that form Special Ecological Formations

7.7.2(d)i.1 Red Tides

At times, red tides (called algal blooms) occur naturally off the sea coast. They are the result of excessive multiplication of unicellular microalgae, diatoms, dinoflagellates and protozoa as a result of which, the sea water is discoloured. The pigments in the microbes in the algal blooms, give it the red or brown colour (**Figure 9.6**). These red tides are caused, due to sudden upwelling of nutrients from the sea floor due to storms. They can also be caused, due to increased nutrient pollution due to human activity. The toxins produced by some of the dinoflagellate members of the red tides, can kill fish, contaminate shell fish and harm humans.

Figure 9.6: Red tide in sea caused by algal bloom. *Consists of a dense population of microalgal species. Picture credit:* https://biolum.eemb.ucsb.edu/organism/redtide.html

7.7.2(d)I.2 Biofilms

Biofilms, colloquially called microbial cities, are a consortium of syntrophic (cross-feeding) microbes, that stick to each other and to a surface. They are ubiquitous and usually found on solid surfaces, submerged or exposed to aqueous solutions, or humid environments. They sometimes occur as floating mats on water surfaces. Every type of microbe, can potentially become a member of a biofilm community. The biofouling of submerged regions of a ship is due to the formation of biofilms. Bacterial infections in humans are established as biofilms. The dental plaque is a biofilm. The pathogenic microbes in sinuses, lungs and intestines of human's form biofilms. It is estimated that 60 to 70 % of the microbial infections in plants and humans establish as biofilms. Since biofilms develop slowly, the effects of bacterial infection are not immediately apparent.

Biofilms can be formed by a single microbial species, or many microbial types like bacteria, archaea, protozoa, algae and fungi. In such biofilms with biodiversity, each microbe type performs specialized metabolic functions. The relationship between the microbes in biofilm, whether it is one of competition or cooperation, depends on the species within these eco-microbial communities.

A biofilm is formed when a single microbial cell which is initiated by such factors as availability of a surface, nutritional cues, or presence of antibiotics, first adheres to a surface. Then it starts multiplication and cell to cell communication – a phenomenon termed 'quorum sensing' resulting in colonization. Other microbial species in the surrounding environment, can sense the formation of a biofilm through quorum sensing and recruit themselves in the biofilm. Colony increases in size through division and recruitment of new microbes and their multiplication. The members in the colony get embedded in slimy matrix composed of carbohydrates, proteins and lipids and DNA is released to the outside, by the component cells. The extracellular matrix of polymeric substances, can also contain components from the surrounding environments, like soil particles, minerals, blood etc. The biofilm becomes a three-dimensional macroscopic structure appearing as a hydrogel. A fully developed biofilm contains channels for circulation of nutrients. Cells in different regions of a biofilm, show different patterns of gene expression, due to quorum sensing. This structural and physiological complexity of biofilms with the coordinated and cooperative group of microbes, is similar to a multicellular organism. A biofilm grows slowly. A well-formed biofilm can undergo dispersion when some part of it is broken and the microbial components are dispersed (**Figure 9.7**).

Microbes in a biofilm have significantly different properties from individually living microbes of the same species. The protected environment of the film allows the microbial components in the biofilm to cooperate and interact in various ways. Microbes in biofilms are not affected by antibiotics, because of the thick matrix surrounding the cells. The extracellular DNA in the matrix of the biofilms, is used for lateral exchange of genes between the microbial members of the biofilm.

The emergent properties of the eco-microbial communities in the biofilm bestow high chances of their survival due to protection from antibiotics (**Figure 9.8**).

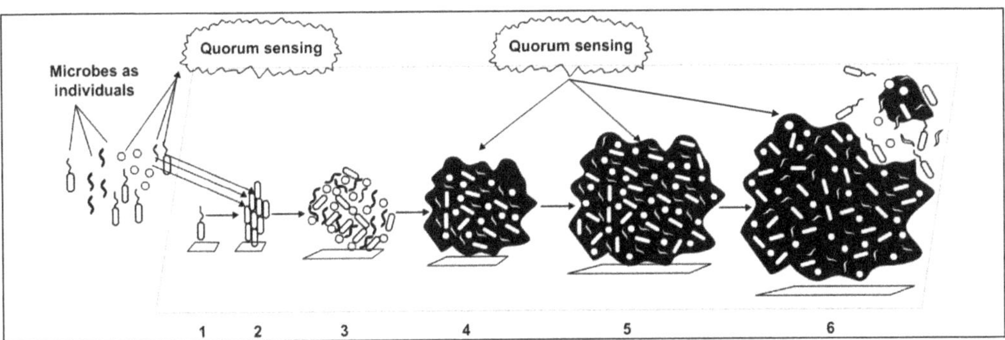

Figure 9.7: Formation of biofilm. 1. *A single microbe attaches to a surface.* **2.** *The microbe starts dividing and the resulting cells attach to each other and to the surface. Other microbes in the surrounding environment sense the microbial cell population set to form a biofilm through quorum sensing and recruit themselves.* **3.** *All the participating microbes reproduce and the new cells formed, get attached to each other and to the surface.* **4&5.** *The increasing microbial cell mass is embedded in an extracellular matrix secreted by them; thus, forming a biofilm. The microbial cells continue to multiply and communicate through quorum sensing; to synthesize specific substances and to assess the cell density.* **6.** *When the cell density in the biofilm is sensed high (through quorum sensing), it starts dispersing by fragmenting and releasing of cells.*

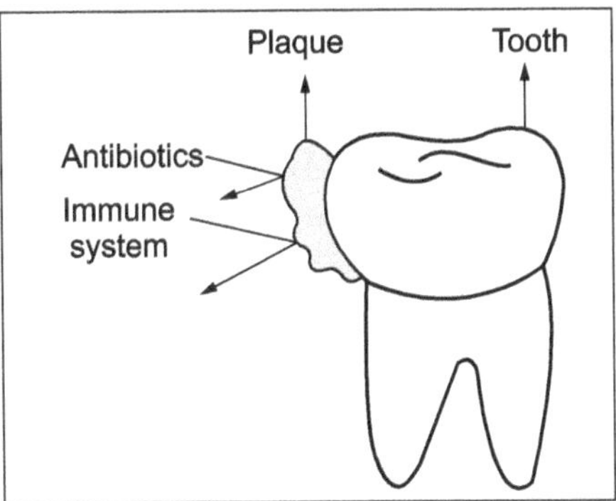

Figure 9.8: Plaque on tooth – a biofilm resistant to antibiotics and not recognized by immune system.

7.7.2(d)i.2(a) Quorum Sensing in Biofilms

Quorum sensing is the ability to detect and respond to the density of cell population. It can occur between diverse microbial species, or between members of a single bacterial species. Quorum sensing is a communication system between the microbes, wherein, they produce signals (which can be different types of molecules in different species and conditions), and respond to the signals in: **1.** a cell density dependant manner, **2.** by regulating gene activity, to produce substances, that can enhance their effect (like virulence, bioluminescence etc). It is thus, a chemical vocabulary, that is used in quorum sensing. It allows the bacterium to monitor the environment and assess the numbers of other microbes and alter its behaviour on a population wide scale in response to change in the number and of species present in the community. Some of the common phenotypes expressed due to quorum sensing are biofilm formation and dispersion, virulence factor expression (in biofilms of pathogenic microbes), bioluminescence, nitrogen fixation and sporulation.

Quorum sensing by bacteria can both facilitate biofilm formation and dispersion. A single celled bacterium, through quorum sensing, can sense if it is surrounded by a dense population of other microbes, and it becomes more inclined to join, and contribute to the formation of a biofilm. Also, when the biofilm becomes large, quorum sensing, enables the component bacteria, to sense the high numbers and initiate dispersal from the biofilm (**Figure 9.7**)

In biofilms of pathogenic microbes, the production of virulence substances, is initiated through quorum sensing, when sufficient individuals are present, so that, enough of the substance is produced, to overtake the host immune system, and utilize the host tissue. Through quorum sensing, these bacteria remain relatively harmless (by formation of biofilm), until population sizes are sufficient, and overwhelm host defences, by producing the virulence substances, when they are in large enough numbers.

The group communication and behaviour synchronization achieved through quorum sensing, allow bacterial populations to develop more quickly, gain access to more resources, and secure better chances of survival.

9.8 MICROBIAL CULTURE

9.8.1 Culture Media and Sterilization

Microbes are cultured in the laboratory for various purposes. Culturing a microbe is mandatory for characterizing and identifying it. Bacteria and yeast are cultured for genetic engineering experiments in the laboratory. On a large scale, microbes are cultured in fermenters for various products. Not all microbes can be cultured artificially in the laboratory.

To successfully culture a microbe, it should be provided with a nutrient medium for growth and suitable physical conditions like air, light and temperature.

9.8.1(a) Physical Conditions for Culture

Suitable physical conditions are provided by setting up bacterial cultures in incubator ovens, or culture rooms set to suitable temperature and light conditions. Liquid cultures are usually grown in culture flasks or tubes in incubator shakers set to suitable temperature and speed of shaking (rotations per min).

9.8.1(a)i Culture Medium

The culture medium is the growth medium that caters to the nutritional requirement of the microbe. For bacteria, basically the growth medium contains a sugar (for energy), a nitrogenous compound and minerals. There are several different kinds of culture media with different constituents. Choice of the medium depends on the type of microbe to be cultured and the purpose of the experiment.

9.8.1(a)ii Types of Culture Medium

9.8.1(a)ii.1 Based on Consistency: Liquid, Solid, Semi-solid

The culture medium is prepared as a liquid with its components dissolved in water. The medium is solidified usually by adding an inert gelling agent like gelatin or agar. In solid media, a concentration of 1.5 to 2% by weight of the gelling agent is used while semi-solid media are prepared with 0.5% of the gelling agent.

Solid medium is poured in Petri dishes or test tubes as a slant. Solid medium is useful for isolating bacteria, or, for examining the colony characteristics of the isolate. Semi-solid medium has a soft custard-like consistency and is useful for determining the bacterial motility and for culture of microaerophilic (requiring lower concentration of O_2 than in the atmosphere) bacteria. Liquid broth media are useful for large scale culture of the microbe.

9.8.1(a)ii.2 Based on Composition: Nutrient, Minimal, Selective, Differential

A nutrient medium is one which contains all the elements required for growth of most types of bacteria. It is used for general cultivation and maintenance of bacteria kept in culture collections. The nutrient medium is described as defined medium when the chemical composition of all its constituents is known. In some nutrient media, yeast, peptone or beef is used as a source of nitrogen and amino acids. Since the exact chemical composition of yeast or peptone (complex proteins) is not known, nutrient medium with these components is termed undefined.

A minimal medium does not contain amino acids and minimum type and amount of nutrients are used which support bacterial growth. A knowledge of the nutritional requirements of the microbe to be cultured is necessary to devise a minimal medium for it.

Selective media are used to encourage the growth of a particular microbe of interest, with the exclusion of the other microbes that may be present along with it.

A medium, devised based on the knowledge of biochemical characteristics of the microbes to be cultured, is called a differential medium. It is also called an indicator medium. Sometimes specific substances like neutral red, phenol red, eosin or methylene blue are added to serve as an indicator to distinguish particular microbial species. This type of medium is used for the detection and identification of microorganisms.

9.8.1(b) Sterilization Techniques

Microbes are present everywhere. To artificially culture a particular type of microbe, the microbial inoculum, should be transferred to the culture medium, in a sterile (where microbes are eliminated) environment. The culture medium and the culture container (flask, tube or Petri dish) should be sterile.

The culture containers are sterilized by heat (dry or moist) or radiation. Dry heat involves heating in the absence of water. Glass rods and metal loops used for inoculation of the bacterial cultures are heated over the flame of a Bunsen burner (red heat). Dry heat sterilization requires exposure to 160°C for 120 min. The cotton plugs, used for closing the mouths of test tubes and the rims of the mouth of the culture containers, are sterilized in this way. The culture medium and the culture vessels are sterilized by autoclaving (moist heat). The autoclave works on the same principle of pressure cooker. The usual temperature and pressure that can be achieved in an autoclave is 121°C at a pressure of 15 psi (pounds for square inch).

Sterilization is complete when kept in these conditions for 15 mins. The culture medium constituents like antibiotics, vitamins etc. that do not withstand high temperature in the autoclave, are filter sterilized by passing the solutions through membrane filters of very small pore size.

While inoculating the microbial inoculum into the culture medium, the surrounding environment should be sterile. A laminar flow unit, which is a cabinet which provides sterile environment, is generally used for inoculations (**Figure 9.9**). The laminar flow cabinet has a continuous flow of air, blown through it after passing through a HEPA (High Efficiency Particulate Air) filter. The surface of the cabinet is sterilized by wiping with alcohol. A Bunsen burner is available in the cabinet for red heat sterilization of the inoculation rod or loop and to sterilize the rim and mouth of the culture container (**Figure 9.10**). The laminar flow cabinet usually has a UV light which is switched on for 10 to 15 mins after placing all the culture containers and media in the cabinet. Through this process, the outer surface of the culture containers is sterilized. The bacterial inoculum is spread on the solid medium in a Petri dish, either with a glass rod with triangularly bent end (for uniform spread of the inoculum), or, streaked with a loop (to obtain single cell cultures) (**Figure 9.11**). Slant cultures in test tubes are inoculated with a glass rod with bent end. Liquid cultures are inoculated with an aqueous solution of the bacterial inoculum using a pipette – either a sterile glass one, or a micropipette with a sterile plastic tip.

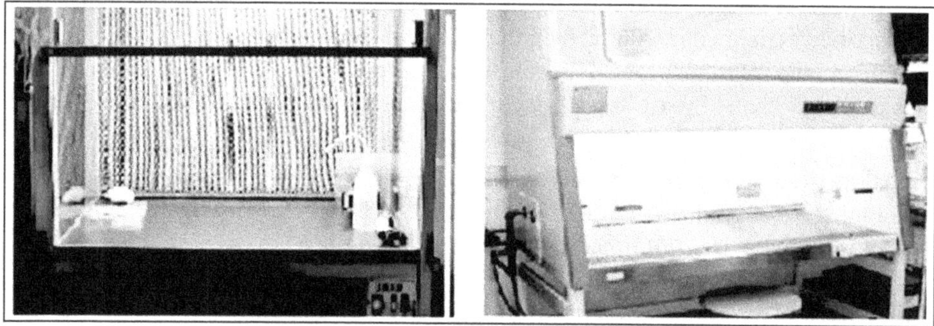

Figure 9.9: Laminar flow cabinet with HEPA filter providing a sterile environment for inoculation of microbes into culture medium.

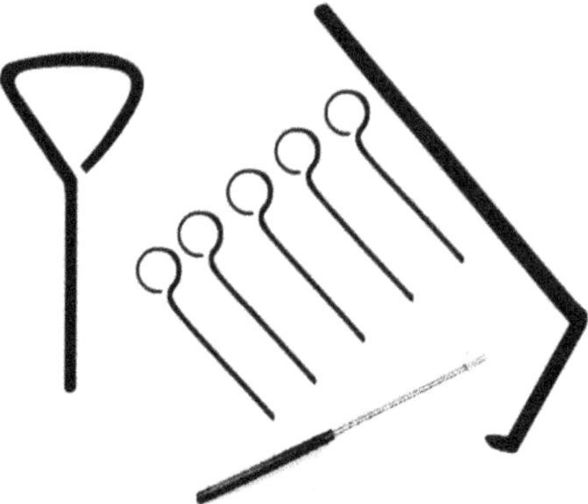

Figure 9.10: Bacterial culture spreading glass rods (with triangular end or bent end like a hockey stick) and metal loops.

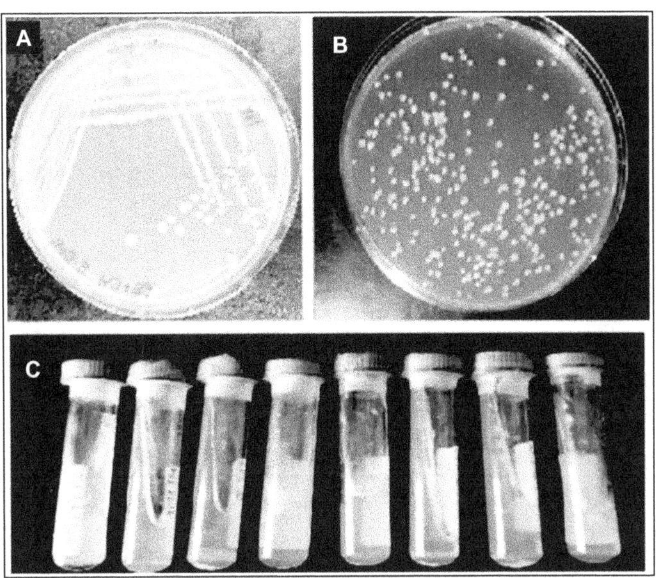

Figure 9.11: Bacterial cultures on solid medium in Petri dishes and agar slants in tubes. A & B. *Streak and spread plate cultures.* **C.** *Slant cultures in tubes.*

9.9 GROWTH KINETICS OF A MICROBE

Microbes multiply at a very fast rate when sufficient nutrients are available and suitable environmental conditions prevail. An *E. coli* cell, can grow and multiply within 30 mins, under ideal growth conditions. The time taken for a microbial cell to divide into two is called its generation time. In a liquid culture medium, the time taken for the cell mass or number to double is called cell doubling time. Growth in microbial cultures is monitored by cell count (haemocytometer counts under microscope), increase in biomass (by measuring dry weight of cells), or, by measuring the optical density of the liquid with a culture with a spectrophotometer, at different times after inoculation of culture.

When a liquid medium is inoculated with a bacterial inoculum, the bacterial cells begin to grow and multiply. When the physical conditions of the culture are kept constant, the rate of growth is directly dependant on the substrate concentration, being higher, at higher substrate concentration (**Figure 9.11**). However, the pattern of growth of the microbial culture, at any given substrate concentration is the same. The growth pattern at a constant substrate concentration in cultures where substrate is supplied in one go at the beginning (batch culture), or it is supplied constantly (fed batch culture), is the same (**Figure 9.12**). Under optimal culture conditions, bacterial growth passes through six phases in both batch and fed batch cultures. They are the lag phase, when the cell number does not increase, acceleration phase, a brief stage when the cell number begins to increase slowly, log phase where the cell number increases rapidly, a very brief deceleration phase where the rate of growth decreases slightly over the log phase, stationary phase when the cell numbers stay constant and finally a death phase when cell numbers start decreasing exponentially (**Figure 9.13**). The duration of the lag phase depends on the type of microbe, its physiological state when inoculated, and the nature of the culture medium. The cells reach a stationary phase because all or some nutrients in the substrate are depleted and because of the metabolic products produced by the cells which inhibit growth. The death phase sets in because of the halt of metabolic activities in the bacterial cells. In fermentation processes, the culture is terminated at the end of log phase.

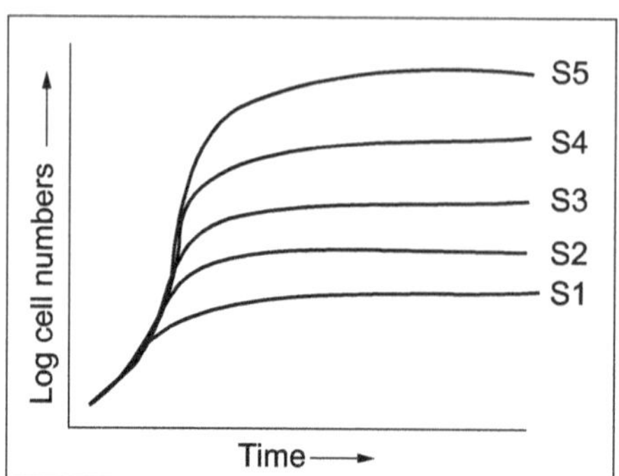

Figure 9.11: Relationship of bacterial growth to the concentration of the substrate. *The concentration of the substrate increases from S1 through S2, S3, S4 to S5. The maximum growth (expressed as log cell numbers) attained, increases with increasing substrate concentration.*

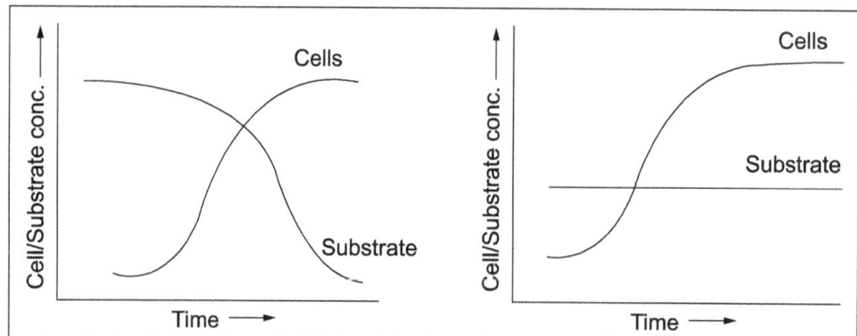

Figure 9.12: Bacterial growth in batch (substrate is supplied one time at the beginning of the culture) (left) and fed batch (substrate is supplied at intervals during culture (right). *The maximum growth achieved and the pattern of growth are similar in both culture conditions. In batch culture, the substrate concentration decreases as the cell culture grows while its concentration remains constant in fed batch culture.*

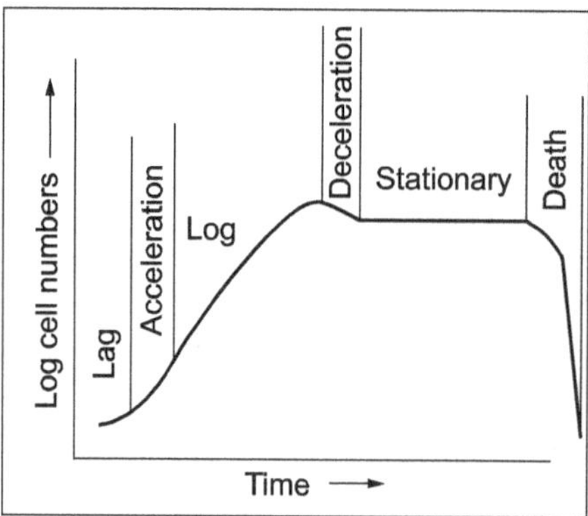

Figure 9.13: Pattern of growth of a bacterial culture in liquid medium. *The growth line shows six phases: lag, acceleration, log, deceleration, stationary and death.*

Growth kinetics is the relationship between specific growth rate and the concentration of a substrate. It follows the classical Monad's equation during the log phase of growth. Monad's equation relates microbial growth rates in an aqueous environment to the concentration of a limiting nutrient in the culture medium.

Monad's equation is similar to Michael-Menton equation of enzyme kinetics, the difference being that it is empirical while the latter is theoretical (**Figure 9.14**).

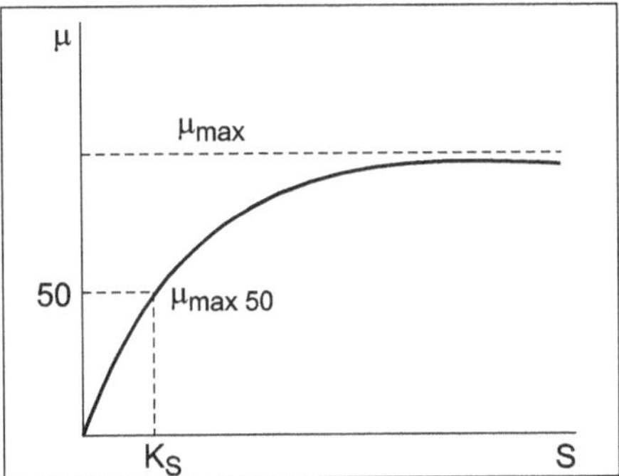

Figure 9.14: Growth kinetics of microbial culture during log phase. Relation of specific growth rate μ **to concentration of the limiting factor in the substrate (culture medium(S)).** *Monad's equation for calculating specific growth rate is derived from this empirical graph.* μ_{max} *is the maximum specific growth rate of the microbe,* K_s *is the value of S when* $\mu/\mu_{max} = 0.5$

Monad equation is given by

$$\mu = \frac{\mu_{max}[S]}{K_S + [S]}$$

μ is the specific growth rate of the microbe.

μ_{max} is the maximum specific growth rate of the microbe.

[S] is the concentration of the limiting substrate for growth.

K_s is the "half-velocity constant" — the value of S when $\mu/\mu_{max} = 0.5$

μ_{max} and K_s are empirical coefficients to the Monod's equation. They will differ with different microbial species and will also depend on the ambient environmental conditions.

9.10 SUMMARY

1. Microbes are organisms that are too small to be seen individually by the naked eye. Though invisible, they are very abundant and ubiquitous, inhabiting even such extreme environments, like hot springs, ocean vents, frozen lands (Arctic zone and Antarctica) and extremely salty water bodies (dead sea). Microbes though abundant are less diverse than the macro-organisms. They do not lead a solitary life, but, live as a population, in the form of colonies or communities. The entire prokaryotic world, some protists and a sub-group of the sac fungi consisting of yeasts, are microbes. Microbes are characterized by very rapid multiplication rates.

2. They are estimated to constitute, more than 50% of the total protoplasmic biomass, on earth. Their existence became evident only after the discovery of the microscope. They are mostly single celled organisms. However, they perform many of the complex activities carried out by large multicellular organisms. They respire, feed, digest, excrete, and reproduce like a typical multicellular animal. Microbes are a strong evidence for the cell theory.

3. A reliable classification of prokaryotes based on external characters is difficult, because they have simple structures. Berge's manual of "Determinative Microbiology (9th edition, 1994)", follows the classification of bacteria based on Gram staining. It was essentially designed to help identification of prokaryotic microbes. In this classification, the prokaryotes (eu- and archaea-

bacteria) are classified into four major categories and 35 groups. The four major groups are: Gram positive-; Gram negative-; Gram variable- and archae- bacteria. In 1987, Carl Woese divided Eubacteria, into 11 divisions, based on the gene sequence coding for 16S ribosomal RNA of ribosomal SSU (small subunit of the ribosome).

4. Protists (organisms which do not belong to plantae, animalia or fungi groups of eukaryotes) have some microbial species. They are classified into three groups: Plant like (e.g. microalgae), animal like (protozoans, e.g. amoeba) and fungi like (e.g. slime and water molds). Among fungi, yeast species are unicellular. Yeast group of fungi are called sac fungi; they belong to sub class Hemi Ascomycetidae of the class Ascomycetes.

5. Identification of bacteria, is to determine to which group within a classification scheme, a bacterium of interest belongs, and what genus/species/strain it is. Culture of the microbe in different media are set up as a first step for identification. The microbe in culture, is sometimes recognized, based on certain end products of its metabolic reactions, either because they are originally coloured, or, an indicator is added in the culture medium, that stains the end product. From culture results, a hint is obtained about of the type of microbe. An authentic identification is then carried out through the following methods: **a.** Traditional: phenotypic and biochemical; **b.** Immunochemical: serological; **c.** Genetic: single strand DNA-DNA hybridization (SS DNA hybridization), % G+C content and 16S rRNA gene sequence. Identification of eukaryotic microbial species is comparatively much easier than prokaryotes, because of the larger cell size. Examination of the cell morphology under the microscope, is sufficient to identify eukaryotic microbes up to species level. To find out their phylogenetic relationship etc., 18SrRNA sequencing is used.

6. In prokaryotes, defining species boundaries based on reproductive compatibility (ability to breed), is not possible because they do not reproduce sexually. Species in higher organisms are usually delimited based on reproductive compatibility (ability to interbreed/mate and produce fertile progeny). In prokaryote taxonomy, a species is considered to be *'a group of strains that show a high degree of overall similarity and differ considerably from related strain groups with respect to many independent characteristics'*. Different strains with close resemblance together form a species. A strain is a genetic variant, a sub-type, or a culture variant within a biological species, which can be distinguished from other organisms within the species, based on certain characteristics. A strain comprises of descendants from a single spore/cell culture. To determine which strains, belong to a species, a "type strain" is used for reference. A type strain is one of the strains, that serves as the representative of a species.

7. A microscope is a must, to observe microbes, as they are too small to be seen with naked eyes. There are basically two types of microscopes: optical and electron microscopes. Optical microscope, also called light microscope, uses light (photons) and has glass lenses for magnification. Electron microscope uses fast moving electrons and has electro-magnetic lenses. Various types of light microscopes, to suit different needs are used: brightfield, darkfield, fluorescent, phase contrast, Nomarski optics, confocal etc. Phase contrast and fluorescent microscopes are useful to observe live cells. Nomarski and confocal microscope images are 3D. Electron microscopes are of two types: transmission and scanning (TEM and SEM). Elaborate preparations of the specimens are required for electron microscopy. SEM gives details of only the surfaces. TEM magnifies an object ~100,000 times and therefore gives internal details of a microbial cell.

8. The tremendous diversity in microbes endows them with the competence to thrive in every square inch on earth, however hostile it might be. Microbes are an important, rather essential,

component of all ecosystems on earth. They play a vital role in the complex nutrient exchange system (biogeochemical cycles) in an ecosystem. Microbial ecology involves the study of the numerous interrelationships of the microbes with the living and non-living components of the environment around them. Microbes live in water, soil, snow covered Tundra (Arctic), Taiga (sub-Arctic) and Antarctic regions.

9. Microbes make associations with plants, animals and other microbes with the relationship being symbiosis or parasitism. Gut bacteria, live in the guts of some insects and vertebrate animals including humans. *Buchnera,* a bacterium lives as an obligate symbiont in the guts of aphids (a kind of insects). The human microbiota far exceeds the total number of cells in a human. Zooxanthellae, a group of dinoflagellates, lead a symbiotic (facultative) life, with marine invertebrates like reef building corals, sponges, jelly fish, some bivalve molluscs and flat worms. Chemosynthetic bacteria inhabit the giant tube worms that live in ocean vents (a very hostile environment); a case of obligate symbiosis. Other cases of symbiotic relationships of microbes are: methanotrophic bacteria inhabiting peat moss, nitrogen fixing bacteria in root nodules. Cyanobacteria and microalgae associate with fungi to form lichens. The lichen is a composite organism, wherein neither of the partners of the association, appear as they do when they live separately. Lichens are in many forms and ubiquitous, being present in most ecosystems. Microbes can associate with other microbes. Red tides in seas and biofilms are examples of microbial associations.

10. Biofilms, are a consortium of syntrophic (cross-feeding) microbes, that stick to each other and to a surface. They are ubiquitous and usually found on solid surfaces, submerged or exposed to aqueous solutions, or humid environments. Biofilms can be formed by a single microbial species, or many microbial types like bacteria, archaea, protozoa, algae and fungi. Microbes in a biofilm have significantly different properties from individually living microbes of the same species. The bacterial infections in sinuses and on the dental plaques are in the form of biofilms. Microbes in a biofilm have quorum sensing. It is a communication system between the microbes, wherein, the participating microbes produce signals (which can be different types of molecules in different species and conditions), and respond to the signals in a cell density dependant manner. They achieve the response through regulating gene activity, to produce substances, that can enhance their effect (like virulence, bioluminescence etc). Quorum sensing also enables a biofilm to manage its size.

11. Microbes are cultured in the laboratory for various purposes. Culturing a microbe is mandatory for characterizing and identifying it. Bacteria and yeast are cultured for genetic engineering experiments in the laboratory. On a large scale, microbes are cultured in fermenters for various products. Not all microbes can be cultured artificially in the laboratory. To successfully culture a microbe, it should be provided with a nutrient medium for growth and suitable physical conditions like air, light and temperature. Different types of culture media are used based on the organism and purpose of culturing. To culture microbes, sterile conditions are required. A laminar flow unit with a HEPA filter is used for inoculation of microbes into culture medium.

12. Microbes multiply at a very fast rate when sufficient nutrients are available and suitable environmental conditions prevail. An *E. coli* cell, can grow and multiply within 30 mins, under ideal growth conditions. The time taken for a microbial cell to divide into two is called its generation time. In a liquid culture medium, the time taken for the cell mass or number to double is called cell doubling time. When physical conditions are suitable for microbial growth,

growth rate of the bacterium is dependent on the substrate (nutrient) concentration, being higher when more nutrients are supplied. Under optimal culture conditions, bacterial growth passes through six phases: the lag phase, when the cell number does not increase, acceleration phase, a brief stage when the cell number begins to increase slowly, log phase where the cell number increases rapidly in logarithmic proportions, a very brief deceleration phase where the rate of growth decreases slightly over the log phase, stationary phase when the cell numbers stay constant and finally a death phase when cell numbers start decreasing exponentially. Growth kinetics is the relationship between specific growth rate and the concentration of a substrate. They follow the classical Monad's equation during the log phase of growth.

9.11 SAMPLE QUESTIONS

9.11(a) Subjective Questions

Q.1. Why and how are bacteria cultured? Describe the physical conditions and nutrients used for their culture and methods of setting up cultures and supply of nutrients.

Q.2. Describe growth kinetics of a bacterial culture. How does it follow Monad's equation?

Q.3. Describe with suitable examples the associations that microbes form with other living beings adding a note on the type of relationship in each of the associations.

Q.4. Why is classification of bacteria difficult? How many categories are bacteria classified in Berge's classification? What is the criterion taken in modern times to classify bacteria? How are microbial species defined and identified?

Q.5. What are biofilms? How are they formed? Give some examples of biofilms. How does quorum sensing operate in biofilms?

9.11(b) Objective Questions

Q.1. Microbes were discovered because of

(*a*) Infections.

(*b*) Their opaque cultures on culture plates.

(*c*) Optical Microscope.

(*d*) Electron microscope.

Q.2. Microscope that is useful in study of live microbial cells

(*a*) Darkfield microscope.

(*b*) Phase contrast microscope.

(*c*) Confocal microscope.

(*d*) Brightfield microscope.

Q.3. Protists are

(*a*) Prokaryotes.

(*b*) A kind of eubacteria.

(*c*) Archaebacteria.

(*d*) Eukaryotes hat do not belong to any of the major groups in Eukaryota.

Q.4. Prokaryotic microbes can be reliably identified from

(*a*) 16S rRNA sequencing.

(*b*) 18S rRNA sequencing.

(*c*) Gram staining.

(*d*) Microscopy.

Q.5. A type strain

(*a*) Is a typical representative of a strain.

(*b*) Is useful for identification of a strain.

(*c*) Is the most abundant type among members of a strain.

(*d*) None of the above.

Q.6. Lichens are

(*a*) An association of a group of microbial species.

(*b*) An association between a macro alga and a fungus.

(*c*) A kind of moss.

(*d*) An association between a microalga or cyanobacterium and a fungus.

Q.7. The phase where the microbial culture grows exponentially is

(*a*) Log phage. (*b*) Lag phase.

(*c*) Acceleration phase. (*d*) Growth phase.

Q.8. Qurom sensing

(*a*) Happens in biofilms.

(*b*) Affords a protection to the microbes.

(*c*) Helps in increasing the virulence of microbes.

(*d*) Helps in managing the size of the biofilm.

(*e*) All the above are true.

ANSWERS

1. (*c*)   **2.** (*b*)   **3.** (*d*)   **4.** (*a*)   **5.** (*b*)   **6.** (*d*)   **7.** (*a*)   **8.** (*e*)

10 Biotechnology

Genetic engineering is to traditional crossbreeding what the nuclear bomb was to the sword.

~ **Andrew Kimbrell**

Cloning is great. If god made the original, then, making copies is fine.   ~ **Douglas Coupland**

Fermentation and civilization are inseparable.   ~ **John Clardi**

10.1 INTRODUCTION

Biology was basically a descriptive science until 1900. With the advent of the field of genetics in 1900, due to the rediscovery of principles of heredity elucidated by Mendel thirty-five years before (1865), biology developed into an inference-oriented science. Following the pioneering work in plants, several significant and exciting discoveries were consecutively made in genetics using maize, fruit fly, bacteria and viruses unravelling the mechanism of inheritance, the chemical constitution of genes and their physical location in the cells, and the basis for the extensive variation observed among individuals of the same species. The exciting ride of discoveries continued, and in early 1950's, the exact molecular structure of the genetic material (DNA) was proposed, using circumstantial evidence, without ever having seen the wonder macro molecule! The era of understanding biology at molecular level, was thus initiated in mid-1900. In quick succession, the process by which genes are expressed, the universal genetic code that all living organisms use, and the customary method in which all living organisms replicate their genetic material – all became obvious. It dawned that the language of life is the same in all living organisms starting from simple bacteria to plants to the most complex human. The understanding that life is the result of chemical behaviour of genes became common knowledge and biological sciences gained the status afforded to physical sciences – physics and chemistry. Soon, the application of this knowledge to develop technologies useful for human welfare began. The era of genetic engineering called recombinant DNA technology (rDNA technology), commenced in mid-1970's with creation of transgenics followed by animal cloning, gene therapy, prenatal diagnosis of genetic diseases, DNA fingerprinting, whole genome sequencing with the scope for personalised medicine, and the list goes on. The gamut of these modern techniques, based on knowledge in biology at molecular level, come under the broad umbrella of modern biotechnology, which has gained recognition and importance on par with information technology. What is extraordinary is that, the time gap between understanding of a natural phenomenon, and use of this knowledge to develop a technology, useful in agriculture, medicine or environment, has become immensely short. The most valuable contributions have been in medicine, with some beneficial results in agriculture. The technology has not yet been used to its full potential, because of social, ethical and environmental concerns.

10.2 RECOMBINANT DNA (rDNA) TECHNOLOGY

Recombinant (r)DNA, is a hybrid DNA, created artificially (in vitro), through joining DNA molecules, from two different sources, be it a human and a bacterium, a bacterium and a plant, a virus and a plant, or a plant and an animal. The technology of making rDNA became possible, after the discovery of the tools required to make it – molecular scissors, to cut DNA, and, glue, to join DNA molecules. The scissors and the glue are both enzymes – the enzyme that is used to cut DNA is called a restriction enzyme, and that which is used to join two DNA molecules, is the ligase. The process of cutting and joining two DNA molecules is called splicing.

10.2.1 Restriction Enzymes

Restriction enzymes, also called restriction endonucleases, were discovered in bacteria in the early 1970's. There are several types of restriction enzymes, with more than 600 of them, available commercially. They are conventionally named, based on the bacterium from which they are isolated. The name is three lettered, with the first letter in capital, which is the first letter of the genus name, and the next two letters being, the first two letters of species name of the bacterium. Sometimes, a capital letter is added next to the three alphabets of the name, which indicates the strain of the bacterium from which it is isolated. Some of the enzymes have a roman numeral in their name. When several restriction enzymes are isolated from the same bacterial strain, the enzymes are assigned roman numerals in the sequence of their discovery (**Table 10.1**). For example, the first restriction enzyme isolated from the bacterium *Escherichia coli* strain R, is called EcoRI. Restriction enzymes differ from each other, in the DNA sequence that they recognize and cut. They are classified into five major types: I to V. The restriction enzymes of type II are used to create rDNA. These enzymes recognize a specific short sequence of DNA – usually a palindromic sequence and make double stranded cuts in the DNA molecule. The recognition sequence of these enzymes is four to eight nucleotides long (**Table 10.1**). The restriction enzymes nick the DNA strands to create a staggering (overhanging), or a blunt end (**Table 10.1**).

Table 10.1: Some type II restriction enzymes used in rDNA technology

Bacterium isolated from	Enzyme		End of the cut DNA molecule		
	Name	Recognition sequence[1]			
Escherichia coli	EcoRI	5'G*AATTC 3'CTTAA*G	5'G 3'CTTAA		Overhanging
Thermus aquaticus	TaqI	5'T*CGA 3'AGC*T	5'T 3'AGCT		Overhanging
Bacillus amyloliquefaciens	BamHI	5'G*GATCC 3'CCTAG*G	5'G 3'CCTAG		Overhanging
Haemophilus influenzae	HinDIII	5'A*AGCTT 3'TTCGA*A	5'A 3'TTCGA		Overhanging
Haemophilus aegyptius	HaeIII	5'GG*CC 3'CC*GG	5'GG 3'CC		Blunt

[1] Asterix in the recognition sequence indicates the site at which the DNA strand is cut. The 5' and 3' indicate the ends of antiparallel strands of the DNA molecule. The recognition sequences are palindromic (read the same in opposite strands of DNA in a given direction (like for e.g. the word – MADAM).

10.2.2 Methodology of Synthesising rDNA

The recombinant DNA (rDNA) molecule, is made by joining two DNA molecules from different organisms (occasionally from different regions within the genome of an organism) and hence termed, recombinant. The restriction enzymes that create overhanging ends through cutting, are more convenient in synthesizing an rDNA molecule, than those which produce blunt ends. The staggering ends, also called sticky ends, of two DNA molecules cut with the same restriction enzyme, are complementary to each other. They join through bonding between complementary nucleotides with the staggering ends of the two DNA molecules overlapping each other (**Figure 10.1**).

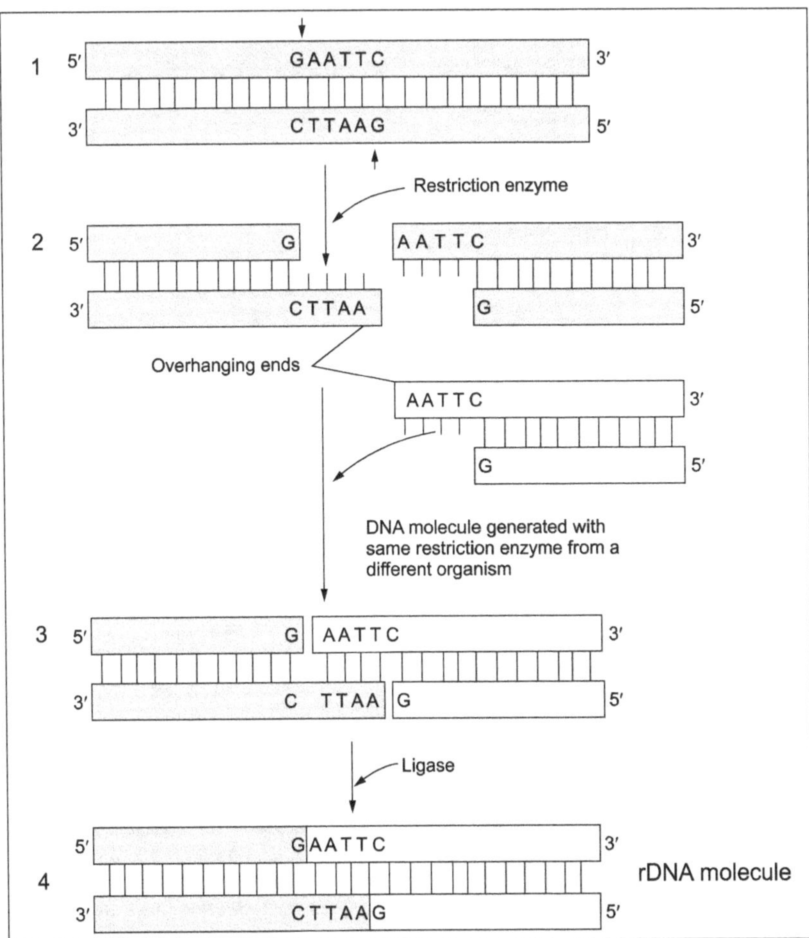

Figure 10.1: Creation of recombinant (r) DNA molecule from DNA molecules with overhanging ends: *Two DNA molecules from different sources, created by cutting with the same restriction enzyme, which generates over hanging (staggering) ends are combined. Due to complementarity of the nucleotides in the overhanging ends, they pair by forming bonds. The sugar phosphate backbone of the two DNA molecules is joined with the aid of ligase which catalyses formation of the phosphodiester bond between the phosphate group of one nucleotide with the hydroxyl group of its neighbouring nucleotide.*

To link a DNA molecule with a blunt end, with a DNA molecule with an overhanging end, to one strand of the DNA with the blunt end, nucleotides which are complementary to the overhanging end of the other DNA molecule are added, to create an overhanging end. The nucleotides on the overhanging ends of the two DNA molecules being now complementary, can bond through overlapping of the strands (**Figure 10.2**)

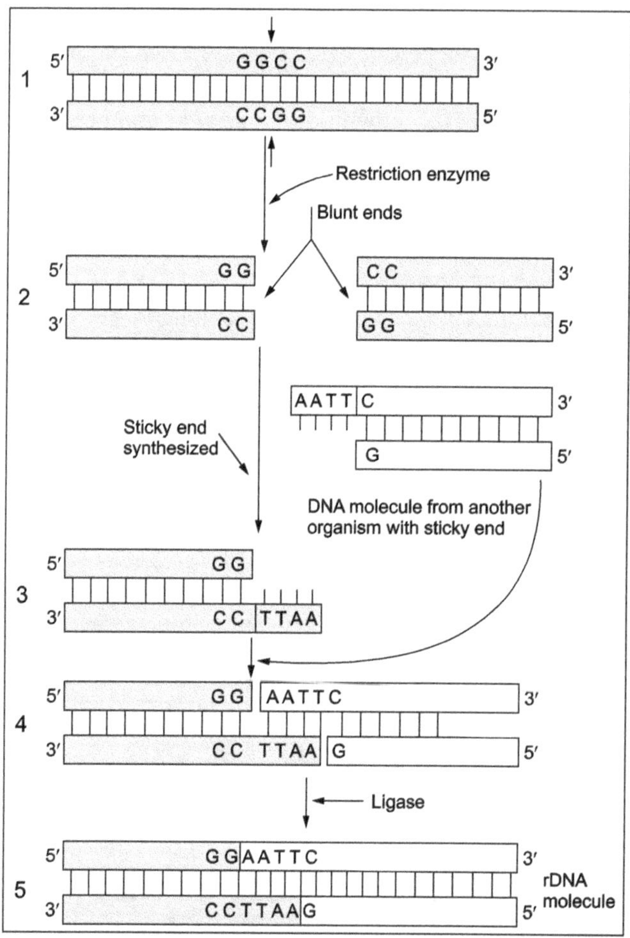

Figure 10.2: Creation of recombinant (r) DNA molecule from DNA molecules one with a blunt end and another with staggering end. *Two DNA molecules from different sources created by cutting with different restriction enzymes, one with blunt end, and the other with staggering end, is joined by adding to one strand of the DNA molecule with blunt end, nucleotides that are complementary to the overhanging strand of the other DNA molecule. The now complementary nucleotides in the single strand extensions of the two DNA molecules pair. The sugar phosphate backbone of the two DNA molecules is joined with the aid of ligase through formation of the phosphodiester bond between the phosphate group of one nucleotide with the hydroxyl group of its neighbouring nucleotide.*

For joining DNA molecules both with blunt (non-sticky) ends, to one of the strand of the blunt end of each DNA molecule, a short length of nucleotide extension is added, to create a staggering or overhanging end. For this, homopolymer tailing with mutually complementary nucleotides is done (**Figure 10.3**).

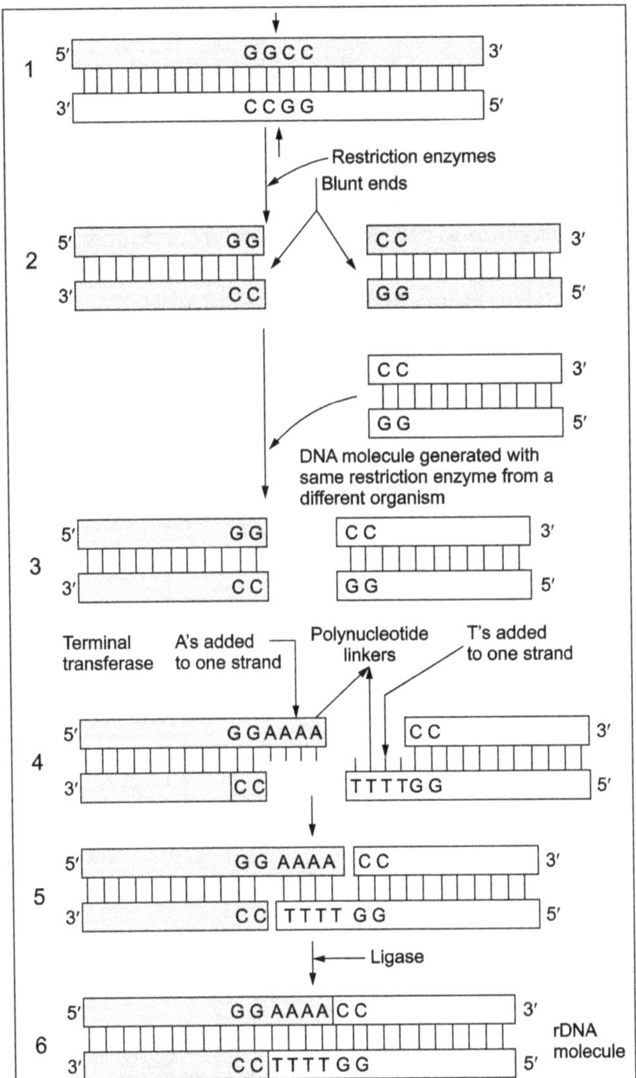

Figure 10.3: Creation of recombinant(r) DNA molecule from DNA molecules with blunt ends. *Two DNA molecules from different sources, created by cutting with a restriction enzyme which creates blunt ends, are joined by adding poly nucleotide linkers (homopolymer tailing). To one of the DNA molecule deoxy adenine triphosphate nucleotides(dATP) are added, and to the other, deoxy thymidine triphosphate nucleotides (dTTP) are added, using the enzyme terminal transferase. The two DNA molecules now link together because of the complementary polynucleotide linker extensions. The sugar phosphate backbone of the two DNA molecules is joined with the aid of ligase through formation of phosphodiester bond between the phosphate group of one nucleotide with the hydroxyl group of its neighbouring nucleotide.*

After the joining of two DNA molecules through bonding between complementary nucleotides in their overhanging strands, a gap remains in the sugar phosphate backbone of the rDNA molecule. This gap is glued with ligase. Ligase catalyses the formation of phosphodiester bond between the phosphate of one nucleotide with the hydroxyl group of the neighbouring nucleotide of the recombinant DNA molecule (**Figures 10.1, 10.2, 10.3**).

10.2.3 Vectors for rDNA

An rDNA molecule, or any DNA sequence of interest, that is to be introduced into another organism, has to be cloned to make more copies of it, for it to be utilized in multiple experiments, and for various purposes. To this end, it is integrated into another DNA molecule which is called vector DNA. There are several kinds of vector DNAs available: plasmids, bacteriophage λ DNA (DNA of λ strain of bacterial virus), cosmids, bacterial artificial chromosome (BAC) and yeast artificial chromosome (YAC). The size of DNA molecule that can be accommodated differs with different vectors. A plasmid can accommodate up to 15kb of DNA, a λ viral DNA can take up to 25 kb DNA, a cosmid, up to 45 kb, up to 350 kb in a BAC and 1000 kb in YAC. The vectors are developed from naturally occurring DNA like the extra genomic plasmid of bacteria and yeasts, λ viral DNA, or, artificially constructed, like BACs and YACs. All DNA vectors, have a short DNA sequence – ori (origin of replication), recognition sequences of several of the commercially available restriction enzymes, and one or two marker genes (**Figure 10.4**).

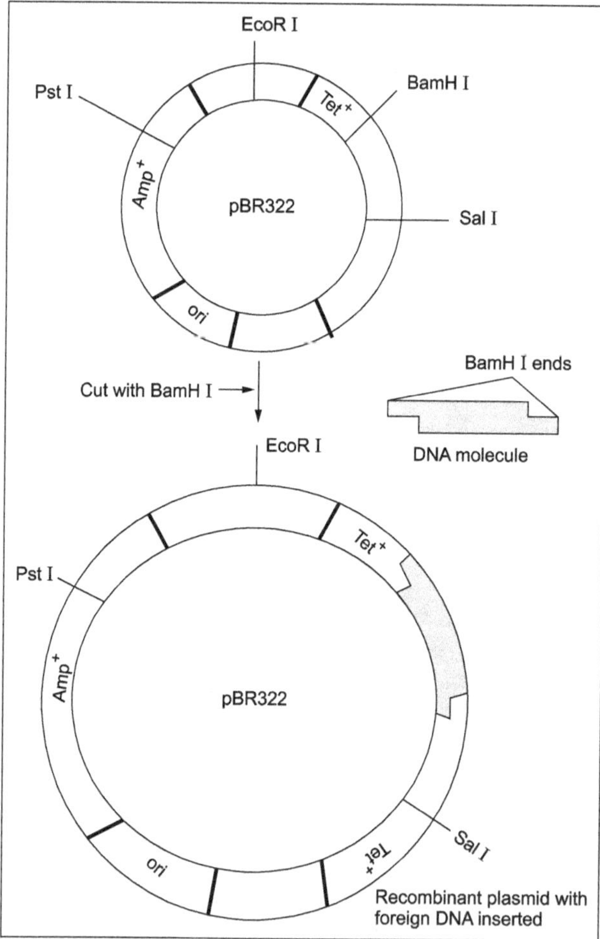

Figure 10.4: Creation of a recombinant plasmid with pBR322. *The pBR322 plasmid has origin of replication (ori), two marker genes – ampicillin and tetracycline resistance genes (Amp⁺, Tet⁺) and recognition sites of several restriction enzymes. The plasmid is cut with BamHI and a DNA molecule with compatible ends for Bam HI cut ends is inserted into it resulting in a recombinant plasmid. Insertion of the DNA molecule in the plasmid results in the inactivation of Tet⁺ gene because the Tet⁺ gene is interrupted with the foreign DNA molecule (insertional inactivation).*

The marker genes can be either coding for resistance to a particular antibiotic like tetracycline, ampicillin, kanamycin or an enzyme (e.g. beta glucuronidase) that catalyses a reaction whose product is coloured. The genetic map with details of the restriction enzyme sites and the marker genes it carries are made available along with each vector. The vector DNA is cut with a restriction enzyme which creates ends complementary to the ends of the DNA molecule that is to be inserted into it. The DNA of interest, gets inserted into the vector DNA through pairing of the complementary nucleotides on the overhanging ends at the two ends of the cut vector. As an example, insertion of DNA into a plasmid vector is shown in **Figure 10.4**. The recombinant vectors with DNA insert are next introduced into their host cells: plasmids into bacteria or yeast; λ viral DNA and cosmids are packed into bacteriophages, which are then used to infect bacteria and thus they enter the bacterial cells; BACs into bacteria and YACs into yeast. To allow the entry of the vector DNA, the bacterial and yeast cells, are treated with chemicals, to bring changes in their cell membrane making them more receptive to outside DNA. Such chemically treated cells are called 'competent'. The cells which take up the vector DNA are called 'transformed' cells. They are recognized and selected based on the expression of the marker gene in the DNA vector (**Figure 10.5**). The vectors are circular and usually extra genomic, i.e., they remain outside the genome of the host cell. They replicate independent of the host DNA, because all vectors have an origin of replication. So, several copies of the vector DNA are made within a single host cell.

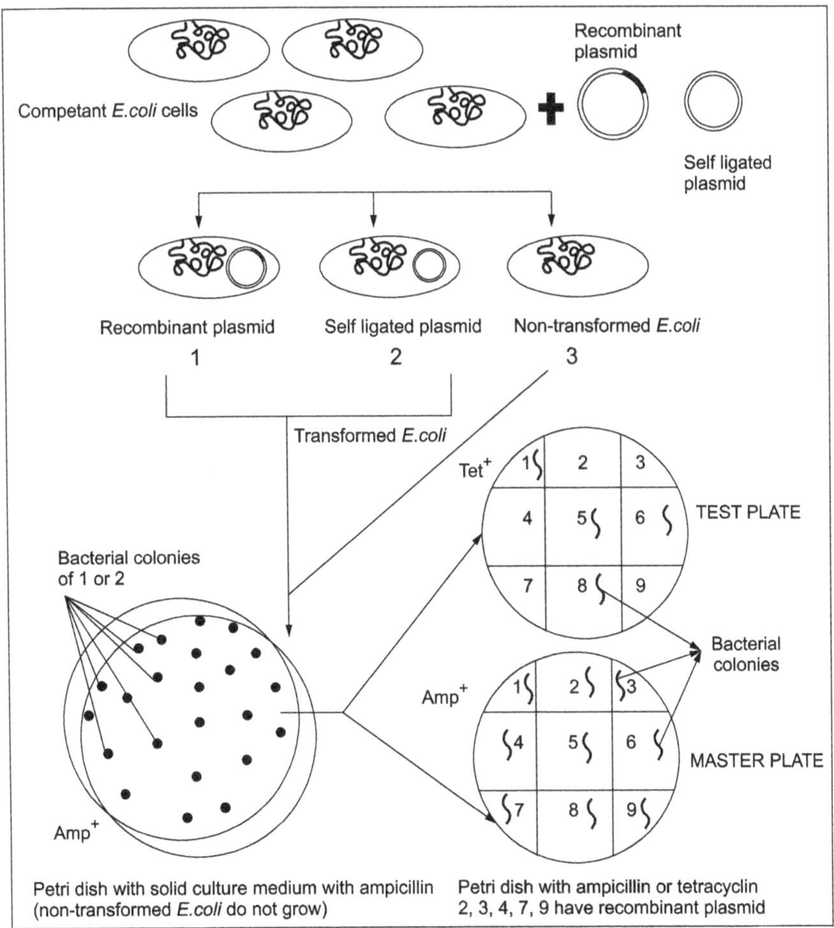

Figure 10.5: Transformation of bacteria (*E. coli*) with a recombinant plasmid.

Competent *E. coli* (*E. coli* treated with chemicals to modify their cell membranes to facilitate entry of plasmid) are incubated with the plasmid recombined with foreign DNA in the EcoRI site knocking out the function of tetracycline resistance gene (from Figure 10.4). The reaction mixture in which the recombinant plasmid is made also contains self-ligated plasmid with no insert of foreign DNA. In the bacterial transformation experiment, three types of bacteria result: non-transformed type with no plasmid, transformed type with a self-ligated plasmid or a recombinant plasmid. The bacterial cells are plated on a medium with ampicillin. Only the transformed bacteria survive because they have plasmids with ampicillin resistance gene. The non-transformed *E.coli* do not survive on medium with ampicillin. The single cell colonies developed on the ampicillin culture are simultaneously inoculated as a wavy line with a tooth pick in the blocks marked and numbered in two Petri dishes one with culture medium containing ampicillin (master plate) and the other containing tetracycline (test plate). The colonies that grow on Amp^+ plate but not on Tet^+ plate have the recombinant plasmid. Transformed *E. coli* with recombinant plasmid cannot grow on a culture medium with tetracycline because the foreign DNA is inserted in the tetracycline resistance gene inactivating it (insertional inactivation). The bacteria with the self-ligated plasmid grow on Tet^+ plate because their tetracycline resistance gene is intact and not interrupted by the foreign DNA. The transformed cells with recombinant plasmid can be picked up from the master plate (**Figure 10.5**).

10.3 TRANSGENIC MICROBES, PLANTS AND ANIMALS

An organism which harbours a DNA sequence of another organism, having been introduced into it artificially, is called a transgenic organism. The sequence that was introduced is termed a transgene and the process of transferring the DNA sequence into an organism is called transgenesis. Expression of the transgene results in a transgenic organism with desired trait.

10.3.1 Methods of Making Transgenics

Different methods are used to introduce a desired DNA sequence into an organism. The DNA molecule of interest is attached to a marker gene. A transformed cell is recognized due to the expression of the marker gene that is transferred along with the DNA of interest.

In bacteria and yeast, the transferred DNA sequence can remain in the plasmid vector which replicates on its own and is thus transmitted to the progeny cells. The DNA sequence introduced through the plasmid into the bacterial cell can express the trait for which it is specific. Plasmids which facilitate the expression of the foreign DNA molecule in the host cell are called expression vectors to distinguish them from plasmids used for cloning DNA (cloning vector).

To create a transgenic multicellular organism like a plant or animal, the transformed cell (with foreign DNA integrated into its genome) should be able to develop into the entire organism such that every cell in the body of the organism has the foreign DNA. Plant cells are totipotent, i.e., a single plant cell, when provided with proper growth medium and physical conditions, can develop into an entire plant. Animal cells do not have totipotency. Therefore, usually a fertilized egg cell is transformed and placed in the womb of the animal, to develop into a transgenic organism. Different strategies are used to introduce DNA into cells of different organisms.

10.3.1(a) Methods of Producing Transgenic Bacteria

For bacteria, the cells are treated with chemicals like magnesium chloride, calcium chloride or calcium sulphate or subjected to electrical treatment. This results in alteration in their cell membranes such that, entry of DNA is allowed. Therefore, such treated bacterial cells are described as competent cells.

The competent cells are frozen and used when required. For transformation, the competent cells are incubated with plasmid DNA in solution and subjected to heat shock at 42°C for two minutes. The transformed cells are identified based on the expression of the marker gene in the plasmid (**Figure 10.5**).

10.3.1(b) Methods of Producing Transgenic Plants

Several methods are used to introduce DNA into plant cells. The most widely and successfully used method is *Agrobacterium* mediated gene transfer. Basically, there are two methods of DNA transfer into plant cells: direct and vector mediated.

10.3.1(b)1 Direct Methods of Transferring DNA

10.3.1(b)1.i Physical Methods

Physical methods involve use of force to introduce genetic material through the cell membrane.

10.3.1(b)1.i(a) Electroporation

This is suitable for plant protoplasts which are obtained by removing the cell wall through physical or chemical means. Also termed electro permeabilization, it involves application of an electrical field to increase the permeability of the cell membrane. To protoplasts in a buffer solution with the foreign DNA, an electric pulse is applied by which the DNA enters the protoplasts (**Figure 10.6**).

10.3.1(b)1.i(b) Microinjection

In this technique, a glass micropipette or microneedle is used to deliver foreign DNA into an immobilised plant protoplast with the aid of a high-resolution microscope. This procedure requires a highly trained technical hand and transformation is limited to one cell at a time (**Figure 10.6**).

10.3.1(b)1.i(c) Microprojectile-Particle Bombardment

The apparatus used is called gene gun. The DNA is coated on tungsten or gold bullets, which are shot through a vacuum, into the plant tissue, from the gene gun. Therefore. it is also referred to as biolistic method (**Figure 10.6**).

10.3.1(b)1.ii Chemical Methods

Natural or synthetic compounds are used to facilitate the transfer of genes into cells

10.3.1(b)1.ii(a) Treatment with Polyethylene Glycol (PEG)

It is suitable for protoplasts. When the protoplasts are incubated in DNA mixed with a solution of PEG 6000, calcium ions and other adjuvants, the DNA enters the protoplasts due to the changes brought about in their membranes with PEG.

10.3.1(b)1.ii(b) Liposome Fusion

Liposomes are vesicular structures with a lipid bilayer. They can be used as a vehicle to transfer DNA. The liposomes with the DNA molecule are induced to fuse with plant protoplasts thereby delivering the DNA into them (**Figure 10.6**).

10.3.1(b)1.iii Vector Based Methods of Transferring DNA

The DNA of interest is inserted in a suitable DNA vector and used for introducing it into the target cell.

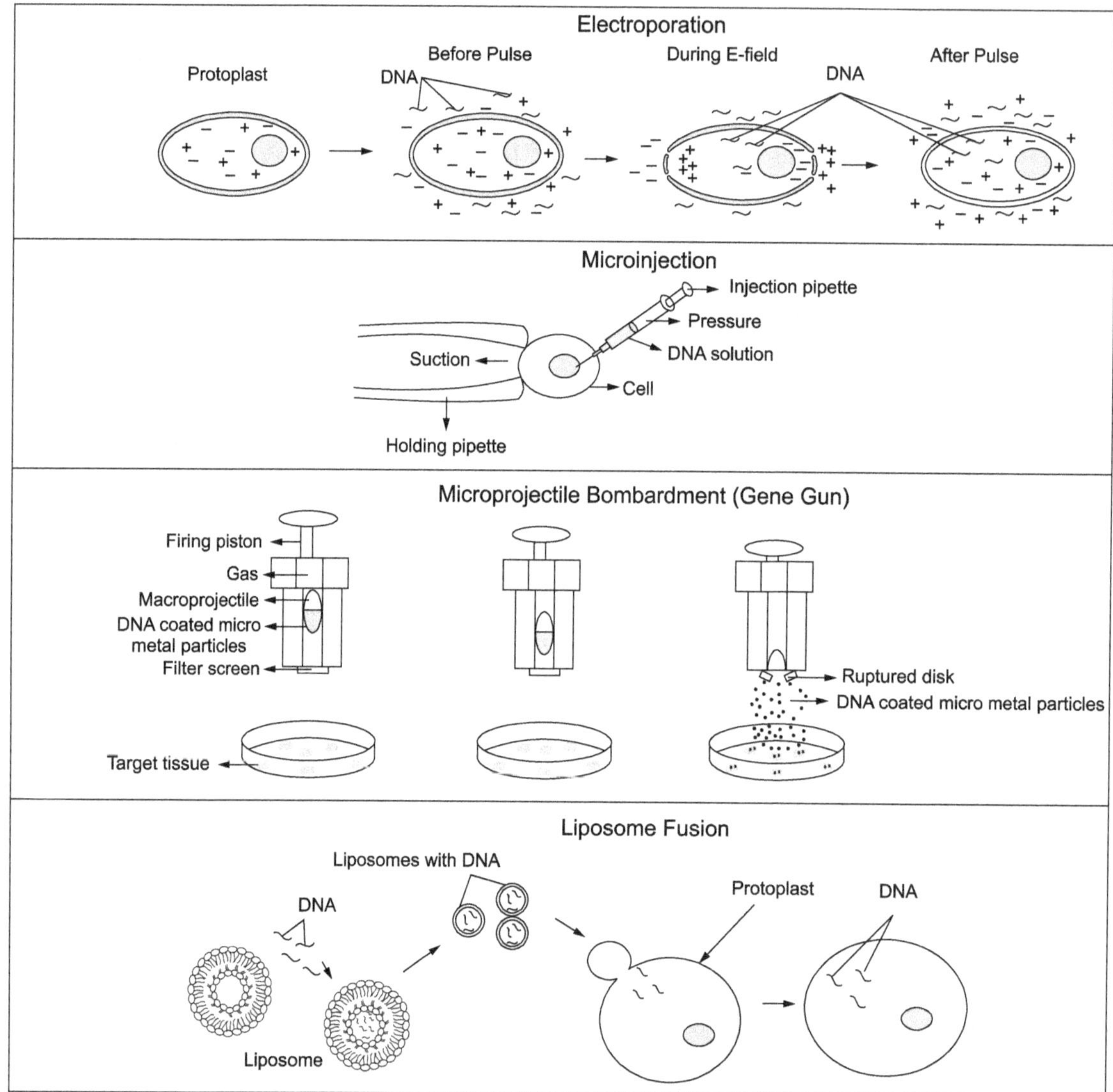

Figure 10.6: Physical and chemical methods of transferring DNA into plant cells. *E.field = Electric field.*

10.3.1(b)1.iii(a) Plant Viral Vectors

The DNA molecule to be transferred, is inserted in a plant viral DNA, in which, virulence genes are knocked out. When plant cells are infected with the virus, its DNA along with the inserted DNA, enters the plant cell. Plant viral vectors are most often used as expression vectors. The avirulent form of viral DNA multiplies and spreads into all cells of the plant and the inserted transgene is expressed in the plant.

10.3.1(b)1.iii(b) Tumour Inducing Plasmid of *Agrobacterium*

Agrobacterium, a soil bacterium that causes tumours in plants is a natural genetic engineer. It has a tumour inducing (Ti) plasmid which carries tumour inducing genes. *Agrobacterium* enters wounded plant cells and the tumour causing genes in the Ti plasmid are integrated into the plant genome, thus engineering its genome to develop tumours. The tumour inducing genes in the Ti plasmid, are bordered on both sides (left and right), with a 25bp sequence, which facilitates the transfer of genes from the plasmid to the plant genome. The tumor inducing genes between the left and right border sequences is termed transfer (T)DNA. The tumour inducing genes in the Ti plasmid are removed, retaining the left and right 25bp border sequences, to create a vector for DNA transfer. The disarmed Ti plasmid (without the tumor inducing genes) used in transgenesis has origin of replication (ori), the left and right border sequences between which there is a multiple cloning site (which contains recognition sequences of several commonly used restriction enzymes), two marker genes one for selecting transformed bacterial cells, and the other for selection of transformed plant cells (**Figure 10.7**). The DNA of interest is inserted in the multiple cloning site. Plant tissues like leaf discs, are co-incubated with *Agrobacterium* with recombinant Ti plasmid. The bacterium infects the plant cells and directly integrates the inserted DNA and the plant marker gene between the left and right border sequences in the Ti plasmid, into the plant genome. *Agrobacterium* Ti plasmid-based vectors are the most convenient, inexpensive and simple way of developing transgenic plants.

The foreign DNA treated protoplasts or tissue are cultured on selective medium to promote the development of whole plants from transformed cells (**Figure 10.8**).

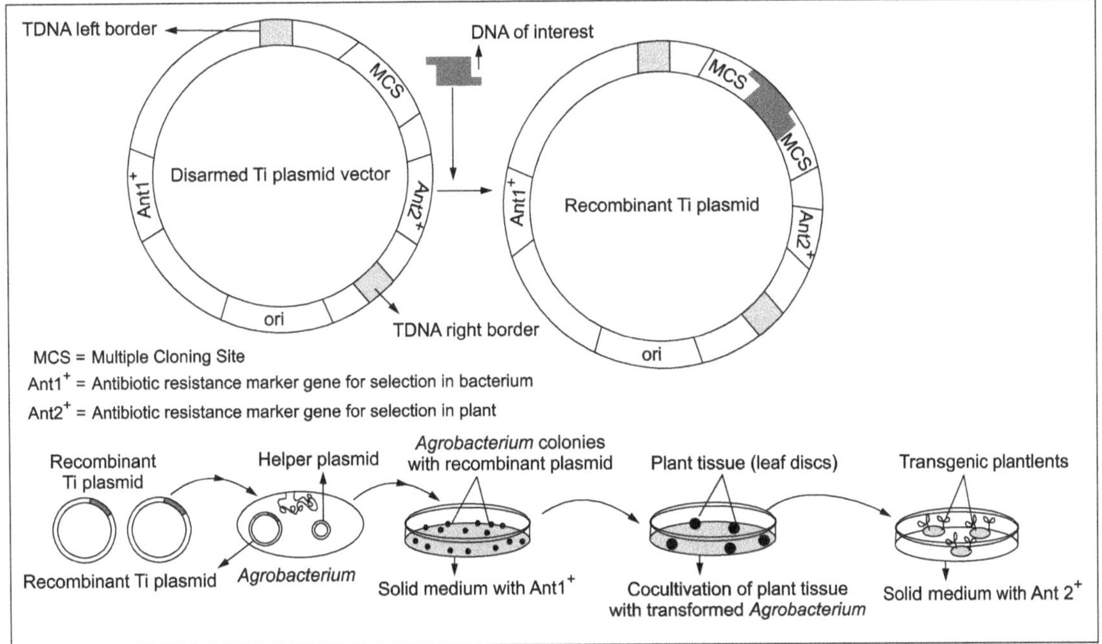

Figure 10.7: *Agrobacterium* **mediated gene transfer into plant cells.** *The DNA of interest is inserted in the multiple cloning site of the Ti plasmid and introduced into Agrobacterium containing a helper plasmid. Agrobacterium transformed with recombinant Ti plasmid is identified based on the expression of the marker gene Ant1⁺(resistance to antibiotic 1). The helper plasmid has genes (Vir genes) which enable the bacterium to infect plant cells. When Agrobacterium transformed with recombinant plasmid is co-incubated with plant tissue like leaf discs, it infects the plant cells and transfers the DNA between the two TDNA border sequences – this includes the DNA of interest and the Ant2⁺ selectable marker gene into plant cells. The plant tissue when cultured on solid medium with Ant2⁺, plantlets emerge from cells that are transformed. These plantlets are later transferred to a rooting medium and finally to soil.*

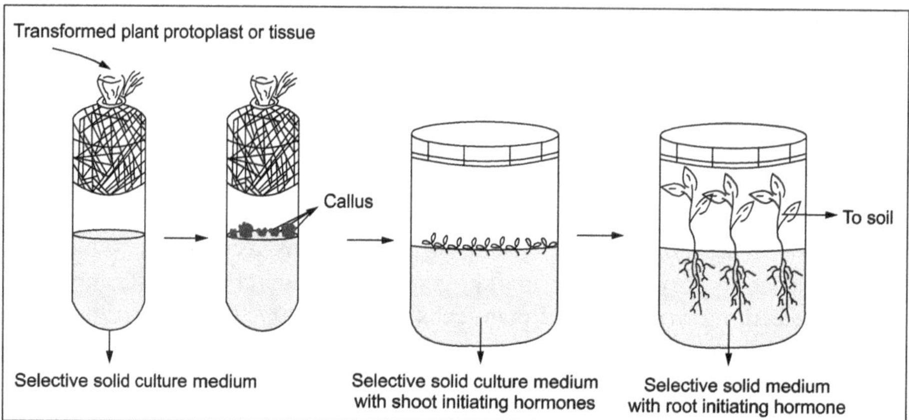

Figure 10.8: Development of transgenic plants in tissue culture using selective (for the transformed cells) medium. *The undifferentiated callus tissue growing from transformed cells on selective medium is transferred to a shoot initiating culture medium. When the shoots differentiate from the callus, they are transferred to culture medium with a hormone composition that induces root development. When the plantlets have sufficiently grown, they are transferred to soil in planters and habituated to grow in a greenhouse and then transferred to soil outdoors.*

10.3.1(b).2 Methods of Producing Transgenic Animals

10.3.1(b)2.i Microinjection

Foreign DNA in aqueous solution is injected into an oocyte/egg/zygote/embryo of animals in vitro with a glass micropipette as described above for the plant protoplasts (**Figure 10.6**). The cell with the DNA injected into it is introduced into the womb of the animal to develop. This is the most commonly used method of producing transgenic animals.

10.3.1(b)2.ii Embryonic Stem Cell-Mediated DNA Transfer

Stem cells are undifferentiated cells that have the potential to differentiate into any type of cell and can give rise to a complete organism. Desired DNA sequence is inserted into embryonic stem cell by homologous recombination. The transformed cells are then incorporated into an embryo at the blastocyst stage resulting in a chimeric animal – i.e., some parts of the animal have the transgene and some don't. This is because the embryo from which it developed had non transformed cells (originally in the blastocyst) and the transformed cells with the transgene that were introduced into it.

10.3.1(b)2.iii Retrovirus Mediated Gene Transfer

An RNA virus with the desired DNA molecule inserted into its genome is used to transfer DNA taking advantage of its ability to infect host cells. Transmission of the transgene is possible only when DNA synthesized from the template retroviral RNA integrates into the genome some of the germ cells.

10.3.2 Transgenic Microbes

The first transgenic organism to have been created in 1978 was a microbe – *E.coli* with human insulin gene. Four years later, it was approved for production of human insulin termed Humulin. Later, genetically modified bacteria have been developed that produce human hormones and proteins for treatment of diseases (**Table 10.2**). Transgenic microbes have also been developed that produce proteins used in food industry (**Table 10.2**). All these are commercially produced today. The transgenic microbes developed for saving crops from frost and those developed for bioremediation (**Table 10.2**), are not widely used, because it involves deliberate release of genetically modified organisms (GMOs) into the environment, where their long-term effects cannot be predicted.

Table 10.2: Products from genetically modified microbes

Product	Used for
Human Health	
Hormones	
Human insulin (Humulin)	Managing diabetes.
Human growth hormone (Somatostatin)	Human dwarfism.
Erythropoietin	Correction of anaemia.
Proteins	
Interferons	Cancer treatment.
Tissue plasminogen activator	Dissolving blood clots in ischemic region of the brain to treat ischemic strokes and myocardial infraction.
Clotting factors	To treat haemophilia.
Epidermal growth factor	Wound healing and skin maintenance.
Lung surfactant protein factor VIII	To treat clinical acute lung injury and acute respiratory distress syndrome.
Interleukin 2	To boost immune system to kill cancer cells.
Viral vaccines (measles, Hepatitis B etc)	Prevention of viral diseases.
Food Industry	
Alpha amylase	Baking.
Chymosin	Cheese making.
Pectin esterase	Fruit juice making.
Agriculture	
Frost minus bacteria (Frostban)	Prevent frosting of leaves and fruits.
Environmental Cleanup	
Super bug	Bioremediation of toxic wastes and oil spills.

10.3.3 Transgenic Plants

The first transgenic plant was released in 1994. It was the 'flavr savr' tomato which had an introduced transgene that delayed fruit ripening. Flavr savr tomato has a long shelf life facilitating its transport to long distances. It was not however extensively cultivated. Since then, several transgenic plants in different crops have been created and tested. The USA approved genetically engineered versions of 19 plant species, but only a few are commercially cultivated (**Tables 10.3, 10.4**). In India, BT cotton has been approved and cultivated. The most popular transgenics are herbicide (glyphosate, trade name – Round up) tolerant and insect resistant crops. The insect resistant transgenic plants carry a gene from a bacterium, *Bacillus thuringiensis* (BT), which codes for a crystal protein that is toxic to insects. BT crops have an environmental benefit of reducing pesticide usage. Some of the transgenics have multiple transgenes (stacked genes). The largest global acreage of genetically modified crops are soybean, corn, cotton, and canola, in that order. The USA has the greatest area under cultivation of transgenic crops – about 40% of the world's total area under cultivation of transgenics. More than 80% of corn, cotton and soybean cultivated in the USA is transgenic. Other large producers include Brazil, Argentina, India, and Canada. There is a large opposition to genetically engineered crops in Europe. Some of the transgenics that have been made but not commercialised are listed in **Table 10.4.** Thus, there is a great deal of unrealized potential in transgenic plants.

Table 10.3: Commercially cultivated transgenic plants

Crop	Conferred character
Corn	Insect resistance, Drought & herbicide tolerance.
Cotton	Insect resistance & herbicide tolerance.
Sugar beet	Herbicide tolerance.
Soybean	Herbicide tolerance.
Canola	Herbicide tolerance.
Alfa alfa	Herbicide tolerance.
Summer squash	Herbicide tolerance.
Papaya	Virus resistance.

Table 10.4: Some transgenic plants created but not commercialised

Transgenic Plant	Trait
Yellow rice	Beta carotene in endosperm (improved nutritional quality).
Nutritionally superior oil containing canola and soybean	Improved nutritional quality.
Virus resistant tomato, potato and tobacco	Plant protection.
Salt tolerant tomato	Can be cultivated in saline soils.
BT tomato, cauliflower, cabbage, rice, potato, brinjal	Insect resistance.
Tearless onions	Less pungent.
Several transgenics that produce antibodies against different viruses or antigenic proteins that can serve as vaccines.	Disease prevention.

10.3.4 Transgenic Animals

Transgenic animals have been made mainly for the following five purposes: **1.** Disease models; **2.** Biological models; **3.** Biopharming or transpharming i.e., production of pharmaceuticals (therapeutic proteins). The transgenic animals practically serve as bioreactors; **4.** To serve in Xenotransplantation i.e., as organ donors for humans; **5.** In animal husbandry for disease resistant animals, animals with improved quality of milk or faster rate of growth (**Table 10.5**). However, none of these transgenics are commercially exploited. The significant contribution of transgenic animals has been from their use as disease and biological models. Disease models have been useful to study the progression of a disease and for drug testing (**Table 10.5**). Among transgenic biological models, knock-out and knock-in mice were used to understand the function of several genes (**Table 10.5**).

Table 10.5: Some animals created through genetic engineering

Transgenic animal	Utility
Disease models	
Alzheimer's mouse	To study the progression of disease, the behaviour of the genes causing disease symptoms, disease pathways and effectivity of drugs.
Onco mouse	
AIDS mouse	
Parkinson's fly	
Biological models to study gene function	
Knock-out and knock-in mice	For study of what happens to an organism when function of a gene is increased (knock-in), decreased, or completely shut off (knock-out).
Smart mouse 'Doogie'	Enhanced memory and learning due to over expression of N-methyl-D-aspartate (NMDA) receptor subunit NR2B.
Transpharming or Biopharming **(The product of the transgene is present in the milk)**	
Transgenic goats with human antithrombin III (hAt) gene (commercial trade name: ATryn®)	Anti-coagulant of blood (for patients with genetic deficiency for antithrombin protein resulting in abnormal blood clots).
Transgenic sheep for ATT gene (alpha-1-antitrypsin)	Treatment of a lung disease – Emphysema.
Transgenic goat with tissue plasminogen activator	Breaks down blood clots for treatment of ischemic stroke.
Transgenic goat and pig with coagulation factor IX gene ATryn®	Treatment of haemophilia.
Transgenic rabbits with C1 esterase inhibitor gene (commercial trade name: Ruconest)	Acute attacks of hereditary angioedema (swelling under the skin in different areas of the body).
Transgenic goats with dragline silk protein from spiders (Bio steel)	For sutures, ballistic (bullet) protection.
Food	
Transgenic salmon 'Aqua Advantage' with growth hormone gene	Fast growth rate reducing cultivation costs.
Transgenic cows with alpha and beta casein genes	Improved efficiency of cheese production (a multibillion-dollar industry).
Transgenic cows with lysostaphin gene	Resistance to bacterial infection induced mastitis (a breast disease).
Xenotransplants (organ donors)	
Pigs with 1,3 galactosyl transferase gene knocked out	Surface proteins on the cells of organs which act as antigens and result in tissue rejection in the recipient are not produced. Therefore, organs can be transplanted without risk.
Companion animals (pets)	
Transgenic zebrafish with glow gene (fluorescent protein gene) from jelly fish and sea anemone (trade name: Glofish™)	Glowing fish as attractive ornamental pets.
Cats with cat allergen gene knocked out	Hypoallergenic cats for humans to pet.

10.3.5 Problems with Transgenics

Compared to the potential of transgenics, the extent of their commercialisation and utilization is very low. This is due to several reasons. Production of transgenic microbes is relatively easy and straightforward. Production of transgenic plants and animals is expensive. Transformation rates are not high, several transgenics are to be screened to pick up individuals with optimum expression of the gene. The transgene can be unstable – it can either get eliminated or lose expression. On top of these biological problems, there is widespread apprehension among public regarding the safety of the products from transgenics. Therefore, it has been made statutory to label products from genetically modified organisms so that consumers are well informed and can choose to use them or not.

10.3.6 Ethical, Environmental, Social and Legal Issues with Transgenics

There have been long debates on various issues related to the transgene in the genetically engineered organisms crossing species boundaries and getting unintentionally inserted into other organisms creating organisms with genes not natural for them, which may turn them to super weeds or monsters. Environmentalists and social activists have mustered a lot of support to rally against the use of genetically modified organisms (GMOs). Public at large are apprehensive of their use. Moratoriums have been imposed several times on GMOs. The approval of use of GMOs is done after stringent examination.

Ethical concerns mainly revolve round the unacceptability of 'playing god' through creation of organisms with altered genomes. With genetically engineered animals, the ethical concerns are even more. They relate to subjecting the animal to hardship (super ovulation, implantation of embryos, surrogacy, miscarriages etc.) for creating a GMO. The suffering that a transgenic animal is subjected to when created to serve as a disease or biological models is disquieting. The side effects of the genetic modification on the health and wellbeing of the animal are also a matter of concern. However, humans have been altering animals in several ways through breeding (exotic animals and pets) for ages. The main religions of Christianity, Islam and Hinduism accept manipulation of animals if it serves human welfare.

Environmental concerns are with respect to the free release of GM microbes into environment like the super bugs (bacteria resistant to many antibiotics) and ice-minus bacteria. The inadvertent effects they might pose to environment are impossible to predict. Similar are the concerns with transgenic plants – the transfer of the engineered gene like herbicide tolerance from a transgenic to a weed plant might wreak havoc. A transgenic with an insect resistance BT gene might affect pollinators like bees and butterflies. Non-governmental organizations (NGOs) such as 'Gene Campaign', 'Centre for Sustainable Agriculture', 'Research Foundation for Science, Technology and Ecology', 'Greenpeace', and 'Friends of the Earth' have also raised concerns relating to genetic manipulation of plants. The food from GMO is branded as "franken food". Another concern is loss of biodiversity due to large scale cultivation and use of transgenics.

Social issues concerns are about takeover of agricultural sector by large agricultural estates, with no place for small land holding farmer, loss of employment to labour involved in weeding, increased import of herbicides, loss of export market of conventional agricultural products etc.

Legal issues relate to patenting of GMOs created from genes isolated from organisms native to developing countries by multinational companies. The disadvantaged farmers in these poor countries need to pay for their own property.

There can be counter arguments for each of these concerns. Less pesticide usage in insect tolerant plants, human health benefits with large scale production of therapeutic proteins, vaccines and effective drugs etc. are some of them. The crux of the matter lies in the philosophy that 'no activity is risk free'. One has to weigh the obvious advantages and the apparent or perceived disadvantages of a GMO to accept or reject the new technology. It can be a case-by-case review but totally banning transgenics is neither wise nor logical when we have come this far.

10.4 RECOMBINANT VACCINES

10.4.1 Mechanism of Immunity to a Pathogen

Vertebrate animals defend themselves from pathogenic (bacterial, viral, plasmodial) infection, in two ways. The first line of defence, is by preventing their entry into cells, while they are circulating in the blood, or lymph, by engulfing them. The second line of defence, when the pathogens have already entered

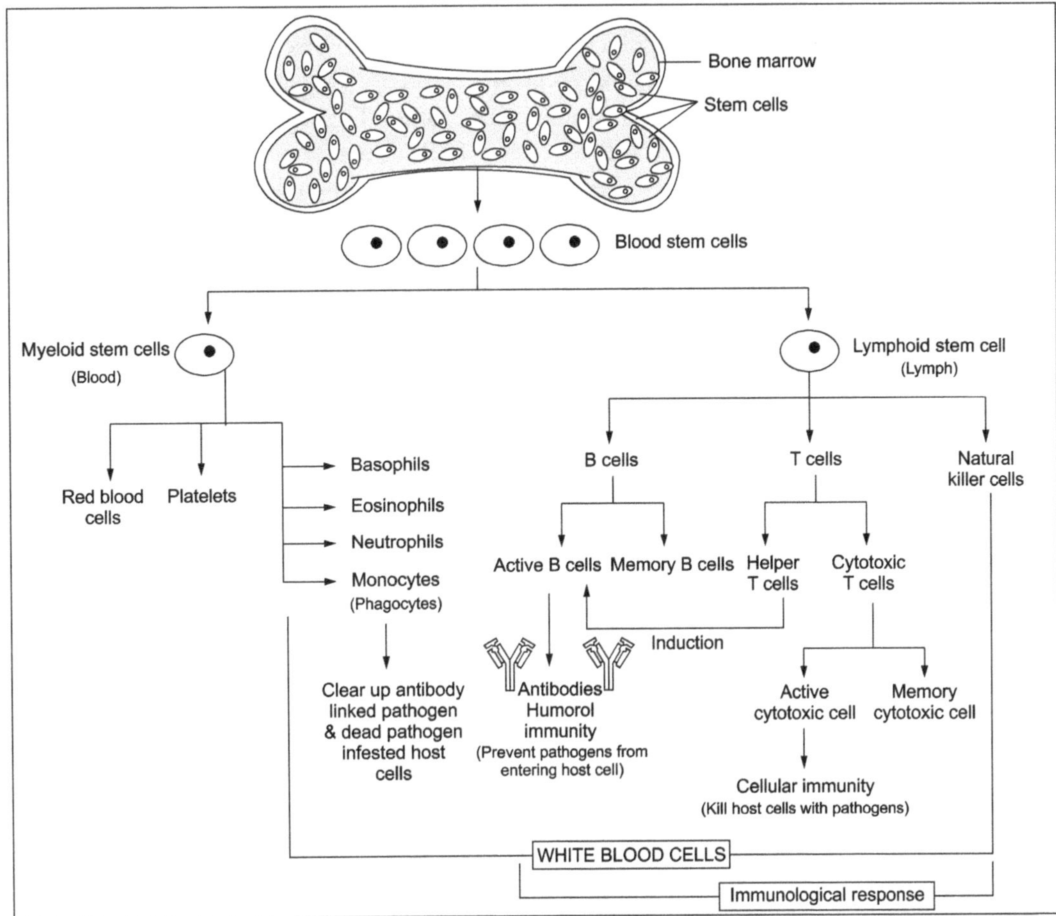

Figure 10.9: Immunological response to a pathogen. *The lymphoid stem cells originating in the bone marrow differentiate into three kinds of cells – B, T and natural killer cells. T cells are of two types: helper and cytotoxic T cells. Both B and T cells differentiate into two versions: active and memory cells. Active cells attack the pathogen. The helper T cells induce active B cells to synthesize antibodies specific for the pathogen. The antibodies bind to the surface of the pathogen when it is circulating in the blood or lymph and prevent it from entering the host cell (humoral immunity). The active cytotoxic T cells kill pathogen infested cells i.e., host cells into which the pathogen has already entered (cellular immunity). The monocytes from the myeloid stem cells engulf (therefore called phagocytes) the dead cells infested with pathogens inactivated by the antibodies thereby cleaning up the debris created by pathogen attack. The B and T memory cells have long life span and give immunity to further infections by the pathogen. Vaccines dupe the pathogen and prime the immunity of the host, so that, the memory T and B cells, can readily attack an invading pathogen and prevent it from causing disease.*

the host cells, is by inducing lysis of the infected cell, killing both the pathogen and the cell. The first line of defence is termed humoral immunity and the latter is called cellular immunity. The defence exhibited by the host is termed immune reaction. Immunological reactions are mediated by white blood cells called lymphocytes that originate from the stem cells in the bone marrow. Three types of lymphocytes: B cells, T cells and killer cells help in defence against a pathogen (**Figure 10.9**). The B cells mature in the bone marrow (therefore named 'B'), and T cells travel to the thymus (in the brain), and mature there.

The T cells are of two kinds: helper T cells and cytotoxic T cells. The helper T cells induce the B cells to produce antibodies. The cytotoxic T cells induce lysis of pathogen infected cells, destroying both the host cell and the pathogen. Phagocytes formed from the myeloid stem cells engulf the lysed cell and the material released from it (**Figure 10.9**).

When an organism is able to produce sufficient antibodies and cytotoxic substances to neutralize the pathogen, the pathogen induced disease cannot progress to severity or become fatal.

When challenged with a pathogen, both B and T cells produce two types of cells – those which act immediately on the pathogen and memory cells (**Figure 10.9**). The memory B and T cells have very long-life spans (decades or life time of the organism) constituting immunological memory. It is due to the immunological memory that a person acquires resistance (is immune) to the pathogen and does not suffer the disease when he is exposed to the same pathogen. The memory cells immediately start producing antibodies and cytotoxic lymphocytes within a short time of the pathogen infection and kill it before disease is initiated.

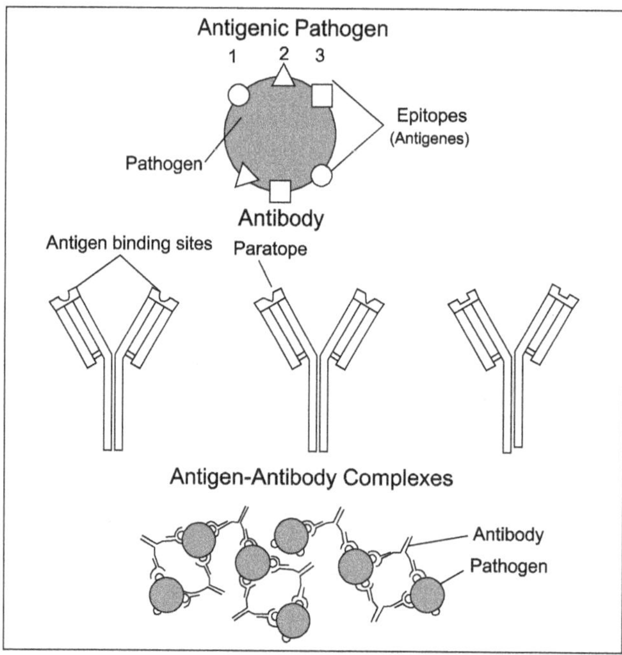

Figure 10.10: Antigens and antibodies: *The surface of the pathogen has several kinds of proteins, that the immune system of an animal, recognizes as foreign. These are called antigens, since they induce the immune system to produce antibodies. The relation between an antigen and its antibody is like a lock and key. The topology of the antigen recognizing stem tips of the antibody termed paratopes match the contours (epitope) of one of the antigenic proteins on the surface of the pathogen. The antibodies bind to the pathogen, their paratope fitting into the epitope. The pathogen – antibody complexes are thus formed and the pathogen is prevented from entering the host cell. These complexes are engulfed by the phagocytes (a type of white blood cells called monocytes) present in the blood.*

B cells are in the first line of defence – they prevent the entry of the pathogen. A protein, or a part of protein, on the surface of the pathogen is recognized by activated (by helper T cells) B cells as foreign (antigen). They produce antibodies against the antigen. The relation between antigen and antibody is highly specific, just as an enzyme and its substrate (**Figure 10.10**). The antibodies bind to the antigenic portion of the pathogen surface. The pathogens are meshed up with antibodies and finally engulfed by the phagocytes that are developed from monocytes which are a type of white blood cells (**Figures 10.9 and 10.10**).

10.4.2 Principle of Immunization Through Vaccination

Preparedness for a threatening pathogen, through developing immunological memory (T and B memory lymphocytes), is the principle on which a vaccine works. Vaccination helps priming the individual for a ready attack on an invading pathogen, in the event of an infection. Almost all current vaccines work through the induction of humoral antibodies: IgG and IgA in serum and mucosa, although it is ideal, if a vaccine induces, cellular immunity as well.

A vaccine is a biological preparation, that provides active acquired immunity, against a certain pathogen induced disease. A vaccine works, by presenting a dead or attenuated pathogen, or pieces of it, to the immune system of a host organism (humans or domestic animals), that sparks an immune reaction – production of T and B memory lymphocytes, resulting in immunological memory of the pathogen. When an immunized human or animal, is exposed to the real pathogen, the memory cells immediately trigger the host defence, and prevent disease due to infection.

The original strategy of vaccinology was to 'isolate, inactivate, and inject' the pathogen. This approach is sometimes limited by the fact that, the material for vaccination is not efficient, not available, or generates deleterious side effects. A possible alternative, is the use of recombinant vaccines. There are several categories of recombinant vaccines, which are based on molecular biological and rDNA technology. Several strategies are used to make them and based on the method used, they are classified into three major types: live genetically modified vaccines, recombinant inactivated subunit acellular vaccines, genetic vaccines.

10.4.2(a) Live Genetically Modified Vaccines

These are of two kinds: attenuated pathogens, vaccine vectors.

10.4.2(a)i Attenuated Pathogens

An attenuated pathogen cannot cause disease, but can serve as an antigen to induce host immunity. The genome of a pathogen, like a virus or bacterium, can be modified in several ways, to create an attenuated form: **a.** by deletion or inactivation of one or more genes responsible for disease (virulence genes); **b.** insertion of a promoter sequence which delays the expression of virulence genes; **c.** over express some genes, like those responsible for development of flagella or cilia, that lead to loss of pathogenicity. Inactivation of disease-causing genes is done through a molecular biology technique called site directed mutagenesis. rDNA technology is used for deletion of virulence gene(s) or insertion of genes or promoter sequences. The attenuated pathogen is multiplied in a suitable host cell culture (like animal cell lines or chicken eggs), and used as a vaccine (**Figure 10.11**).

For dealing with the Corona 19 pandemic, Sinovac (from China), Covaxin (by Bharath Biotech company, India) vaccines, were developed from attenuated virus.

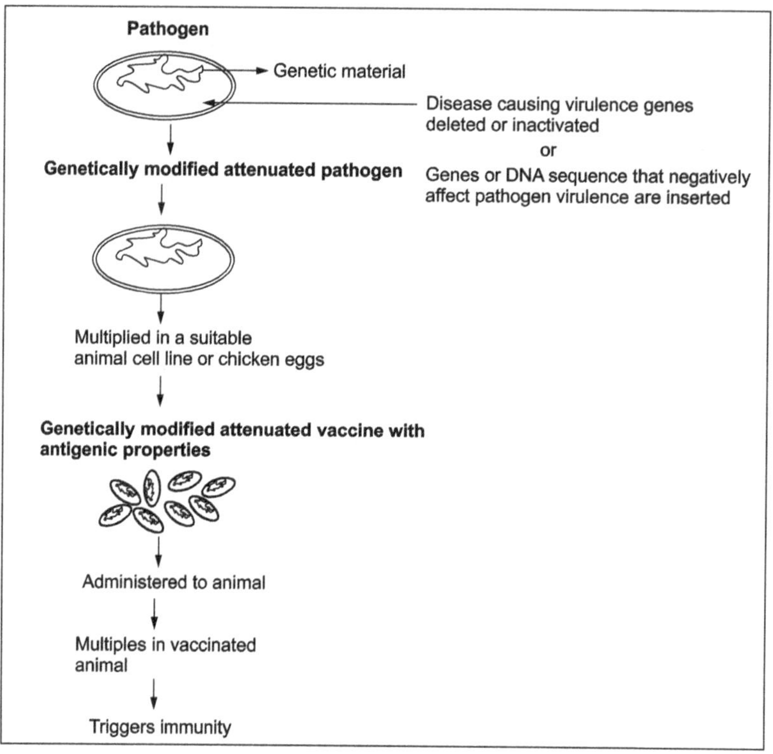

Figure 10.11: Production of live genetically modified attenuated pathogen with antigenic properties that serves as a vaccine.

10.4.2(a)ii Vaccine Vectors

A gene from a pathogen, that can elicit immune response, is inserted into a non-pathogenic microbe. The genes coding for surface proteins of the pathogen (e.g. spikes in viral pathogens), are usually used for creating the recombinant non-pathogenic microbe. The recombinant microbe is multiplied in a suitable cell system and harvested and used as a vaccine (**Figure 10.12**). It effectively infects and replicates in the host (vaccinated animal) and expresses the transgene resulting in production of protein with antigenic properties of the pathogen, thereby triggering the immune system of the host.

Thus, these vaccines with recombinant microbes, serve as vectors to transfer immunity eliciting genes, into an animal or human. Vaccines for veterinary animals have been made using different strains of non-pathogenic vaccinia viruses and adenoviruses using this strategy. The UK based Astrazeneca company and a Chinese company, developed a vaccine marketed as 'Covishield' in India, based on this technology to deal with the Corona 19 virus pandemic.

Live genetically modified vaccines elicit strong humoral and cell-mediated immune responses, resulting in immunological memory. They are relatively inexpensive and easy to transport. However, when a live attenuated virus is used, there is always a risk of reversion to virulence. This is overcome by knocking out more than one virulence genes of the pathogen. Another issue with the vaccines that use adenoviruses as vector is that, they have a potential to transform host cells to a cancerous phenotype.

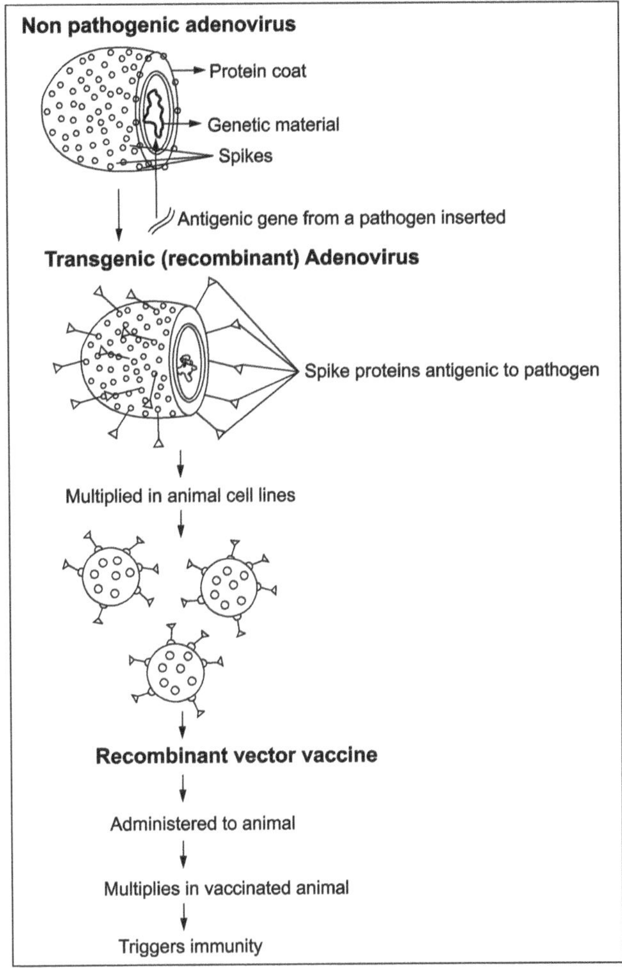

Figure 10.12: Production of live recombinant vector vaccine

10.4.2(b) Recombinant Inactivated/SubUnit/Acellular Vaccines

Subunit vaccines contain usually a protein or a part of the protein of one of the several proteins that are present in the pathogen surface which can serve as an antigen and provoke the host immune system. The gene coding for the protein (or part of it) is introduced into a plasmid or viral vector and the recombinant vector is expressed in suitable eukaryotic host cells like yeast, insect, animal or plant cells resulting in the production of antigenic protein (**Figure 10.13**).

The insect viral (bacculo virus) vector/insect cell system is more commonly used for production of the sub unit vaccines. Some experiments indicate that duck weed and microalgae are promising for synthesis of recombinant vaccines. The major advantage with this system is that it serves as a universal "plug and play" system to produce a variety of vaccines for humans. Yeast (*Pichia pastoris*) was used to synthesize hepatitis B virus vaccine. Novovax, a vaccine for the notorious Corona 19 pandemic was developed based on subunit protein.

Subunit vaccines need an adjuvant – a substance that stimulates the immune system of the animal to respond to the vaccine. Alum (aluminium based salt) is the only approved adjuvant available to date. However, it is a weak adjuvant for antibody induction against recombinant vaccines.

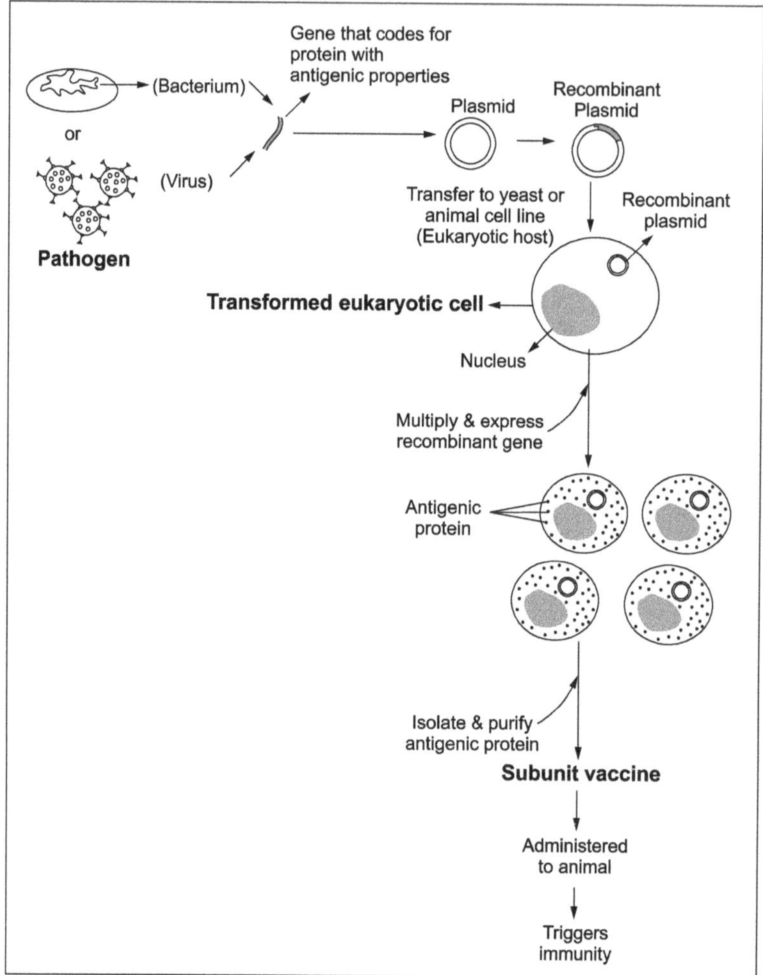

Figure 10.13: Preparation of recombinant subunit vaccine

10.4.2(c) Genetic Vaccines

Genetic vaccines contain no protein or polysaccharide. The active principle in these vaccines is a nucleic acid.

10.4.2(c)i DNA Vaccines

A gene from a pathogen coding for protein or short peptide with antigenic properties, is inserted into a plasmid and the recombinant plasmid is multiplied in a bacterium. The recombinant plasmid is then isolated from the bacterial culture and purified. This 'naked' DNA, is used as a vaccine and injected intramuscularly or intradermally (into the skin). The pathogen gene in the recombinant plasmid is expressed, as a result of which, the antigenic protein is synthesized in the vaccinated animal, resulting in its immunization (**Figure 10.14**). Genetic vaccines also have been designed, to include immune-stimulatory genes along with the gene for antigenic protein. Several DNA vaccines are in clinical trials.

10.4.2(c)ii mRNA Vaccines

These vaccines consist of mRNA encoded by antigen genes of the pathogen. A bacterium is transformed with a recombinant plasmid, with a gene for antigenic protein, where it is transcribed into mRNA. The mRNA is harvested from the transformed bacterial culture (**Figure 10.14**). When it is administered into host cells, it translates and produces protein antigens, that elicit immune response. mRNA vaccines initiate a quicker response compared to DNA vaccines, because one step (transcription) in the synthesis of antigenic protein, is shortened. The instability of mRNA, has been largely overcome with recent technological advances. mRNA vaccines hold great promise, particularly to tackle the emergence of pandemic pathogenic strains. The Moderna Covid 19 vaccine, and Pfizer-BioNTech Covid 19 vaccine based on mRNA for spike protein was widely used to contain the pandemic of 2020–2021.

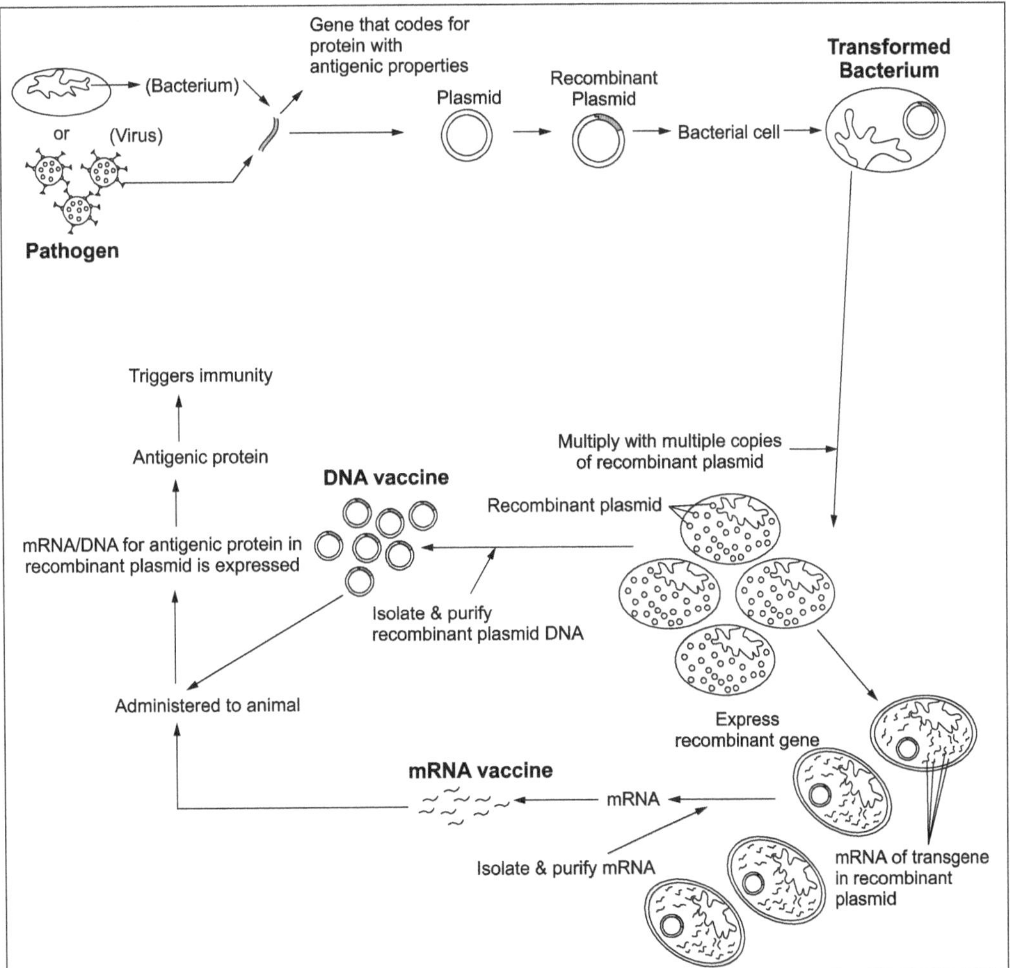

Figure 10.14: Preparation of DNA and mRNA vaccines. *Both of them are expressed in the vaccinated animal/ human as antigenic proteins that elicit immunity. mRNA vaccines have a faster activity because it is just one step of translation for the antigenic protein to be synthesised; the recombinant gene in the DNA vaccines has to be transcribed to mRNA which is to be processed and translated to protein.*

The nucleic acid and subunit recombinant vaccines are highly safe. They are safer than the genetically modified attenuated vaccine types. Recombinant vaccines lack pathogenicity, have defined composition and their production systems have been simplified. The only disadvantage is, multiple doses of vaccine need to be administered and adjuvants are required in several occasions.

10.4.2.(d) Reverse Vaccinology to Identify Candidate Genes which Code for Proteins with Antigenic Property

Identification of genes of the pathogen that can act as antigens is the first step in recombinant vaccine development. The entire genome of the pathogen is sequenced and screened by employing bioinformatics tools to explore all genes of the pathogen that can serve as antigens. The antigen most effective in eliciting immunological response can thus be identified. Effective vaccines against *Meningococcus* B, and antibiotic-resistant *Staphylococcus aureus* and *Streptococcus pneumoniae* have been developed with reverse vaccinology.

Though recombinant vaccines have several advantages over conventional live vaccines, not many have reached the market. There were only two recombinant vaccines available up to 2020 for humans. Several recombinant vaccines were released on war footing in the end of 2020 and during 2021 to deal with the Corona pandemic which shook the entire world. Most of the commercially available recombinant vaccines are for domestic animals (**Table 10.6**).

Table 10.6: Commercially available recombinant vaccines

Vaccine for	Type	Organism
Hepatitis B virus	Subunit	Humans
Human papilloma virus (causes cancer)	Subunit	Humans
Rota virus (causes diarrhea)	Live attenuated	Humans
Dengue virus	Live attenuated	Humans
Corona 19 Virus (causes Covid)	Live attenuated (Covaxin, Sinovac)	Humans
	Subunit (Novavax)	Humans
	Viral vector (Covishield of AstraZeneca)	Humans
	mRNA (Pfizer, Moderna)	Humans
Neisseria meningitidis B strain virus (causes meningitis)	Subunit	Humans
Rabbit haemorrhagic disease virus (RHDV)	Subunit	Rabbits
Poultry New castle disease virus (NDV)	Subunit	Poultry
Poultry Bursal disease virus (IBDV)	Subunit	Poultry
Avian influenza (bird flu) virus	Subunit	Poultry
Laryngotracheitis virus	Subunit	Poultry
Salmonella (bacterium that causes typhoid)	Genetically modified live attenuated	Humans
Canine distemper virus	Subunit	Dogs
Canine distemper virus	Genetically modified live attenuated	Dogs
Rabies virus	Genetically modified live vaccine that expresses an antigenic protein	Cats
Borrelia burgdorferi (bacterium causing lyme disease)	Subunit	Dogs
Equine west Nile virus (WNV) (encephalitis)	Genetically modified live vaccine	Horses
Covid 19	Attenuated, subunit, mRNA	Humans

10.5 ANIMAL CLONING

An identical copy (hereditary facsimile), of a cell or an organism is called its clone. The progeny of cells or organisms, generated through mitotic cell division, all represent, clones of the mother cell /organism. Bacteria reproducing through binary fission, or yeast through budding, or plants that reproduce vegetatively through suckers, stolons or bulbs, result in clonal progeny. Progeny produced through asexual reproduction are clones. In sexual reproduction, the progeny receive genes from both parents and are therefore, only partially similar to their parents and never identical to them. Since each gamete formation is the result of an independent meiotic division, during which recombination between genes takes place, siblings born of the same parents, are not identical, but only similar to each other. Identical twins or triplets, that arise due to splitting of zygotes, are examples of naturally occurring clones.

Creating clones of sexually reproducing animals, was thought to be impossible, unlike in plants. Plant cells are totipotent, and any cell of a plant cultured in an appropriate medium, can develop into a plantlet, that grows into an adult plant, identical to the plant of its donor cell. Animal cells were believed to lose their totipotency, once they differentiate into specialised cells. Cell and tissue culture of animals, often resulted in undifferentiated mass of cells, and nothing beyond. Cloning was once in a while successful, in the early experiments of cloning with frogs and salamanders. In mammals, the most complex among animal kingdom, the first success with cloning, was achieved in mid-1990's. Attempts of animal cloning, through a technique called 'Somatic Cell Nuclear Transfer' (SCNT), resulted in the first successful production of an animal clone – a sheep which was named 'Dolly'. Dolly was created using a cell from the udder (breast). It meant that, a nucleus from an adult cell, that has specialised for a specific function, could be reprogrammed to develop into an embryo. Cloning a mammal defied the scientific beliefs of that time. Since then, clones have been created in pigs, goats, rabbits, horses, deer, cows, dogs, cats, mice, wolves, ferrets, camels, mules and even monkeys.

Clones are essentially identical genetic copies. While identical twins represent naturally originating clones born at the same time, clones created artificially, though are identical genetic copies, are born at different times. The parent contributes to its identical clone much later in life. There are advantages in producing animal clones, but there are serious moral issues raised, with this artificial way of duplicating animal life, specially in humans. Cloning in humans is banned. Though possible, animal cloning is a difficult and expensive process, with a very low success rate and therefore, can only be adopted for special needs.

10.5.1 Methodology of Cloning

Animal cloning is an extension of the 'Assisted Reproductive Technologies'(ART), where fertilization between the egg and sperm, is done in vitro (in a test tube so to say), and the zygote is allowed to undergo a few cycles of division to produce a multicellular embryo. The embryo is then, implanted in the uterus of the biological (egg donor) or surrogate (one who serves to gestate the embryo to full term), mother. In cloning, fertilization is bypassed. Instead, a somatic cell (from any body part), of an animal to be cloned, is used to generate a clone. An egg cell of the same, or another animal, is used for harbouring the nucleus from the somatic cell. For this purpose, the egg cell is first enucleated (removing of nucleus). The somatic cell is then allowed to fuse with the enucleated egg cell. The resulting cell, is allowed to undergo a few cycles of division in invitro culture, to form a multicellular embryo. The process involves, dedifferentiation/reversion of a nucleus from a differentiated cell, into a pluripotent cell, that can develop into a whole organism, with multiple type of cells, tissue and organ systems. The multicellular embryo is then, implanted in the uterus of an animal which serves as the surrogate mother.

After successful gestation, the clone is born. The strategy – somatic cell nucleus transfer (SCNT), used in creating Dolly, the first successfully cloned animal is described in **Figure 10.15**. The somatic cell, egg donors and the animal which carries the embryo to term, can be one and the same individual or different, depending on feasibility.

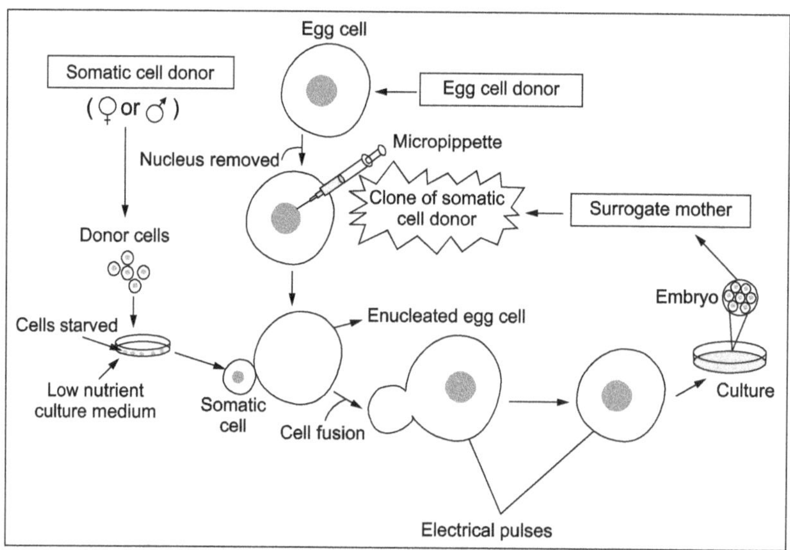

Figure 10.15: Animal cloning through somatic cell nucleus transfer (SCNT) technique. *The process depicted was followed for creation of the first cloned mammal, a sheep named Dolly. The somatic cell donor, egg donor and surrogate mother can be one and the same, or any two of them can be the same or all three may be different depending on feasibility.*

The procedure of animal cloning as figured out above, is a complex and cumbersome process, that can go wrong, at any step. The reprogramming of the nucleus, from a mature differentiated somatic cell, to an embryonic stage, is a multistep event, where a single mistake at any of the steps, can lead to disastrous outcome – either the embryos may not be formed, or embryos may have defects, which lead to abortions mid gestation or death after birth. Successfully born clones may also have several abnormalities. Thus, the frequency of success of animal cloning is very low – with a maximum of 10 to 15%. Given this success rate, animal cloning is expensive and can be resorted only when the benefits are overwhelming.

10.5.1 Problems in Creating Animal Clones

Cloning is such a delicate procedure that, it is really remarkable, it works at times! Given the numerous stages and steps it can go wrong, it is in fact, very challenging to figure out, how it is even sometimes, successful. Associated with the low success rate, are the moral issues, that arise from considerations of animal welfare – the heath, trauma, and fate of the surrogate mother. Enlisted below are some of the issues of animal cloning:

10.5.1(a) Pre-natal Failures

Only a small percentage (less than 10%) of pregnancies with cloned embryos, result in live births.

10.5.1(b) Surrogate (Host) Suffering

The surrogate mother undergoes grave suffering, much of which is caused by, the inordinately high rates of spontaneous abortions. Cloning often results in 'large offspring syndrome', due to which, early term stressful caesarean deliveries become necessary. There is also substantial percentage of deaths of surrogate mothers, during pregnancy.

10.5.1(c) Post-natal Health of Cloned Animal

Cloned animals born on a farm, without the help of veterinary services, have very little chance of surviving. Cloned baby animals look like premature born animals, with not yet fully developed lungs, improperly functioning hearts, and fatty livers. Cloned animals have been reported to have, a wide range of health defects and deformities like brain abnormalities, heart and lung damage, flattened faces, large tongues, diabetes, intestinal blockages and kidney failures. Some studies report that, cloned animals have a short life span – they die young.

10.5.2 Applications of Animal Cloning

The first success of mammal cloning with the birth of Dolly, led to bizarre predictions like – humans would be cloned; diseases would be prevented, and lost children rebirthed. None of these have materialised. The potential of cloned animals is in the following areas.

10.5.2(a) Transgenic Animals

Through cloning by SCNT of transformed animal cells, it is possible to obtain transgenic animals, at relatively lesser cost, and with more success rate, than other methods. Efficiency of production of transgenic animals can be increased through nuclear transfer from cultured transgenic cells to the enucleated egg for SCNT cloning. The DNA sequence of interest can be introduced into cultured animal cells, which can then be screened to identify those cells that have successfully incorporated the foreign DNA. The identified transgenic cells, can be used as the nuclear donor in the SCNT process, to create cloned transgenic animals. Cloning could also be used to create multiple copies of an existing transgenic animal.

10.5.2(b) Preserve Endangered Species and Prevent Their Extinction

Cloning can be very valuable to preserve endangered animal species or breeds. Enderby island cattle breed of New Zealand, was saved from extinction, by cloning the last surviving cow. Cloning animals of the yester eras, like dinosaurs or woolly mammoth, remain still in the realm of science fiction, with a distant possibility.

10.5.2.(c) Duplicating Elite (Prize) Animals

Animal cloning is useful, to replicate animals with favourable genotypes, even when they are male or infertile e.g. cloning of prizewinning animals like Texas long horn steer and racing mules. Cloning of working dogs like police K9s, search-and-rescue dogs, and other service dogs would be very useful. The fine service abilities of a dog, are realized much later to their routine neutering. Therefore, cloning is a practical solution to sire animals with such fine capabilities.

10.5.2(d) Cloning Best Breeds of Livestock

Cloning can ensure superior livestock genetic stocks, so that quantitatively more productive herds of the highest quality, can be maintained. Cloned animals are almost always used as breeding animals, and not directly for food. Most countries legally accept cloning of farm animals. The benefits of cloning in livestock are imminent. For example, if a cow which gives double the yield of milk than the rest in the herd is spotted, it can be cloned. With such high milk yielding cows, the number of cows to be maintained in the barnyard, can be reduced. Cloning disease resistant animals, will serve many goals. Farmers do not need to use antibiotics in raising such disease resistant animals, and a supply of safe meat and other food products from them, is possible. With a constant supply of such healthy animals, worry of possible disease due to animal consumption, can be done away with. The expenditure of farmers on treatment of diseased animals, and for prevention of disease and loss due to death of diseased animals, can all be avoided making animal husbandry, a more profitable engagement. Most often, the food that is derived from the animals, is not directly from the cloned animal but its offspring. Therefore, negative impact of cloning if any, is not an issue. Thus, cloning of livestock is considered safe and profitable and made legal in many countries.

10.5.2(e) Cloned Animals as Disease Models

The SCNT method of cloning has also been used to produce animal models for diseases like cystic fibrosis, diabetes, retinitis pigmentosa, cancer, amyotrophic lateral sclerosis etc. These models are useful, to gain more insight about the diseases, that might lead to development of new therapeutic treatments.

10.5.2(f) Animal Cloning for Therapeutics

Therapeutic cloning is used to produce cloned embryos, instead of cloned animals. Embryos developed from cells of diseased people through SCNT cloning, serve as source of stem cells, genetically identical to the donor. The stem cell like cells of these embryos, can be stimulated to differentiate into any of the more than 200 cell types in humans, and used to transplant into the patient, to replace diseased or damaged cells, without the risk of rejection (since they are cloned from their own cells). Such cells can be used to treat diseases like Alzheimer's, Parkinson's, diabetes and spinal cord injury. Because they are created artificially and can be induced to develop into a variety of different cell types, they are called induced pluripotent stem (iPS) cells. Earlier, sexually produced embryonic cells were used for this purpose which raised ethical concerns. Even therapeutic cloning is distasteful to people, who cannot accept the idea of creating and expending human life for a required purpose. Those who support this approach hold the view that, it is a moral imperative to heal the sick.

10.5.2(g) Cloning of Companion Animals (Animal Pets)

Most pet owners are very emotionally attached to them and consider the pet (him/her) as family. Because of the short life span of the pet, given a chance, a pet owner would prefer a clone of it, to have his once in a lifetime companion, all through his life time. Cloning facility of pets was initially available only in Asia, but is now possible in USA. Texas based ViaGen livestock cloning company, began pet cloning from 2015 onwards. Given the complexity and low success rate of cloning, a pet clone is expensive. It was reported to be 50,000 US $ for a dog and 25,000 US $ for a cat.

The whole idea of pet cloning is that, one gets a replica of the original. This is not always true. The pet is more loved for its behaviour and attitude, which are shaped in its early age, due to nurture and surrounding conditions. The circumstances in which a clone is born, are not the same as that of its donor, which was born long ago, to a different mother. The pet owner gets the puppy or kitten, only after more than three months after its birth, during which its personality is moulded. Therefore, its behaviour may not be similar to the original pet. It has been an observation with the many pet clones created, that behaviour cannot be duplicated. Thus, though genetically identical to the original, the cloned pet will grow up, to have its own behavioural traits and attitudes. Also, because of the differences in diet, during pregnancy between the original parent of the pet and the surrogate mother of the clone, physical attributes, like shades of coat colour, may also vary. Thus, a clone is no way of getting the original favourite pet, back.

Another perspective about pet cloning is that, it amounts to arrogance and insult to the pet. Assuming one is rich, it is as if, challenging the pet, that I can have one like you! It is also an unfair expectation on the cloned (duplicated) pet, to perform just as its original!

Human cloning is universally condemned, since it is incompatible with human dignity. Laws are in place to prohibit all forms of human cloning.

10.6 BIOSENSORS

Development of biosensors for analytical purpose, be it human health, or environmental monitoring, is a very promising field, with a vast, yet to be tapped market. Well-known biosensors are: the pregnancy test strip kit, blood glucose meter and cholesterol meter.

10.6.1 What is a Biosensor?

Biosensors (short for biological sensors), are small analytical devises, with an ingenious amalgamation of, a biological element and microelectronics, for rapid, easy and straight forward detection and estimation of specific chemicals, or their activity, or pathogens. Biosensors have high selectivity and sensitivity for the element they are devised to analyse.

10.6.2 Components of a Biosensor

Biosensor is a blend of two building blocks: **1.** a sensor or receptor, which is a biological component and **2.** a sensing element or transducer, which is an electronic component, that detects and transmits the sensed signal. The signal is amplified, processed and displayed by an electronic devise, connected to the transducer (**Figure 10.16**)

10.6.2(a) Biological Component

The biological component in a biosensor can be a biomolecule (enzyme, antibody, DNA, plant protein, lectin), organelle, cell, tissue or microorganism. Receptors modeled after biological systems (biomimetic substances), like aptamers e.g. oligo nucleotides (short nucleotide sequences) and peptides, are also used as the biological component in a biosensor (**Figure 10.17**).

The biological component interacts (catalytic), or binds with (affinity based) the element that is to be detected (**Figure 10.17**). A biosensor detects, only one particular element, for which it is specific, even when it is present in very low concentration, surrounded by many other elements (**Figure 10.16**). To ensure the specificity of the biological component, it is immobilised through several ways (**Figure 10.17**), on a solid support like glass, or polymer, or a metal oxide nano material, that is in close contact with a transducer, which is the sensing element of the biosensor. To prevent non-specific attachment of substances other than the recognition element of the biosensor, the surface on which the bio-element is immobilized, is coated with an appropriate material. Thus, fabricating the surface on which the sensing biological component is immobilized, is a skilful step in the design of a biosensor.

There are two types of bio-elements based on their mode of action: catalytic and affinity based.

10.6.2(a)i Catalytic Bio-elements

The bio-elements that work on catalytic principle are enzymes, whole cells, tissues and microorganisms (**Figure 10.17**). The target molecule to be detected, is identified, by any of the following reactions of the target molecule, with the catalytic properties of the enzyme: **a.** result of a product from the target molecule due to enzyme action, **b.** activation or inhibition of the enzyme activity, **c.** modification of enzyme properties.

10.6.2(a)ii Affinity Based Bio-elements

Antibodies, cell receptors and DNA are affinity-based bio-elements. Biosensors with antibodies as the biological element are called immunosensors. Artificially synthesized peptides and oligonucleotides (short nucleotide sequences), described as biomimetic, and called aptamers, are also used as bio-elements. Aptamer peptides are small – about 100 amino acids in length. Depending on their nature, biomimetic peptides can either detect due to catalytic properties, or recognition can be affinity based. Since these peptides are much smaller than antibodies, they are more suitable as affinity-based bio-elements. The DNA and oligonucleotide biosensors are based on complementary strand hybridization. Single-strand nucleic acid molecule is used in the biological element. If a sequence complementary to the nucleic acid strand in the bio-element of the biosensor, is present in the analysed sample, it hybridizes with it. In principle, complementary strand hybridization can also be described as affinity based.

10.6.2(b) Transducer

The transducer is the detecting element of the biosensor, that translates the physical change accompanying the interaction of the target molecule with the bio-element, into a signal. The transducer is made up of semi-conducting or nano-material and it is attached to an electronic system, that includes a signal amplifier, processor and a display unit (**Figure 10.16**). The display unit is the most expensive part of a biosensor. The working principle of the biosensor is depicted in **Figure 10.16**.

10.6.3 Types of Biosensors

Depending on the change that is detected in the sensing of the target by the bio-element, biosensors are of different types: **a.** optical (light), **b.** caloric/thermo/pyro-metric (heat), **c.** potentiometric (creation of electric potential due to redistribution of charges), **d.** piezoelectric (change in mass), **e.** electrochemical (detection of release or consumption of electrons due to the activity of redox enzymes of the biosensor) (**Figure 10.17**).

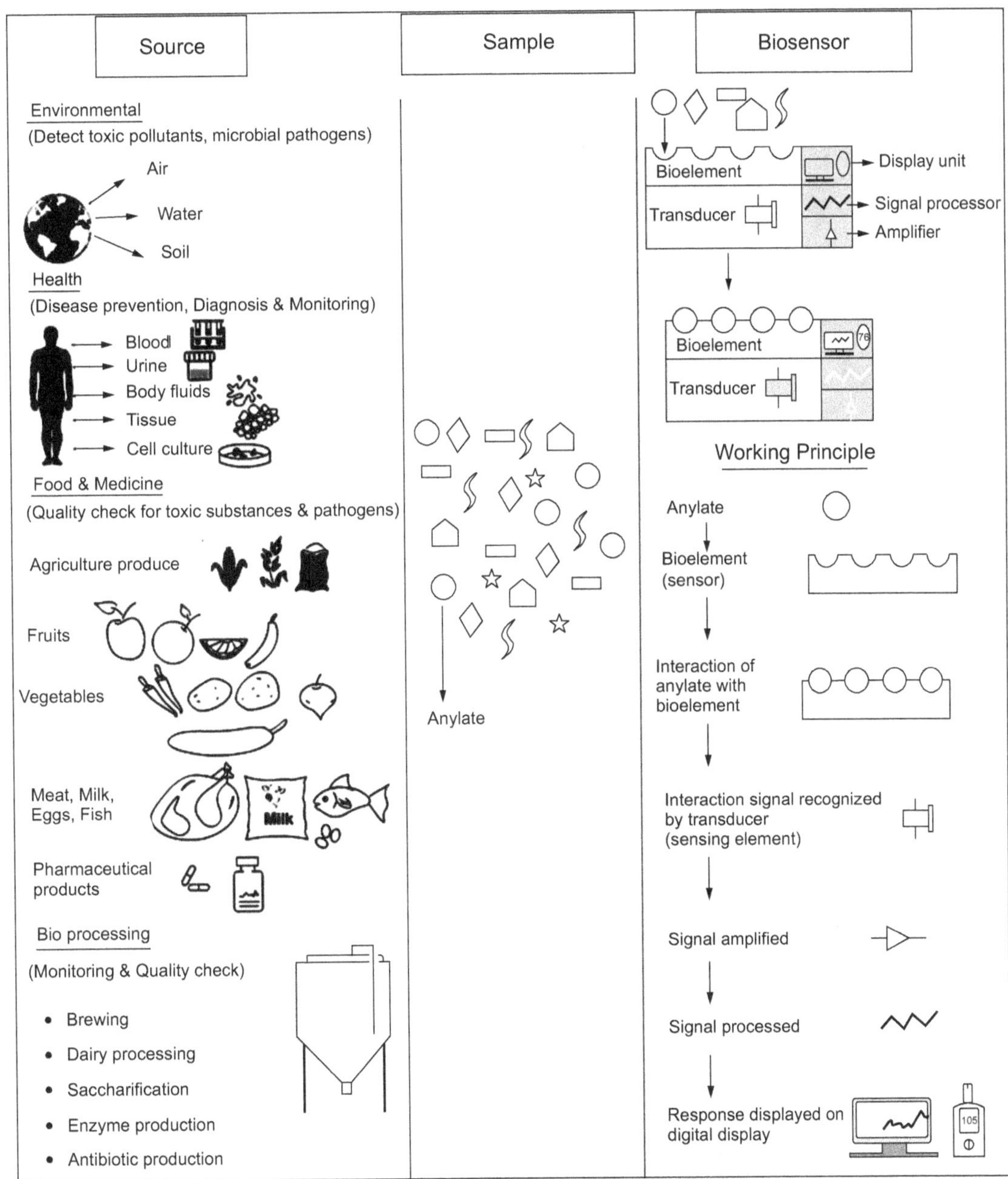

Figure 10.16: Biosensor: sources of sample, sample with anylate and working principle. *Only the anylate if present in the sample will be detected by the biosensor.*

Based on the mode of transmission of the response of the bio-element, biosensors are described as first, second or third generation types. In the first-generation biosensors, the product resulting from the reaction of the target molecule with the bio-element, diffuses to the transducer generating an electrical signal. The second-generation biosensors have 'mediators', that relay the occurrence of the reaction in the bio-element, to the transducer, resulting in improvement in the display of response. In the third-generation biosensors, the reaction at the bio-element itself produces the signal.

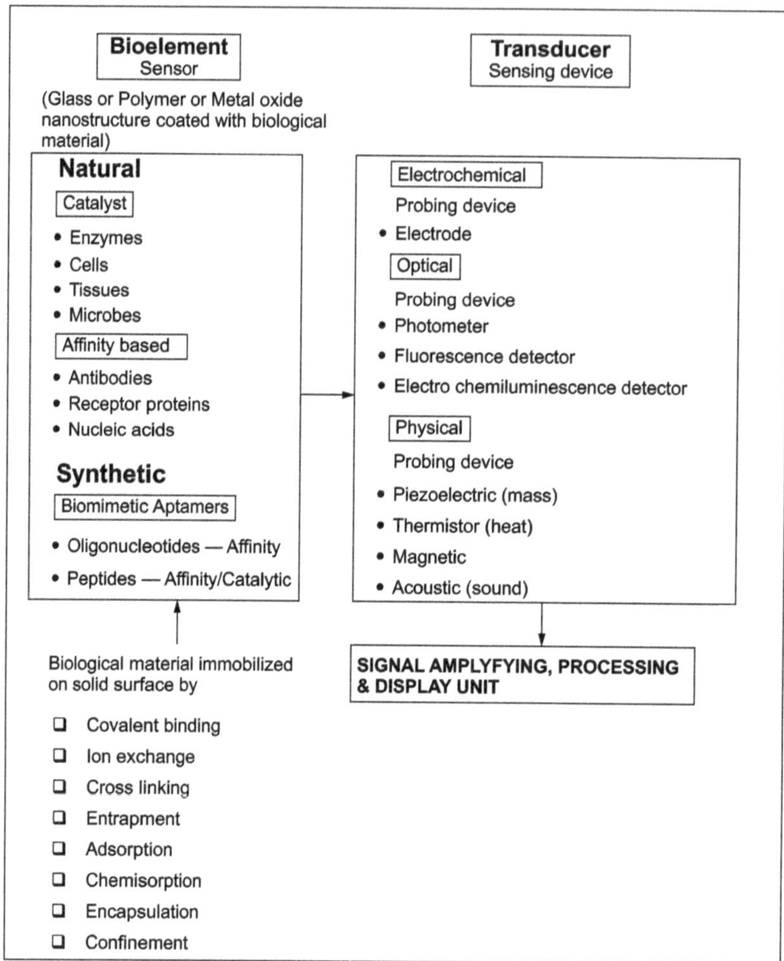

Figure 10.17: Biosensor components and types.

A biosensor should be cheap, small, portable, sterilizable, reusable and capable of being used by a semi-skilled operator. When it is used in invasive monitoring in clinical situations, it should be biocompatible and non-toxic.

10.6.4 *Applications of Biosensors*

Biosensors are highly valuable devices, for measuring a wide spectrum of analytes, including organic and inorganic compounds, heavy metals, gases, ions and bacteria. They have a wide spectrum of applications (**Figure 10.18**).

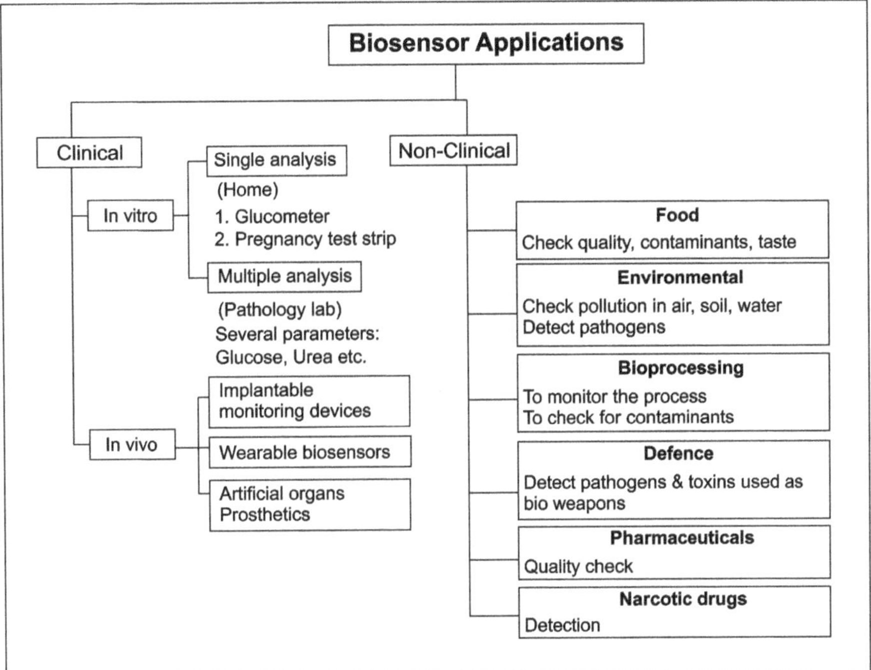

Figure 10.18: Applications of biosensors

10.6.4(a) Food Industry

Biosensors are useful to assess the quality, authenticity (distinguish between natural and artificial) and safety of food. Immuno and ligand binding biosensors, with fluorescence detection system, are used to detect small water-soluble vitamins, drug residues like sulphonamides, biological toxins and allergens. Biosensors can be used in detection of pathogenic organisms and antibiotics in meat, fish, poultry and eggs. Catalytic biosensors (based on inhibition of choline esterase enzyme), are used to detect traces of chemical pesticides, like organophosphates and carbamates, in farm produce.

Another trending biosensors are 'E noses', used to detect odd volatile compounds that affect the taste of food. E noses are particularly useful in tea, coffee, wine and spice industries. E tongue is a biosensor that has many applications. It is used to assess particular taste attributes of food like ageing, flavour and taste of wines, quality of taste masking substances used in medicines, to identify different varieties of tomatoes, coffee etc.

10.6.4(b) Bioprocess Monitoring and Analysis

Biosensors are used to analyze, the chemical composition of culture broth in fermentation, to detect the level of glucose, different types of amino acids, proportion of alcohol (in brewing industry), contaminants etc. Biosensors can be used in-line, on-line, at-line and off-line. The first three types, are useful to monitor real time during bioprocessing. In-line sensors are used directly in the fermentation broth, without the need to take a sample, from the bioprocessing stream. The broth is diverted from the manufacturing process for on-line measurements. For at-line analysis, the sample is taken from the fermenter and analyzed in close proximity to the process stream. Off-line biosensors analyze the samples at the end of bioprocessing. Biosensors are used to monitor quantity of lactose in real time, in dairy processing plants.

10.6.4(c) Environmental Monitoring

Biosensors are used to detect pollutants like chemical pesticides and heavy metal ions, organic and other pollutants in river water, remote sensing of quality of coastal waters, assessing the level of toxic substances before and after bioremediation, detection of airborne pathogenic microbes etc.

10.6.4(d) Narcotic Drugs and Explosives Detection

A biosensor based on a multi-sensing platform - BIOSENS 600 was developed to detect drugs like methamphetamine, cocaine, heroin, MDMA (ecstasy), cannabis, and common explosive chemicals, like trinitrotoluene and nitro glycerine.

10.6.4(e) Health and Pharmaceuticals

Biosensors are used to prevent, diagnose and monitor diseases. Biosensors have been developed for accurate and early diagnosis of cancer, heart disease, diabetes, autoimmune diseases and infectious diseases like dengue. Implantable biosensors and wearable biosensors which help in real time monitoring are available and several are under development. The wearable biosensor is integrated with smart watches, smart shirts or tattoos to monitor the levels of blood glucose, BP and the rate of heartbeat. User friendly biosensors that can be linked to mobile phone are also in the pipeline.

In pharmaceutical industry, biosensors are useful for quality check and to detect contaminants in pharmaceutical products.

10.6.4(f) Forensics

Biosensors are used to detect biomolecules and biological materials found in the crime site for identification of criminals. Biosensors are even used for lie detection; they are based on thermal detection and more accurate than the conventionally used lie detectors.

10.6.4(g) Military Operations

Biosensors that detect multiple types of pathogenic bacteria and viruses that are used as bio-weapons and chemicals like neurological toxins, that can wipe away human populations, have been developed and are available.

The applications of biosensors in a wide array of fields are endless. Extensive research is ongoing to develop refined and sensitive biosensors through development of new technologies for their design and construction. Biosensors that have been developed as prototypes, for proof of concept at present, will soon become commercial products, with a significant impact, specially in medicine.

10.7 BIOCHIPS

Biochips are the outcome, of an elegant confluence of electronics (semiconductors), computer science and biology. Biochip, is the name given to a multitude of computer chip alike devises, that function in a microfluid environment, or on RFID (Radio frequency identification) technology (**Figure 10.19**). Biochips, encompass a family of microprocessor devises (chips), that have a biological recognition component.

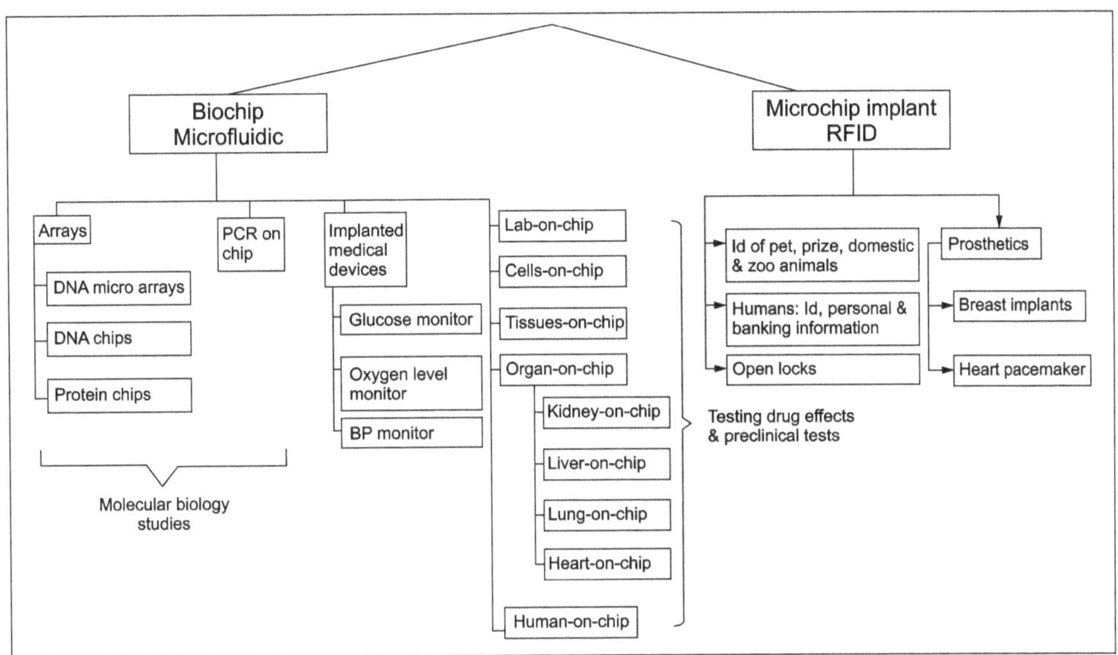

Figure 10.19: Different types of biochips

10.7.1 Microchips Based on RFID that are Used for Biochipping

Microchip is an electronic devise – an integrated circuit device, that serves as an RFID transponder (Radio Frequency Identification and responding unit), encased in biocompatible silicate glass capsule, that can be implanted in animals and humans. It is no bigger than the size of an uncooked rice grain. It is implanted, with the help of a hypodermic syringe, at the neck region, between the shoulder blades for pets and domestic animals and, in humans, on the back side of the palm, between the thumb and index finger (**Figure 10.20**). The process of introducing microchips inside a living organism is termed biochipping.

A micro-biochip has two components – a transponder and a reader (**Figure 10.20**). The transponder is a small glass capsule, in which is enclosed, the computer microchip, along with its tuning capacitor and antenna. It needs no charge and hence, has a very long (~99 years) life time. The reader can be a hand-held devise, or within an electronic door lock, or integrated in a scanner. The reader has an exciter and a receiving coil, to communicate with the transponder and a display unit. The reader can read the information on the implanted microchip, within a distance of 2 to 12".

The microchips are being used as identification tags for pets, domestic and prize animals and animals in zoos. In some corporate offices, microchips are used as ids for the employees. Using these biochips for mentally ill and Alzheimer's patients, can be handy in tracing them, when they go missing. They can be used for opening electronic locks, or as e tickets (railway in Sweden is totally microchip enabled). Biochip systems can, serve as a combination of credit card, driving licence, passport, medical & other records.

There are concerns regarding the use of these micro-biochips in humans, as they infringe the personal space. However, proponents of biochipping argue that, for the extreme convenience it offers, it is much less intrusive, than the smart phones that we now use. They contend that, biochips are inert and passive; they pose fewer privacy risks than smartphones, which continuously transmit our whereabouts. The future, with our life on one chip, seems imminent.

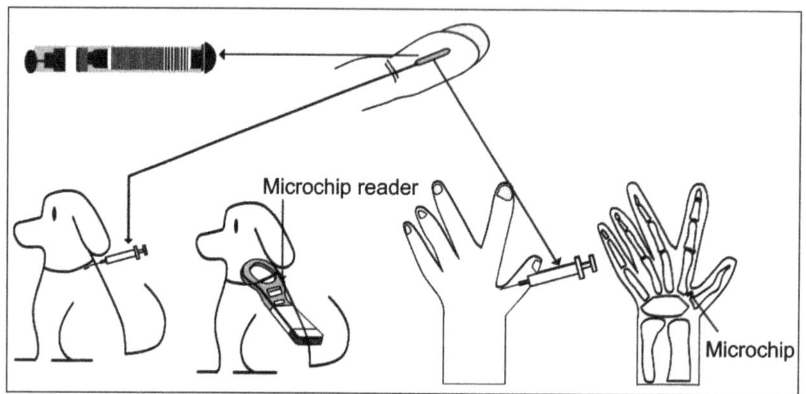

Figure 10.20: Microchips for biochipping organisms. *The microchip is no bigger than an uncooked rice grain with an information printed on it. It is inserted with a suitable syringe within the body – at the back of a neck in animals and between the index finger and thumb in humans. The microchip can have an identification number or other information like code for opening an electronic lock, or accessing a bank account, an electronic ticket for a public transport, details of medical records, passport, driving licence etc. A reader or scanner, placed within a distance of 2 to 12" of the implanted chip, can read the information on it, and display the details on a display unit, or open the doors, or carry out other needful activities.*

10.7.2 Microfluidic Based Biochips

The underlying mechanism in these chips, utilizes microfluidic micro-electromechanical systems (MEMS) technology. These devices are used, to analyse biological elements such as DNA, RNA, proteins, chemicals and drugs. The microfluidic biochips have four components – the microarray containing the biosensing elements, a transducer, signal processer and display unit.

10.7.2(a) DNA Microarrays and DNA Chips

These are miniature chips, that contain multitude (up to thousands) of DNA molecules, which serve as biosensors, arranged in rows, on a two-dimensional microgrid, about one-centimetre square. The DNA molecules are arranged in arrays in the grid, either through photolithography, or spotting (micro inkjet spot deposition). The chips with DNA arrayed through photolithography are called DNA chips and the ones with DNA spotted are termed DNA microarrays (**Figure 10.21**).

The DNA microarrays and chips, are used in functional genomics, to study gene expression. The thousands of genes, expressed in a particular tissue, under specific conditions, can be identified in one go, using a DNA microarray. The identification is based on, complementary strand hybridization, using fluorescent dyes for labelling. The image of the hybridization sites is displayed on the computer screen (**Figure 10.21**). A dedicated software is used, to analyse the reactions of the various DNA molecules on the chip, visible as colour spots on the image (**Figure 10.21**). DNA chips contain short sequences of DNA (oligonucleotides). They are usually used, to locate single nucleotide mutations in a gene sequence of interest.

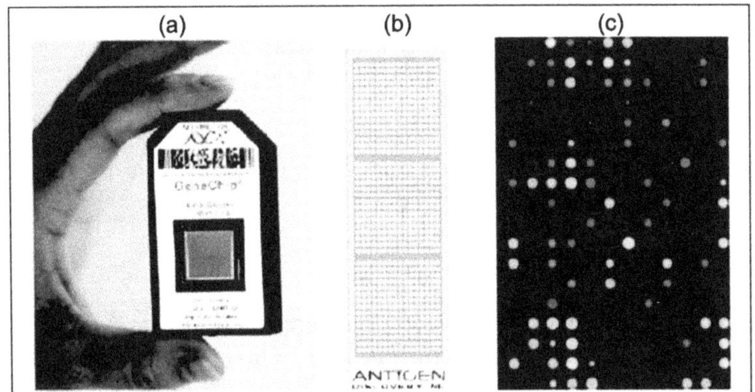

Figure 10.21: DNA and protein microarray chips and the image displayed after reaction in the arrays (a, b, c respectively). *The different coloured fluorescent spots in* c *indicate the different expression levels of the probes on the chip.*

10.7.2(b) Protein Microarrays

Similar to DNA microarrays, the protein microarrays are made, by spotting hundreds of proteins or peptides, on the grid of the microchip (**Figure 10.21**). Protein microarrays are also used in functional genomics, to identify all the genes, that are expressed in a particular tissue, under particular conditions. The proteins are labelled with fluorescent dyes, and identification is based on affinity binding, between the proteins in the sample, with those on the chip.

The DNA and protein microarrays are thus, high throughput devises, which enable to get a wholistic picture of the several hundreds of genes, that are expressed, in a snap shot. Like a computer chip, that can carry out millions of mathematical operations in seconds, a microarray can perform, thousands of biological reactions in a few seconds.

The DNA microarrays gauge gene expression, through identification of mRNAs in a sample; i.e., they give a picture of gene expression, at transcriptional level. The protein microarrays assess gene expression at translational level. The DNA chips enable high throughput screening of mutations in hundreds of DNA sequences, in one go. Identification of such single nucleotide mutations, which forebode the development of cancer at a later stage, can thus serve in molecular diagnostics. Such diagnosis can shift health care, from a focus on detection and treatment, to a process of prediction and prevention. The DNA or protein molecules in these arrays, are termed probes, since they are used to search for genes that are expressed, or look for gene mutations.

10.7.2(c) PCR Biochips

These chips facilitate carrying out multitude PCRs (Polymerase Chain Reaction for synthesis of DNA) in nano volumes in one go.

10.7.3 Biochips in Health

10.7.3(a) Lab-on-Chip

A lab-on-chip is a biochip, with reagents for several diagnostic tests, arrayed on a solid support, to serve as a miniaturised laboratory (**Figure 10.22**), for performing multiple tests simultaneously on a single sample resulting in generation of a patient's health profile at high speed. Lab-on-chips for various biochemical and microbiological studies are being widely used.

10.7.3(b) Tissue-on-Chip

These are miniaturized units, with a tissue on it, that can be implanted to replace a damaged tissue, or some part of it, to enable an organ to work normally.

10.7.3(c) Organ-on-Chip

These are biochips on which human organs are mimicked. The functional elements of an organ, are organized in a miniaturised form on a chip (**Figure 10.22**). As in the body, the organ-on-chip is supplied with oxygen and food via circulating fluids, through the channels in the chip. The organ-on-chips are useful to study human physiology, with respect to each organ. They can also be used as in vitro disease models. A very valuable application for these chips is that, they can potentially serve as replacements, for animals used in drug development and toxin testing.

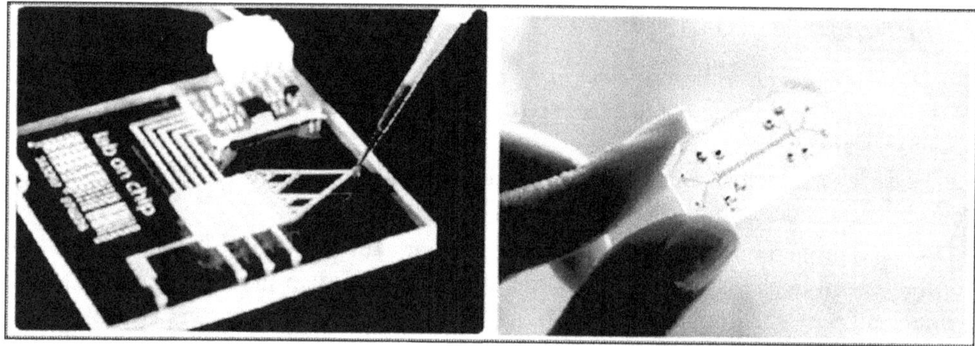

Figure 10.22: Microfluidic biochips that serve as Lab-on-Chip and Organ-on Chip.

Kidney-on-chip is used for checking the toxicity of a drug; liver-on-chip for the study of the metabolism and the toxicity levels of a drug, and pathology of chronic liver disease and other infections. Lung-on-chip is used for nanotoxicological studies of various nanoparticles which are introduced into the air channels. Heart- on-chip is handy to gauge the effect of physiological factors or test drugs for cardiotoxicity.

10.7.3(d) Human-on-Chip

The human-on-chip is an ingenious design of human organ constructs of heart, liver, lung and the circulatory system in communication with each other on a microfluidic biochip. It is under development and will be capable of accurately predicting drug and vaccine efficacy in preclinical testing. With this chip, it will be possible to assess effectiveness and toxicity of drugs in a way relevant to humans and their ability to process these drugs in place of animal testing.

10.7.3(e) Biochip Implants to Monitor Health

Biochip implants, to measure blood glucose levels, heart murmurs, oxygen levels and blood pressure, are being used for better health care (**Figure 10.23**). Microchips with a reservoir arrays of potent drugs, are being implanted, to deliver drugs when needed, or on a predetermined schedule, to precisely control drug release (**Figure 10.23**).

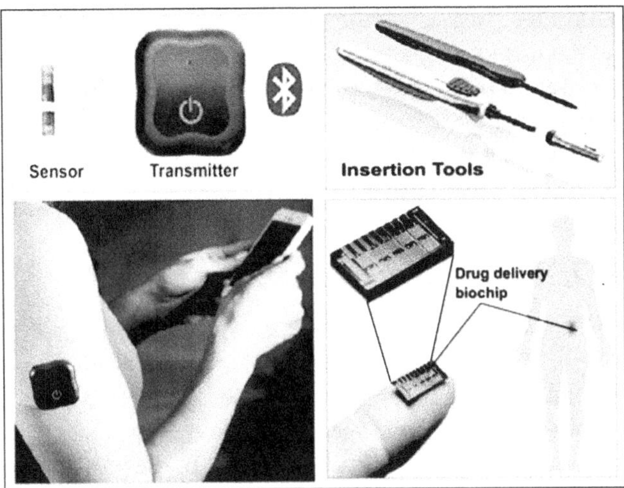

Figure 10.23: Implantable glucose monitor and drug delivery biochip. *The sensor (biochip) of the glucose monitor is implanted in the arm with an insertion needle. The glucose level can be measured by securing the transmitter on the arm and the readings can be seen on the mobile phone. The drug delivery biochip with several wells in rows filled with the drug is inserted in the body. The drug in each well is discharged into the body at a pre-set time regularly.*

10.8 INDUSTRIAL APPLICATIONS OF ENZYMES

Enzymes, the biocatalysts that drive biochemical reactions at phenomenal speed, and get life going, are used in industries, for their catalytic functions. They are used in two ways: to drive industrial processes and; as components of products like detergents, cosmetics and pharmaceuticals. Enzymes are used to make and improve nearly 400 everyday consumer and commercial products (**Figure 10.24**).

Enzymes are being used as industrial biocatalysts because, they make production process a green technology, with a higher process efficiency, requiring milder reaction conditions, such as lower temperature and pressure, non-toxic solvents, and lower volume and less toxic by-products and effluents, compared to chemical synthetic processes. Enzymes have exceptional chiral and positional specificity – attributes that traditional chemical processes lack. Thus, enzymes help to provide environmentally friendly products to consumers, which are manufactured using less energy, water and raw materials, and generating less waste. Employing enzymes in industrial production, also simplifies the machinery required, and makes process control easy.

Enzymes can be obtained from plants, animals and microbes. The microbial enzymes are more active and stable, than plant and animal enzymes. Microbes can be cultured in large quantities, in a short time by fermentation. Separation and purification of enzymes, from the fermentation broth, is easy. Hence, it is economical to produce enzymes from them, for industrial use. To illustrate this point: rennet, an enzyme used in cheese making, was initially obtained from calf stomachs, with an average yield of, 10 kg rennet per calf. Several months of intensive farming is required to obtain a calf. In comparison, a 1000-litre fermenter of recombinant *Bacillus subtilis,* can produce 20 kg of rennet enzyme within 12 h.

Of the ~4000 known enzymes, ~200 types of microbial enzymes are being commercially used at present. Of these, only 20 are produced on truly industrial scale, through fermentation. The large-scale production of microbial enzymes through fermentation processes, in a very short time, in a small production facility, enabled the availability of enzymes, at commercially feasible prices for the consumers.

Proteases, lipases, amylases and cellulases are the most abundantly used enzymes. They are supplied as liquid concentrates, soluble powders or granules. Two thirds of global enzyme market is catered to by the companies – Novazyme and DuPont. The projected market demand for enzymes is in billions of dollars. Enzymes are used in many industries like the food, textile, paper, brewing, baking, diary, tanning, pharmaceutical, biofuel etc. (**Figure 10.24**).

10.8.1 Food Industry

10.8.1(a) Making Fructose Syrups from Starch

Fructose is used in food manufacture as a sweetener. It is sweeter than glucose. Fructose is made usually from corn starch, in two steps. First the starch is converted to glucose using amylase enzyme. Amylase is inexpensive, being easily obtained from fungi. Next, glucose is converted to its isomer – fructose, using the enzyme glucose isomerase. This enzyme is more expensive to produce, so it is used in immobilised form, so as to be easily recovered and reused. Fructose is used in the preparation of fructose syrup.

Figure 10.24: Industrial uses of enzymes – as components of products and in driving the production process in industries.

10.8.1(b) Other Uses in Food Industry

Trypsin, a protease enzyme is used to predigest proteins during the manufacture of baby foods. Cellulases and pectinases are used to clarify fruit juices during their processing. Papain is used to tenderize meat for cooking. Lipases play a major role in the fermentative steps during preparation of sausages. Lipases are also used for refining rice flour and modifying soybean milk.

10.8.2 Baking Industry

α amylases are used, to break down starch in the flour to sugars, which are later fermented by yeast, resulting in raising of the dough. α amylases also improve bread quality – its taste, crust colour and toasting property and delay staling. Proteases are used by biscuit manufacturers to lower the protein level of flour.

An emerging trend, is the use of enzyme catalysis, for commercial-scale production of, probiotics, artificial sweeteners, and rare sugars. Probiotics are non-digestible food additives, that stimulate growth of gut bacteria and reportedly improve digestion. Galacto oligosaccharide, a probiotic and a low-calorie sweetener, is commercially produced, through enzyme catalysed process.

10.8.3 Diary Industry

Rennet, a complex of proteolytic enzymes – chymosin, pepsin and lipase, obtained from calves, was used to hydrolyse casein protein, during manufacture of cheese. Later, fermentation-produced chymosin from plant, fungal, and microbial sources, was used for industrial cheese-making. In recent years, microbial *(from the fungus Mucor michei)* rennet, which is acid aspartate protease is being used in cheese production. Lipase enzymes are used, to enhance the ripening of blue mould, during the production of Roquefort cheese. Lactases are used to break down lactose to glucose and galactose.

10.8.4 Brewing Industry

Industrially produced barley enzymes from *Bacillus amyloliquefaciens* – α amylase, β glucanase and protease, are used in the brewing process. Amylase, glucanases, and proteases are used, to split polysaccharides and proteins in the malt, which are later fermented by yeast, to beer. Beta glucanases and arabino xylanases are used, to improve the wort and beer filtration characteristics. Proteases are used, to remove cloudiness produced during storage of beers.

Pectinases, proteases and glucoamylase enzymes, are used in wine industry to speed up fermentation, and to obtain high yield of wine with enhanced aroma.

10.8.5 Paper Industry

For paper making, lignin needs to be removed from plant material. Ligninolytic enzymes (xylanase), obtained from fungi, are used in pre-treatment of wood pulp, to remove lignin (enzymatic bleaching), which is a milder and cleaner strategy of removing lignin, compared to chemical bleaching. Amylases and cellulases are used for the modification of fibre in wood pulp, deinking and removal of pitch to obtain quality paper.

10.8.6 Textile Industry

Enzymes are used in textile processing and fabric finishing (e.g. to get stone washed effect). Amylase is used, for warp sizing of textile fibres, that prevents the breakage of yarn in the weaving machine. Lipases are used, for the removal of oil and grease stains from the fabric, that are formed from the lubricants used in the machinery of the textile industry. The absorbance ability of the fabric, is thus increased, which imparts improved levelness in dyeing.

10.8.7 Pharmaceutical Industry

Penicillin G/V acylase obtained from *E. coli* is used for the production of semisynthetic Penicillin. Sitagliptin, a drug marketed by Merck, for type II diabetes, is synthesized through enzyme catalysis. Amylases are used as a digestive aid in pharmaceutical products. Microbial enzymes play an important role in the diagnosis, cure and monitoring of several diseases. Several different kinds of enzymes are used to make different kinds of biosensors.

10.8.8 Biofuel Industry

Lipases are used in the production of biodiesel from waste palm oils and waste edible oils. Enzymatic biodiesel production does away with the chemical stripping of the free fatty acids, generated during biodiesel production, saving a lot of steps required to obtain pure biodiesel. Amylases and cellulases, are used in ethanol production from vegetable sources.

10.8.9 Enzymes in Detergents

Proteases, lipases and amylases are used, in household products like, detergent powders and dish washing liquids, for helping in the removal of protein and oil stains from clothes and grease from dishes used in cooking and serving. Subtilisin and other proteases have been engineered and modified, to work alongside with bleaches and other chemicals, in detergents. The alkaline serine protease, obtained from *Bacillus licheniformis,* is mostly used in detergents. In addition, the serine protease of *B. amyloliquefaciens* is also used for this purpose.

10.8.10 Enzyme Immobilization

In industrial processes using bio-catalysis, the enzymes can be added in soluble form, or immobilized, on a solid but porous support. Enzymes are immobilized through several techniques like, adsorption, covalent binding, affinity, and entrapment in jelly like beads made from sodium alginate. The beads are packed in a container, and a solution of the substrate is allowed to flow over them, to be converted into the product. Immobilization of enzymes, greatly simplifies the recovery process of the enzyme, which can be reused, enhances process control, and reduces operational costs. The immobilized enzymes remain very stable under varying conditions of temperature and pH.

10.9 FERMENTATION AND ITS INDUSTRIAL APPLICATIONS

Fermentation was traditionally practiced by humans for centuries, in brewing and baking, using yeast, to convert sugars to alcohol and carbon di oxide, under anerobic conditions. The process of fermentation, has now been improvised, to carry out on an industrial scale, for commercial production of several quality and safety assured products. For this, normal, mutated, or genetically engineered, bacteria, yeast, fungi, microalgae, insect and mammalian (CHO) cells are used. Technological improvement in industrial fermentation processes, was partly driven by, the two world wars – I and II. The production of acetone through fermentation, provided the explosives used in first world war. In the second world war, the necessity to heal wounds, to save lives of the soldiers, through antibiotics, led to the advancement of fermentation techniques, for the large-scale fermentative production of Penicillin.

Industrial fermentation is a sensitive process, and requires the application of knowledge and principles of, biology (microbiology, biochemistry, genetics), chemistry, engineering (chemical, material and bioprocess (fluid dynamics) engineering), mathematics and computer science. Fermentation is performed in large vessels called fermenters or bioreactors, often of several thousand litres in volume (**Figure 10.25**).

10.9.1 Types of Fermentation

Based on the state of the medium used in fermentation, fermentation is described as solid state and liquid or submerged state. In solid state fermentation, nutrient rich grains or wastes like bran, bagasse or paper pulp, are used as the substrates, which are slowly and steadily, used by the microbe, usually a fungus. Many industrial bioactive metabolites such as cellulases, pectinases, and proteases are produced through solid-state fermentation. Fungal spores that are used as biopesticides are obtained through solid-state fermentation.

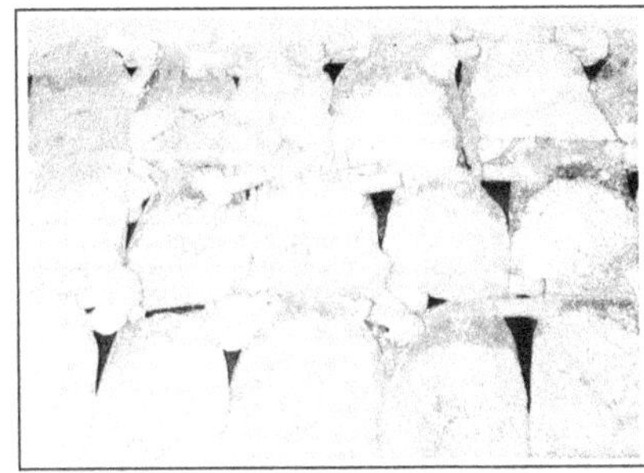

Figure 10.25: Solid state fermentation of a fungus to produce spores that are used as bioinsecticide. *(From author's laboratory)*

Submerged fermentation systems are more commonly used. Here, liquid medium is used. The microbial or eukaryotic cells, inoculated into the medium, utilize the dissolved substrates rapidly and grow. They secrete bioactive metabolites into the fermentation broth. This system is best suited, for bacterial cells that require high moisture content and whose secreted metabolites are in liquid form.

Fermentation of cocoa beans and coffee cherries is done without any medium. The cocoa beans are just left covered with banana leaves, or kept in a closed bucket. The coffee cherries are floated in water. The microbes grow due to heat and moisture and ferment them, by which the cocoa beans attain a chocolate flavour and the coffee beans can be easily harvested from the berries.

Industrial fermentation is used, for production of diverse products, such as antibiotics, cheese, pickles, wine, beer, biofuels, vitamins, enzymes, amino acids, solvents, bioinsecticides and pesticides, commodity chemicals such as acetic acid, citric acid, and ethanol.

10.9.2 Fermentation Process

10.9.2(a) Bioreactor

The heart of bioprocessing through liquid fermentation, is the vessel in which it is carried out. It is called the bioreactor (**Figure 10.26**). Bioreactors are a sophisticated version of shake flasks, with computer-controlled supply of gases, adjustment of pH regimes, and feeding of liquid broth with nutrients. The fermentation process is well monitored, to ensure optimal growth of the cells used for fermentation and production of the desired product.

There are several types of bioreactors, each designed for a specific microbial fermentation process. Some of the types of bioreactors are – continuous stirred tank bioreactor, bubble column and airlift loop bioreactor, photobioreactor, fluidized bed bioreactor and packed-bed bioreactor.

Figure 10.26: *Bioreactors used in fermentation*

10.9.2(b) Liquid Medium for Submerged Fermentation

Different media (fermentation broths), are used based on the organism used in fermentation, and the end product required from the process. All media invariably contain, a carbon source, a nitrogen source, water, salts and micronutrients.

Carbon sources are typically sugars or carbohydrates. For sensitive fermentations, purified sugars like glucose, sucrose and glycerol are used, to minimize variation between different fermentation batches, and to ensure the purity of the final product. Starch is provided as carbon source, for organisms meant to produce enzymes like beta galactosidase, invertase or other amylases. Grape is the carbon source in the production of wine, and rice or barley malt, for beer production. For bioethanol, inexpensive carbon sources such as molasses, corn steep liquor, sugar cane, or sugar beet juice, are used to minimize costs. In substrate transformations such as the production of vinegar for example, the carbon source may be an alcohol.

Type of nitrogen source provided in the medium, depends on the organism used for fermentation. Soy meal, pre-digested polypeptides such as peptone or tryptone, or chemicals like ammonia or nitrate salts, are used as nitrogen source. Cost is also an important factor in the choice of a nitrogen source. Phosphates are added to some culture broths, depending on the nature of the fermentation product, and the organism used.

Growth factors, trace nutrients and vitamins are included in the fermentation broth. Yeast extract is a common source of micronutrients and vitamins for fermentation media. Inorganic nutrients, including trace elements, such as, iron, zinc, copper, manganese, molybdenum and cobalt are typically present in unrefined carbon and nitrogen sources, but may have to be added, when purified carbon and nitrogen sources are used.

Gases are released during fermentation, or added, for the process to go on. Due to this, froth is formed in the culture broth. To prevent froth formation, anti-foaming agents are added. To maintain an optimal pH in the culture broth, mineral buffering salts, such as, carbonates and phosphates are used.

10.9.2(c) Fermentation Modes

There are mainly, three fermentation modes of providing culture broth for fermentation, depending on the fermentation product. They are the batch, fed-batch, and continuous cultivation systems. These modes are monitored and regulated based on mathematical models for optimal fermentation.

Batch fermentation mode is the simplest operation system. It is carried out in a closed vessel, where all the nutrient requirements in the culture medium, are added at the beginning of fermentation process.

The fed-batch fermentation mode, starts off in the batch fermentation mode, until, one or more substrates in the medium, are exhausted. Fresh medium is added, at different time regimes, to compensate for the exhausted nutrients in the medium. This feeding regime requires, accurate monitoring and operation control, to prevent the repressive effects of high substrate concentrations and avoid catabolic repression. Th fed-batch mode of fermentation, prolongs the exponential growth phase, of the fermenting cells. The fermentation product is harvested from the bioreactor at specific time intervals.

In continuous mode, the culture medium is continuously made to flow through the bioreactor, all through the fermentation process.

In industrial fermentation, to ensure the desired output, it is very important to meticulously monitor and regulate, all the culture parameters such as, temperature, pH, cell density, oxygen concentration, level of nutrients and product concentration. Computers are often used for this purpose.

10.9.2(d) Organisms/Cells Used in Fermentation

Yeast (*Saccharomyces*) and bacteria (*E. coli*), are the most commonly used organisms in fermentation. Recently, insect cells and mammalian cells – Chinese hamster ovary (CHO) cells, are also being used in fermentations. Several enzymes and chemicals are obtained as fermentation products using fungi.

10.9.2(e) Steps of Fermentation Process

Fermentation involves the following six steps:

A. Formulation of media: For each fermentation process, composition of two media needs to be formulated – one, that is used to initially culture the fermenting organism, to use as the inoculum in the bioreactor; the other, that is to be used, in the bioreactor.

B. Sterilization of the medium, fermenters and ancillary equipment to prevent contamination.

C. Production of an active pure culture, in sufficient quantity, to serve as the inoculum into the bioreactor.

D. Providing optimal conditions for fermentation to proceed and result in good yield of product.

E. Harvesting of the biomass, or extraction of the product, and its purification.

F. Disposal of effluents produced by the process.

10.9.3 Products from Industrial Fermentation

Industrial fermentation is used for production of diverse products such as baker's yeast, *Lactobacillus*, antibiotics, cheese, pickles, wine, beer, biofuels, vitamins, enzymes, proteins, amino acids, vinegar, solvents, bioinsecticides and pesticides, commodity chemicals such as acetic acid, citric acid and ethanol.

The primary and largest sector of industrial fermentation is the food industry. Cheese, sour cream, kefir, and yogurt are fermented dairy products. Many products of industrial fermentation are used as, food additives, flavours, vitamins, colours, preservatives and antioxidants. Thiamine (vitamin B1), riboflavin (vitamin B2), cobalamin (vitamin B12), and ascorbic acid (vitamin C, an antioxidant) are produced through microbial fermentations. Carotenoids are used as a natural food colour for butter and ice cream. Citric acid and lactic acid, produced through industrial fermentation, are extensively used as food preservatives. Nisin, a protein produced through fermentation, is used in the production of processed cheese. Bread making, involves using yeast produced through industrial fermentation. Fermentation plays a big role in the meat industry. Cured sausages called salami are made through fermentation.

Beer brewing and wine making, are huge agricultural product based industrial fermentation processes.

Production of ethyl alcohol from corn and other agricultural products and wastes, is one of the most successful commercial applications of fermentation. Ethyl alcohol is used in making gasohol fuel, which is a mixture of gasoline (~90%) and 10% alcohol. Acetone-Butanol- Ethanol (ABE) fermentation of glucose by an anaerobic bacterium *Clostridium,* is another major source of ethanol. Fermentative production of ethanol is an industry of national interest in Brazil.

Fermentation technology, is the heart of rapidly growing biopharmaceutical industry. Penicillin is produced by the fermentation of corn by the fungus *Penicillium chrysogenum.* New generation fermentation products include, anti-viral drugs, therapeutic recombinant proteins, bacterial vaccines, steroids, gene therapy vectors and monoclonal antibodies. Therapeutic proteins include, growth hormone, insulin, wound-healing factors and interferon.

Vinegar is produced via industrial fermentation of alcohol (ethanol) by acetic acid bacteria. Amino acids produced through fermentation are used in food, animal feed, pharmaceutical, cosmetic and chemical industries. Most of the industrial enzymes are produced through fermentation using *Bacillus* species.

The product of fermentation may be: **a.** biomass of the organism used in fermentation; **b.** substance secreted by the fermenting organism into the medium; **c.** primary or secondary metabolite of the fermenting organism and **d.** transformed product of the substrate used in fermentation.

10.9.3(a) Cell Biomass

Single cell protein (cyanobacterial /blue green algal cells), baker's yeast, *Lactobacillus* (for use in cheese making) and *E. coli,* are harvested, as the biomass of the organism used in fermentation. One major product harvested as cell biomass is baker's yeast. Baker's yeast is required for making bread, bakery products, beer, wine, ethanol, microbial media, vitamins, animal feed and biochemicals for research.

10.9.3(b) Extracellular Metabolites

These are harvested from the fermentation broth. Metabolites produced during the growth phase of the organism, are termed primary metabolites, while, those produced during the stationary phase are called secondary metabolites. Ethanol, lactic acid, citric acid, amino acids like glutamic acid, threonine, tryptophan and lysine, vitamins and polysaccharides, are some examples of primary metabolites. Penicillin, cyclosporin A, gramicidin(S), gibberellin, lovastatin and fungicides like griseofulvin, are some examples of secondary metabolites.

10.9.3(c) Intracellular Substances

The fermentation cells have to be lysed, to obtain these fermentation products. Further, a purification step is required, to separate the substance of interest, from other substances, in the lysed cells. Some such products are microbial enzymes: catalase, amylase, protease, pectinase, cellulase, hemicellulase, lipase, lactase and streptokinase. Recombinant proteins, such as, human insulin, hepatitis B vaccine, interferon, streptokinase, are also produced in the cells of fermenting organisms.

10.9.3(d) Substrate Transformation

Some fermentation processes result in the transformation of a substance into a useful product, e.g. benzaldehyde, is transformed by yeast to L phenylacetyl carbinol, which is used in the synthesis of a pharmaceutical – pseudo ephedrine.

10.10 SUMMARY

1. Genetic engineering involves creation of a recombinant DNA (rDNA) molecule and therefore also termed rDNA technology. Recombinant(r)DNA, is a hybrid DNA, created artificially (in vitro), through splicing DNA molecules from different sources, using restriction enzyme and ligase. A DNA molecule of interest can be inserted into another DNA called the vector thus forming a rDNA molecule. The recombinant vector is then cloned to get multiple copies. A vector is a DNA molecule that has an origin of replication, marker genes and recognition sites for several commercially available restriction enzymes to insert the DNA to be cloned. There are several kinds of vectors: plasmids, bacteriophage λ DNA (DNA of λ strain of bacterial virus), cosmids, bacterial/ yeast artificial chromosome (BAC/YAC). A plasmid can accommodate up to 15kb of DNA, a λ viral DNA can take up to 25 kb DNA, a cosmid, up to 45 kb, up to 350 kb in a BAC and 1000 kb in YAC. The vectors are transferred to a suitable host cell. The transformed cells are identified from the expression of the marker gene in the vector.

2. Transgenic organisms are created through transfer of a gene (a DNA sequence) of interest into an organism. This process is called transgenesis, the gene transferred is the transgene. Transgenic bacteria are made by chemical treatment of cells to make them receptive to the transgene. Transgenic plants are made using physical methods like electroporation, microinjection, micro projecticle particle bombardment (gene gun), or chemical methods and use of liposomes. The most popular and easy way of producing transgenic plants is using the plasmid vectors in *Agrobacterium*. The transformed plant cells are favoured to grow into plantlets in tissue culture using appropriate selective media. Making transgenic animals is more challenging, as animal cells are not totipotent like plant cells, and the fertilized egg cell or stem cells need to be transformed and placed in utero to develop into the animal. For transgenesis of animals, microinjection and sometimes retroviruses are used.

3. The potential of transgenic organisms has not been fully harnessed. Most have been produced as a proof-of-concept, with few examples of successful applications. Transgenic microbes are being used for production of human insulin (humulin), human growth hormone and some therapeutic proteins and enzymes. Among plants, herbicide and insect tolerant crops are being cultivated. Transgenic animals of disease models are very helpful in studying disease prognosis. Transpharming i.e., using transgenic animals for production of drugs is yet another promising activity. The untapped potential of transgenics is due to the several social, ethical, legal and environmental concerns raised regarding them.

4. Vertebrate animals combat pathogen infection with their white blood cells: monocytes/ phagocytes and the lymphoid cells – T, B and killer cells. During this process, memory T and

B cells with a long life span, are also formed which confer immunity against further infections, with the same pathogen. Vaccines dupe the pathogen and prime the immunity of the host, so that, the memory T and B cells, can readily attack an invading pathogen and prevent it from causing disease. Presently several kinds of recombinant vaccine technologies are available for synthesizing vaccines. They are live genetically modified vaccines – either attenuated or vaccine vectors and recombinant acellular vaccines like subunit, DNA and RNA types. The corona 19 pandemic is being dealt with using recombinant vaccine technologies.

5. An identical copy (hereditary facsimile), of a cell or an organism is called its clone. Clones are created artificially. Though they are identical genetic copies, they are born at different times. The parent contributes to its identical clone much later in life. Animal cloning is an extension of the 'Assisted Reproductive Technologies'(ART). In cloning, fertilization is bypassed. Instead, a somatic cell (from any body part), of an animal to be cloned, is used to generate a clone. An egg cell of the same, or another animal, is used for harbouring the nucleus from the somatic cell. The strategy used in animal cloning is called somatic cell nucleus transfer (SCNT). Animal cloning is a complex technique with very low success rate. There are several issues in animal cloning: prenatal failures, surrogate suffering and postnatal health of cloned animals. Animal cloning is mainly useful: to create transgenic animals, to preserve endangered species and prevent their extinction, duplicating elite (prized) animals, cloning best breeds of livestock, cloning animals as disease models, cloning pet animals and for synthesizing therapeutics. Human cloning is banned.

6. Biosensors are small analytical devises, with an ingenious amalgamation of, a biological element and microelectronics, for rapid, easy and straight forward detection and estimation of specific chemicals, or their activity, or pathogens. Biosensors have high selectivity and sensitivity for the element they are devised to analyse. The biological component in a biosensor can be a biomolecule (enzyme, antibody, DNA, plant protein, lectin), organelle, cell, tissue or microorganism. Receptors modeled after biological systems called aptamers e.g. oligo nucleotides and peptides, are also used as the biological component in a biosensor. The biological component interacts (catalytic), or binds with (affinity based) the element that is to be detected. Biosensors have a wide spectrum of applications: to assess the quality, authenticity and safety of food, in fermentation for monitoring bioprocessing and checking for contaminants, in environmental monitoring, detecting narcotic drugs and pathogens used as bio-weapons, in pharmaceutical industry for checking for contaminants. Biosensors have several medical applications : as medical devises-glucometer, pregnancy test strips, for estimation of urea, glucose and as implantable devises to treat diabetes, wearable biosensors to check diabetes, BP, heart beat and in forensics.

7. Biochip, is the name given to a multitude of computer chip alike devises, that have a biological recognition component. They function in a microfluid environment, or on RFID (Radio frequency identification) technology. Microchips based on RFID that are used for biochipping are used as identification tags for pets, domestic and prize animals and animals in zoos, as ids for employees, in tracing mentally ill and Alzheimer's patients, for opening electronic locks, or as e tickets. Biochip systems can, serve as a combination of credit card, driving licence, passport medical & other records. Microfluidic based biochips are used, to analyse biological elements such as DNA, RNA, proteins, chemicals and drugs. The DNA microarrays and chips, are used in functional genomics, to study gene expression. The thousands of genes, expressed (at mRNA level) in a particular tissue, under specific conditions, can be identified in one go, using a DNA microarray. Protein microarrays are used in functional genomics, to identify all the genes, that

are expressed (as proteins) in a particular tissue, under particular conditions. The DNA and protein microarrays are thus, high throughput devises. There are several kinds of biochips: PCR biochips, lab/tissue/organ/organism/human-on chip for carrying out experiments. Biochip implants, to measure blood glucose levels, heart murmurs, oxygen levels and blood pressure are being used for better health care . Microchips with a reservoir arrays of potent drugs, are being implanted, to deliver drugs when needed, or on a predetermined schedule, to precisely control drug release.

8. Enzymes have several industrial applications. They are being used as industrial biocatalysts because, they make production process a green technology, with a higher process efficiency, requiring milder reaction conditions, such as lower temperature and pressure, non-toxic solvents, and lower volume and less toxic by-products and effluents, compared to chemical synthetic processes. Enzymes are used in two ways: to drive industrial processes and; as components of products like detergents, cosmetics and pharmaceuticals. Enzymes are used to make and improve nearly 400 everyday consumer and commercial products. Enzymes can be obtained from plants, animals and microbes. The microbial enzymes are more active and stable, than plant and animal enzymes. Microbes can be cultured in large quantities, in a short time by fermentation. Separation and purification of enzymes, from the fermentation broth, is easy. Hence, it is economical to produce enzymes from them, for industrial use. Enzymes are used in food industry, baking and brewing industries, diary, paper, textile, pharmaceutical industries as drugs and in detergents. Enzymes are immbolized on solid support when they are used in driving industrial processes.This enables multiple time use of the enzyme.

9. The process of fermentation has now been improvised, to carry out on an industrial scale, for commercial production of several quality and safety assured products. Based on the state of the medium used in fermentation, fermentation is described as solid state and liquid or submerged state. The heart of bioprocessing through liquid fermentation, is the vessel in which it is carried out. It is called the bioreactor. Bioreactors are a sophisticated version of shake flasks, with computer-controlled supply of gases, adjustment of pH regimes and feeding of liquid broth with nutrients. There are mainly, three fermentation modes of providing culture broth for fermentation, depending on the fermentation product. They are the: batch, fed-batch, and continuous cultivation systems. Industrial fermentation is used for production of diverse products such as baker's yeast, *Lactobacillus*, antibiotics, cheese, pickles, wine, beer, biofuels, vitamins, enzymes, proteins, amino acids, vinegar, solvents, bioinsecticides and pesticides, commodity chemicals such as acetic acid, citric acid and ethanol. Production of ethyl alcohol from corn and other agricultural products and wastes, is one of the most successful commercial applications of fermentation The product of fermentation may be: **a.** biomass of the organism, used in fermentation; **b.** substance secreted by the fermenting organism into the medium; **c.** primary or secondary metabolite of the fermenting organism and **d.** transformed product of the substrate used in fermentation.

10.11 SAMPLE QUESTIONS

10.11(a) Subjective Questions

Q.1. What is rDNA technology? Describe the methodology of creation of a rDNA molecule and vectors used for cloning.

Q.2. What are biosensors? Describe the components, working principle and applications of biosensors.

Q.3. What are transgenics? How are transgenic microbes, plant and animals created? Give some examples of transgenic organisms. What are the concerns regarding transgenic organisms?

Q.4. What is fermentation? How is it done? What are the products obtained through fermentation?

Q.5. Describe the types and applications of biochips.

Q.6. Why and how are enzymes used in industries? Describe the various industrial applications of enzymes.

10.11(b) Objective Questions

Q.1. Recombinant DNA technology involves

(*a*) Use of restriction enzyme and ligase.

(*b*) Creation of a hybrid DNA molecule using two DNA sequences.

(*c*) Recombining genes.　　　　　(*d*) (*a*) & (*c*) are true.

(*e*) (*a*) & (*b*) are true.

Q.2. DNA vectors are used for

(*a*) Only cloning DNA.　　　　　(*b*) Cloning and sometimes expression of DNA.

(*c*) Only to transfer DNA to organisms.　(*d*) As vaccines.

Q.3. Splicing involves

(*a*) Cutting DNA.　　　　　(*b*) Cutting nucleic acids.

(*c*) Cutting and joining DNA.　　　　　(*d*) Transferring a gene for genetic engineering.

Q.4. Xenotransplants are

(*a*) Transplanted organs.

(*b*) Transgenic animals which serve as organ donors.

(*c*) Artificial devises implanted in humans.

(*d*) None of the above.

Q.5. Immunity is due to

(*a*) White blood cells.　　　　　(*b*) Lymphoid cells.

(*c*) Monocytes, B &T cells and killer cells. (*d*) Stem cells.

Q.6. Covaxin used for Covid is

(*a*) Vaccine vector.　　　　　(*b*) Attenuated virus.

(*c*) Accelular vaccine.　　　　　(*d*) mRNA vaccine.

Q.7. Animal cloning requires

(*a*) an egg cell, a donor nucleus and surrogate mother.

(*b*) An enucleated egg cell, nucleus from somatic cell and surrogate mother.

(*c*) A somatic cell and surrogate mother.

(*d*) A somatic cell and an egg.

Q.8. Biosensors

(*a*) Detect biological materials.　　　　　(*b*) Electronic devise to remove biological material.

(*c*) Have a biological component and a transducer.

(*d*) Used to remove environmental pollutants.

Q.9. Microchips are

(*a*) DNA chips.
(*b*) Microfluidic chips.

(*c*) Are inserted into the organism.
(*d*) High throughput protein chips.

Q.10. Fermentation is

(*a*) Culture of yeast only.

(*b*) Large scale culture of microorganisms for a product.

(*c*) Laboratory culture of cells.
(*d*) Baking industry.

ANSWERS

1. (*e*)   **2.** (*b*)   **3.** (*c*)   **4.** (*b*)   **5.** (*c*)   **6.** (*b*)   **7.** (*b*)   **8.** (*c*)

9. (*c*)   **10.** (*b*)

References

1. Wikipedia *https://en.wikipedia.org/*
2. https://bio.libretexts.org/
3. https://commons.wikimedia.org/
4. http://textbookofbacteriology.net/Todar's online Text Book of Microbiology
5. https://www.yourgenome.org/
6. https://www.nih.gov/
7. https://mycbseguide.com/
8. https://www.cienotes.com/a-level-biology-notes-9700/
9. https://biologyaipmt.com/2016/06/01/chapter-9-biomolecules/
10. https://byjus.com/
11. https://www.britannica.com/
12. https://www.khanacademy.org/
13. Molecular Cell Biology, 4th edition
    Harvey Lodish, Arnold Berk, S Lawrence Zipursky, Paul Matsudaira, David Baltimore, and James Darnell.
14. https://www.library.ucdavis.edu/database/merlot-ii-multimedia-educational-resources-learning-online-teaching/
15. The *Cell. A Molecular Approach.* Sixth Edition. Geoffrey M. *Cooper* and Robert E. Hausman.
16. https://biologydictionary.net/
17. https://courses.lumenlearning.com/
18. https://www.ifst.org/resources/information-statements/insight-animal-cloning
19. The future and prospects of bio-chips. Neeta Shivakumar, Poornima S, Sukanya Raghu and Pratibha. Innovare Journal of Education, Vol 1, Issue 1, 2013, 5-9
20. https://www.elprocus.com/what-is-a-biosensor-types-of-biosensors-and-applications/
21. https://www.pharmanewsonline.com/applications-of-enzymes-in-industry
22. https://www.lensrentals.com/
23. Fred Levine. "Basic Genetic Principles", Elsevier BV, 2017
24. opengenetics.net
25. www.omicsonline.org
26. https://www.scribd.com/
27. https://www.coursehero.com/
28. https://www.chgf.vu.lt/
29. https://www.slideshare.net/
30. kids.britannica.com

31. http://www.akubihar.ac.in/

32. www.ncbi.nlm.nih.gov

33. epdf.pub

34. wikimili.com

35. inba.info

36. onlinelibrary.wiley.com

37. https://www.goodreads.com/

38. "Industrial Fermentation", Applied Science, 2012

39. http://www.pharmanewsonline.com/

40. Several books that I read through my four decade avocation in Biology as a researcher and teaching faculty in the University.

Index

www.ingramcontent.com/pod-product-compliance
Lightning Source LLC
Chambersburg PA
CBHW080734120726
48001CB00009B/2586